T0327514

Data Analysis and Chemometrics for Metabolomics

Data Analysis and Chemometrics for Metabolomics

Richard G. Brereton

University of Bristol
United Kingdom

WILEY

This edition first published 2024
© 2024 John Wiley and Sons Ltd

All rights reserved, including rights for text and data mining and training of artificial technologies or similar technologies. No part of this publication may be reproduced, stored in a retrieval system, or transmitted, in any form or by any means, electronic, mechanical, photocopying, recording or otherwise, except as permitted by law. Advice on how to obtain permission to reuse material from this title is available at http://www.wiley.com/go/permissions.

The right of Richard G. Brereton to be identified as the author of this work has been asserted in accordance with law.

Registered Offices
John Wiley & Sons, Inc., 111 River Street, Hoboken, NJ 07030, USA
John Wiley & Sons Ltd, The Atrium, Southern Gate, Chichester, West Sussex, PO19 8SQ, UK

For details of our global editorial offices, customer services, and more information about Wiley products visit us at www.wiley.com.

Wiley also publishes its books in a variety of electronic formats and by print-on-demand. Some content that appears in standard print versions of this book may not be available in other formats.

Trademarks: Wiley and the Wiley logo are trademarks or registered trademarks of John Wiley & Sons, Inc. and/or its affiliates in the United States and other countries and may not be used without written permission.

All other trademarks are the property of their respective owners. John Wiley & Sons, Inc. is not associated with any product or vendor mentioned in this book.

Limit of Liability/Disclaimer of Warranty
In view of ongoing research, equipment modifications, changes in governmental regulations, and the constant flow of information relating to the use of experimental reagents, equipment, and devices, the reader is urged to review and evaluate the information provided in the package insert or instructions for each chemical, piece of equipment, reagent, or device for, among other things, any changes in the instructions or indication of usage and for added warnings and precautions. While the publisher and authors have used their best efforts in preparing this work, they make no representations or warranties with respect to the accuracy or completeness of the contents of this work and specifically disclaim all warranties, including without limitation any implied warranties of merchantability or fitness for a particular purpose. No warranty may be created or extended by sales representatives, written sales materials or promotional statements for this work. This work is sold with the understanding that the publisher is not engaged in rendering professional services. The advice and strategies contained herein may not be suitable for your situation. You should consult with a specialist where appropriate. The fact that an organization, website, or product is referred to in this work as a citation and/or potential source of further information does not mean that the publisher and authors endorse the information or services the organization, website, or product may provide or recommendations it may make. Further, readers should be aware that websites listed in this work may have changed or disappeared between when this work was written and when it is read. Neither the publisher nor authors shall be liable for any loss of profit or any other commercial damages, including but not limited to special, incidental, consequential, or other damages.

Library of Congress Cataloging-in-Publication Data:

Names: Brereton, Richard G., author.
Title: Data analysis and chemometrics for metabolomics / Richard G. Brereton.
Description: Hoboken, NJ : Wiley, 2024. | Includes index.
Identifiers: LCCN 2024002533 (print) | LCCN 2024002534 (ebook) | ISBN
    9781119639381 (hardback) | ISBN 9781119639374 (adobe pdf) | ISBN
    9781119639404 (epub)
Subjects: LCSH: Metabolites. | Chemometrics.
Classification: LCC QP171 .B74 2024 (print) | LCC QP171 (ebook) | DDC
    572/.4—dc23/eng20240307
LC record available at https://lccn.loc.gov/2024002533
LC ebook record available at https://lccn.loc.gov/2024002534
Hardback: 9781119639381

Cover Design: Wiley
Cover Image: © berCheck/Shutterstock; Courtesy of Prof Richard G. Brereton

Set in 10.5/13pt STIX Two Text by Straive, Chennai, India

Contents

Foreword   xi

Acknowledgements   xv

About the Companion Website   xvii

CHAPTER 1   Introduction   1
            1.1   Chemometrics   1
            1.2   Metabolomics   9
            1.3   Case Studies   14
            1.4   Software   15
            References   20

CHAPTER 2   Instrumental Methods   26
            2.1   Introduction   26
            2.2   Coupled Chromatography Mass Spectrometry 27
                  2.2.1   Chromatography   27
                  2.2.2   Ionisation and Detection   29
                          2.2.2.1   GCMS   29
                          2.2.2.2   LCMS   30
                  2.2.3   Data Matrices and Peak Tables   30
                  2.2.4   Step 1 of Creating a Peak Table: Transforming
                          Chromatographic Data to an Aligned Matrix
                          of Peaks   33
                          2.2.4.1   XCMS   33
                          2.2.4.2   Multivariate Curve Resolution 35
                          2.2.4.3   AMDIS   38
                          2.2.4.4   MZmine   38
                          2.2.4.5   Other Approaches   39
                          2.2.4.6   Which Approach is Most Suitable? 41
                  2.2.5   Step 2 of Creating a Peak Table: Manual
                          Inspection   42
                  2.2.6   Step 3 of Creating a Peak Table: Identifying
                          Metabolites and Annotating the Peaks 43
                          2.2.6.1   Experimental Libraries 45
                          2.2.6.2   Computational Mass Spectral
                                    and Retention Libraries 46
                          2.2.6.3   Expert Systems   47
            2.3   Single Wavelength HPLC   47

2.4	Nuclear Magnetic Resonance	50
	2.4.1 Fourier Transform Techniques	50
	2.4.1.1 FT Principles	50
	2.4.1.2 Resolution and Signal-to-Noise	53
	2.4.2 Preparing the Transformed Spectra	54
	2.4.3 Preparing the Data Table	55
	2.4.3.1 Chemometric Approach	56
	2.4.3.2 Deconvolution and Identification	58
	2.4.4 Identification of Metabolites	60
2.5	Vibrational Spectroscopy	61
	2.5.1 Raman Spectroscopy	62
	2.5.2 Fourier Transform Infrared Spectroscopy	64

CHAPTER 3   Case Studies   66

3.1	Introduction	66
3.2	Case Study 1: Presymptomatic Study of Humans with Rheumatoid Arthritis Using Blood Plasma and LCMS	67
3.3	Case Study 2: Diagnosis of Malaria in Human Blood Plasma of Children Using GCMS	69
3.4	Case Study 3: Measurement of Triglyclerides In Children's Blood Serum Using NMR	70
3.5	Case Study 4: Glucose Intolerance and Diabetes in Humans as Assessed by Blood Serum Using NMR	71
3.6	Case Study 5: Metabolic Changes in Maize Due to Cold as Assessed By NMR	72
3.7	Case Study 6: Effect of Nitrates on Different Parts of Wheat Leaves as Analysed by FTIR	74
3.8	Case Study 7: Rapid Discrimination of Enterococcal Bacteria in Faecal Isolates by Raman Spectroscopy	75
3.9	Case Study 8: Effects of Salinity, Temperature and Hypoxia on *Daphnia Magna* Metabolism as Studied by GCMS	76
3.10	Case Study 9: Bioactivity in a Chinese Herbal Medicine Studies Using HPLC	77
3.11	Case Study 10: Diabetes in Mice Studied by LCMS	78

CHAPTER 4   Principal Component Analysis   80

4.1	A Simple Example: Matrices, Vectors and Scalars	80
4.2	Visualising the Data Direct	81
4.3	Principal Components Analysis: Scores, Loadings and Eigenvalues	84
	4.3.1 PCA	84
	4.3.2 Scores	84
	4.3.3 Loadings	86
	4.3.4 Relationship Between Scores and Loadings	88
	4.3.5 Eigenvalues	89
	4.3.6 Reducing the Number of PCs	89
4.4	Exploration by PCA of Case Study 5 in Detail: NMR Study of the Effect of Temperature on Maize	92

| | | 4.4.1 | Variable Plots | 92 |

4.4.1   Variable Plots   92
4.4.2   Scores and Loadings Plots of the Whole
         Standardised Data   92
4.4.3   Scores and Loadings of the Low-temperature
         Data   96
4.5   PCA of Different Case Studies   98
4.5.1   Case Study 1: LCMS Studies of Pre-arthritis   98
4.5.2   Case Study 4: NMR of Human Diabetes   100
4.5.3   Case Study 10: LCMS of Diabetes in Mice   103
4.5.4   Case Study 6: FTIR of Effect of Nitrates
         on Wheat   107
4.5.5   Case Study 3: NMR for Triglycerides in Serum   107
4.5.6   Case Study 7: Raman of Bacterial Faecal
         Isolates   109
4.6   Transforming the Data   114
4.6.1   Row Scaling   114
         4.6.1.1   Row Scaling to Constant Total   115
         4.6.1.2   Standard Normal Variates   121
         4.6.1.3   Scaling to Reference Standards   123
4.6.2   Column Centring   125
4.6.3   Column Standardisation   129
4.6.4   Logarithmic Transformation   136
4.7   Common Issues   144
4.7.1   Missing Data   144
4.7.2   Quality Control Samples   147
4.7.3   Variable Reduction   149

CHAPTER 5   Statistical Basics   151
5.1   Use of P Values and Hypothesis Testing   151
5.2   Distributions and Significance   152
5.2.1   Simulated Case Study   152
5.2.2   The Normal (z) Distribution and p Values   153
5.2.3 t-Distribution and Degrees of Freedom   161
5.2.4 χ^2-Distribution   169
5.2.5 F-Distribution and Hotelling's T^2 174
5.3   Multivariate Calculation of P Values and the
      Mahalanobis Distance   182
5.4   Discriminatory Variables   190
5.5   Conclusions   194

CHAPTER 6   Choosing Samples   195
6.1   Motivation   195
6.2   Design of Experiments   196
6.2.1   Factors, Response and Coding   198
6.2.2   Replicates   199
6.2.3   Statistical Designs   200
         6.2.3.1   Fully Crossed Designs   201
         6.2.3.2   Two-level Full Factorial Designs   202
         6.2.3.3   Fractional Factorial Designs   204

6.3	Sampling Designs		206
	6.3.1	Simple Random Sampling	208
	6.3.2	Systematic Sampling	208
	6.3.3	Stratified Sampling	208
	6.3.4	Cluster Sampling	209
	6.3.5	Multi-stage Sampling	209

CHAPTER 7 Determining the Provenance of a Sample — **210**

7.1	Pattern Recognition		210
7.2	Preliminary Processing Prior to Classification		211
7.3	Simulated Case Studies		212
7.4	Two-Class Classifiers		218
	7.4.1	Linear Discriminant Analysis	220
	7.4.2	Partial Least Squares Discriminant Analysis	227
		7.4.2.1 PLSDA for Equal Class Sizes	228
		7.4.2.2 PLSDA for Unequal Class Sizes	234
		7.4.2.3 OPLS	240
7.5	One-Class Classifiers		244
	7.5.1	Quadratic Discriminant Analysis	245
	7.5.2	SIMCA	260
		7.5.2.1 Disjoint PCA	260
		7.5.2.2 D- and Q-statistics	264
		7.5.2.3 Limits and Decisions	266
7.6	Multiclass Classifiers		272
	7.6.1	LDA as a Multiclass Classifier	272
	7.6.2	PLSDA as a Multiclass Classifier	275
		7.6.2.1 One Versus All	275
		7.6.2.2 One Versus One	285
		7.6.2.3 PLS2DA	285
	7.6.3	Multilevel PLSDA	287
7.7	Validation, Optimisation and Performance Indicators		288
	7.7.1	Classification Performance	288
		7.7.1.1 Two Classes	288
		7.7.1.2 Multiclasses	291
		7.7.1.3 One-Class Models	293
	7.7.2	Validation	294
	7.7.3	Optimisation	300

CHAPTER 8 Multivariate Calibration — **305**

8.1	Introduction	305
8.2	Partial Least Squares Regression	306
8.3	Training and Test Sets	310
8.4	Optimisation: Number of PLS Components	317

CHAPTER 9 Selecting the Most Significant Variables and Markers — **320**

9.1	Introduction	320
9.2	Univariate Approaches	320

	9.3	Loadings, Weights and VIP Scores		324
		9.3.1	Principal Component Loadings	324
		9.3.2	PLSDA Loadings and Weights	326
		9.3.3	VIP Scores	331
		9.3.4	P Values	338
		9.3.5	Multilevel PLSDA	341
	9.4	Selectivity Ratios		346
	9.5	Volcano Plots		349

CHAPTER 10 Which Factors are Most Significant   352

	10.1	Introduction		352
	10.2	Terminology and Definitions		353
	10.3	Single Factor (One-Way – One-Factor) ANOVA Test and Regression		357
		10.3.1	Balanced Design at Two Levels	357
			10.3.1.1 Degrees of Freedom	357
			10.3.1.2 ANOVA Test	358
			10.3.1.3 Regression	361
			10.3.1.4 The t-test	363
		10.3.2	Unbalanced Design at Two Levels	364
		10.3.3	Multiple One-Way Design with Two Levels: Multilinear Regression	365
		10.3.4	Multilevel Designs	371
			10.3.4.1 One-Way Multilevel ANOVA Test	371
			10.3.4.2 One-Way Multilevel Regression with Dummy Variables: Unrelated Groups	373
			10.3.4.3 One-Way Multilevel Multilinear Regression: Related Groups	375
			10.3.4.4 Comparison and Interpretation	376
	10.4	Multiple Factor (Multiway) ANOVA Test and Regression		379
		10.4.1	Simulated 2×3 Case Study: ANOVA Test and Regression	379
			10.4.1.1 ANOVA Test	380
			10.4.1.2 Regression with Dummy Variables	382
		10.4.2	Two-level Multiway Factorial Designs	385
	10.5	ASCA		389
		10.5.1	Simulated Dataset	390
		10.5.2	Case Study: Environmental Effect on Daphnia	395

Index		406

Foreword

The term *metabolomics* was first introduced in the early 2000s, when spectroscopic and chromatographic techniques were being developed for profiling of metabolites within cells and tissues. The capabilities of instrumental techniques, primarily NMR, LCMS and GCMS, to obtain large quantities of such data resulted in big datasets. Common to other emerging omics technologies such as genomics, transcriptomics and proteomics, computational methods were required to make sense of this data.

Computational methods for processing data can be applied to two steps of the metabolomics workflow. The first is for resolving and characterising raw instrumental data primarily from coupled chromatography and NMR spectroscopy, to produce peak tables for subsequent statistical processing. We list the main packages currently available in this book and their underlying principles, but this is not the focus of the text. These packages evolve rapidly, and most will change over the lifetime of this text. In addition, many involve proprietary or very complex and ever-evolving algorithms often linked to expanding databases, which would be difficult or impossible to describe in detail. Finally full information is not always easily available from the developers or main users of some of the packages, but there are regularly updated websites with user manuals to which readers should refer. We have tried to gather information on most of these approaches in Chapter 2, all of which are unique, but some have been developed and are maintained by large teams, and it would be unrealistic for a reader of this book to reproduce these methods and not all developers are easily forthcoming about details.

In contrast, approaches for statistical or chemometric processing have a long vintage and are likely to remain available in decades to come in a similar form to now, and all common methods are public domain. A textbook is designed to have a long lifetime, and the focus of this text is on commonly available chemometric methods. Many approaches and concepts used in current metabolomic statistical analysis were first formulated over 100 years ago, such as p values, ANOVA, distributions, least squares regression, PCA and so on. Another set of methods emerged around 50 years ago, such as PLS and SIMCA when there developed a more widespread need to interpret multivariate analytical data. A few are more recent such as ASCA. This text is an aid to modern-day research but not a theoretical text reviewing the latest chemometric methods proposed in the literature.

The choice of methods in this book is based on this author's perception of some of the most widespread in current practice, based on the literature, on talking to practicing colleagues and implemented in widespread packages. Of course there will be many other methods, and a comprehensive description of all chemometric approaches used in metabolomics would be a series of texts: unfortunately such a series of texts would take many years to compile and probably involve several authors, and with the rapid development of this field the first book would almost certainly be dated when the last one appears. Within a single text and a single author, one can only describe the most common, but the advantage is that there is uniformity in presentation, so the methods are described in similar depth using similar notation and datasets. Most users are only exposed to a few of the most widespread methods.

This book can be used at various levels. At the top level, it is a description of the use and basis of the most common chemometric methods, illustrated by case studies. Readers may discover methods they had not been aware of, or interpretations they had not previously appreciated. This book advocates a hypothesis-based approach, as most of metabolomics is hypothesis driven. At a deeper level, some will want to follow the calculations in the book. This will enable understanding of the methods and can enable comparison with in-house software. There is sufficient description of how the methods are implemented and sample output to allow readers to compare with their own calculations; if there are differences, this may lead to changes in software usage or catalyse additional interpretation of data. Sometimes there are several comparable methods, and users may want to look at their results from different angles.

Although this author performed all calculations with in-house software developed using MATLAB, there is no requirement to use this package to reproduce results. Some numerical output has been compared by myself and by colleagues using other approaches, notably R, Excel (with VBA), PLS Toolbox and SIMCA, to both ensure identical results (except for the sign in PCA) and check numerical accuracy. Readers will be using a wide variety of favoured software environments. This author has over many years performed chemometrics calculations in MATLAB, Fortran, SAS, BASIC, Excel (with VBA), C and PL/1 where appropriate and has co-operated with colleagues who have additionally used R, PLS Toolbox, SIMCA, UNSCRAMBLER, Pirouette, Sirius and Minitab. There are, of course, many other packages available suitable for the statistical analysis of metabolomics data. However, unlike the methods for resolution and characterisation of instrumental data, if correctly used, all these packages should come to identical answers. Most low-level programming languages allow all the steps in an algorithm to be coded in, and many high-level environments allow for scripting or macro commands to swiftly develop applications without complex programming. Some users will not want to do any programming and want more or less automated laboratory-based software, although usually some functionality is available by menu commands – approaches such as PCA and PLS should, if steps such as pre-processing are performed as described in this text, result in identical answers, providing there is sufficient flexibility in the software.

The 10 case studies in this book have been carefully chosen to span a range of applications. We have focussed on the main instrumental workhorses, namely NMR (3 case studies), LCMS (2 case studies) and GCMS (2 case studies). Raman (1 case study), FTIR (1 case study) and single wavelength HPLC (1 case study) are used by some investigators who view themselves as working in metabolomics; therefore to satisfy such readers, we have also included data from these sources. The case studies come from human (4), plant (3), animal (2) and microbial (1) experiments reflecting a range of applications, with data sources from Europe (6), Asia (2), North America (1) and Africa (1).

Some case studies are used more frequently to illustrate different methods than others, some being just illustrated by PCA, whereas others are mentioned in three or four chapters. There are excellent articles describing the full analysis of each case study referenced in the text, and it is not the aim of this book to copy the existing literature but to describe the main approaches and illustrate them where appropriate with one or more case studies using one or more steps in the analysis. The case studies can be downloaded from the companion Wiley website and are all in Excel format. The case studies are supplemented by a small number of simulations, with the larger simulations also available for download.

Although chemometric methods are widely recognised as essential to the analysis of metabolomic data and there are many texts on general chemometrics methods mainly aimed at analytical chemists, there is a lack of books focussed on their application to metabolomics. It is hoped that this text will be a useful reference.

In addition to a primary focus on metabolomics, this book will also be of interest to the general user of chemometrics in related fields, covering most of the common methods such as PCA, PLS, calibration, classification, experimental design and so on. It will also be of interest to the applied statistician interested in methods used in chemometrics. For these readers, the choice of and relative importance of methods discussed in the text are oriented towards metabolomics, and case studies are also related to data encountered in this field, but the applicability of the statistical approaches can easily be transferred to other fields.

Richard G. Brereton
Bristol, University of Bristol, UK

December 2023

Acknowledgments

In the course of preparing this book, over a three-year period, I have been helped by a large number of colleagues around the world.

Several have generously suggested datasets, some after significant discussions looking at which would be most suitable for the purposes of this book. It was necessary to span various applications and techniques so the sources have been likewise varied. Some have generously given their time for personal meetings where I discuss their protocols, and many have carefully commented on relevant sections. This has been particularly important to ensure that the book encompasses a range of different applications.

Several colleagues have checked my calculations by running data from this book through other packages, especially where the results obtained appeared unexpected, and to check the results of my programs in MATLAB agree with those from software in other environments, and that my explanations are sufficiently detailed.

Furthermore, many colleagues have read sections of the text often providing detailed corrections and comments. Some have expressed opinions and explanations for different approaches, often from different viewpoints to my own. Having a wide variety of opinions and experiences has helped shape this text.

A list of collaborators, who have shared knowledge, data and expertise is below.

Trygve	Andreassen	(Norwegian University of Science and Technology, Norway)
Paul	Benton	(Scripps, USA)
Elizabeth	Carter	(University of Sydney, Australia)
Olivier	Cloarec	(Sartorius, France)
Catherine	Deborde	(INRAE, France)
David	Duewer	(NIST, USA)
Oliver	Fiehn	(UC Davis, USA)
Roy	Goodacre	(University of Liverpool, UK)
Andris	Jankevics	(University of Birmingham, UK)
Olav M	Kvalheim	(University of Bergen, Norway)
Gavin	Lloyd	(University of Birmingham, UK)
Gary	Mallard	(NIST, USA)
Annick	Moing	(INRAE, France)
Tomas	Pluskal	(IOCB Prague, Czechia)
Alexey	Pomerantsev	(Federal Research Center for Chemical Physics, Russia)
Francesc	Puig	(INSERM, France)
Rich	Sleeman	(Mass Spec Analytical, UK)
Hans	Stenlund	(University of Umeå, Sweden)

Izabella	Surowiec	(Sartorius, Sweden)
Roma	Tauler	(IDAEA-CSIC, Barcelona, Spain)
Johan	Trygg	(University of Umeå, Sweden)
Yulan	Wang	(Nanyang Technological University, Singapore)
Yun	Xu	(University of Liverpool, UK)

Wiley are also thanked for their patience and encouragement throughout and Jenny Cossham for suggesting this text.

About the Companion Website

This book is accompanied by a companion website:

www.wiley.com/go/Brereton/ChemometricsforMetabolomics

This website includes:

- Case Study Data
- Simulated Data

The website hosts the datasets in this book as two downloadable Excel files:

1. The file "case study data" contains the data for all 10 case studies in this book.

2. The file "simulated data" contains data for the larger simulations in this book.

These datasets can be used freely for private study or for use in courses. If in courses, please reference the book. The datasets may also be used if required in publications or presentations. If using case study data, please cite both the book and the original source (as cited in Chapter 1), and for the simulated data, just the book.

The datasets can be used to reproduce numerical and graphical results from relevant chapters of the book or used for further exploration.

It is recommended to export the data to an external package for further processing.

CHAPTER 1

Introduction

The subject matter of this book is a synthesis between chemometrics and metabolomics, both relatively recent scientific disciplines. This chapter describes the background to these disciplines and then introduces the background to the case studies which are used to illustrate the chemometric methods and describes some software packages that can be used to obtain results described in this text.

1.1 CHEMOMETRICS

The name chemometrics was first proposed by Svante Wold in 1972 in the context of spline fitting [1]. Together with Bruce Kowalski, they founded the International Chemometrics Society and the term slowly took off in the 1970s. However, the pioneers did not widely use this term for some years, but a major event that catalysed it was a workshop in Cosenza, Italy, in 1983 [2] where many of the early pioneers met. After this time several initiatives took off, including the main niche journals, *Journal of Chemometrics* (Wiley) [3] and *Chemometrics and Intelligent Laboratory Systems* (Elsevier) [4], together with courses and the first textbooks [5, 6] with regular reviews and ACS (American Chemical Society) symposia starting a few years earlier [7].

However, these events primarily concern name recognition and organisation, and the main seeds for the subject were sown many years earlier.

Applied statistics was one of the main influences on chemometrics, although the two approaches have diverged in recent years. The modern framework for applied statistics was developed in the early 20th century and we still use terminology first defined during these decades. Before that, early academic statistics was mainly mathematical and theoretical, often linked to probability theory, game theory, statistical mechanics, distributions etc. and viewed as a subdiscipline of mathematics. Although many early pioneers had already used approaches previously that we would now regard as the

Data Analysis and Chemometrics for Metabolomics, First Edition. Richard G. Brereton.
© 2024 John Wiley & Sons Ltd. Published 2024 by John Wiley & Sons Ltd.
Companion website: www.wiley.com/go/Brereton/ChemometricsforMetabolomics

forerunners of modern applied statistics, their ideas were not well incorporated into mainstream thinking until the early 20th century.

A problem in the 19th century was partly the division of academic disciplines. Would a mathematician talk to a biologist? They worked in separate institutes and had separate libraries and training. For applied statistics to develop, less insular thinking was required. There also needed to be some level of non-academic contribution as many of the catalysts were at the time linked to industrial, agricultural and medical problems. With core academic disciplines, the application of statistical methods in physics and chemistry, which would eventually progress to quantum mechanics and statistical mechanics, fell outside mainstream applied statistics and has led to specialist statistically-based methods that are largely unrelated to chemometrics.

However, in the first three decades of the 20th century, there was a revolution in thinking. Such changes primarily involved formalising ideas that had been less well established over the previous decades and even centuries. Karl Pearson [8] and William Gossett publishing under the pseudonym 'Student' [9] are recognised as two of the early pioneers. Pearson set up the first statistics department in the world, based in London, and his 1900 paper first introduced the idea of a p value, although historic predecessors can be traced several centuries back [10–12].

It was not until after the First World War that applied statistical methods were properly formalised in their modern incarnation. Ronald Fisher was possibly the most important figure in developing a modern framework for statistical methodology that many people still use today. In 1925 he published *Statistical Methods for Research Workers* [13] and established the concepts of p values, significance tests and ANOVA (analysis of variance). Ten years later he wrote à book that described the basis of almost all statistical experimental designs [14] used even now, and his paper on classification of irises (the plants) [15] is an essential introduction to multivariate classification techniques, with this dataset used even now for demonstrating and comparing new approaches. Other important workers over that period, included Harold Hotelling, who among others was attributed with progressing the widespread use and recognition of PCA (principal components analysis) [16, 17] and Jerzy Neymar and Ergon Pearson who developed alternative approaches to hypothesis tests to those proposed by Fisher [18].

During the interwar period, many of the cornerstones of modern applied statistics were developed, and we continue to use methods first introduced during this era; many approaches used in chemometrics have a hundred-year vintage. However, there were some significant differences from modern practice. There was no capacity to perform intensive computations or generate large quantities of analytical data, so applications were more limited. Agriculture was at the forefront. During this era, the old land-owning classes had to modernise to survive: many farm labourers left for the cities and agriculture became more automated. The relationship between landowners and tenants weakened and larger farms were viewed more as an industry rather than the birthright of aristocratic classes. This required a significant change in production, and agricultural statistics was very important, especially to improve the economies of Western Nations. Other important driving forces came from the use of psychology to interpret test scores, and from economics. Common to all these types of data is that experiments involved considerable investment in time, so it was reasonable to spend substantial effort analysing the results, some required weeks of manual calculations, as data was expensive and precious. In modern days, spectra, in contrast, can be obtained relatively rapidly and quickly, so spending days or weeks performing statistical calculations would be an unbalanced use of resources.

Furthermore, without the aid of computers many of the multivariate methods we now take for granted would involve a large amount of time. Salsburg [19] claims as follows: 'To get some idea of the physical effort involved, consider Table VII that appears on page 123 of *Studies in Crop Variation. I.* [20]. If it took about one minute to complete a single large-digit multiplication, I estimate that Fisher needed about 185 hours of work to generate that table. There are fifteen tables of similar complexity and four large complicated graphs in the article. In terms of physical labor alone, it must have taken at least eight months of 12-hour days to prepare the tables for this article.' Of course, Fisher would have had many assistants to perform calculations, and he would have been very well resourced compared to most workers of the time. Hence, only quite limited statistical studies could be performed routinely. Some algorithms and designs such as Yates' algorithm [21] were developed with simplicity of calculation in mind as the data had special mathematical properties and although still reported in some textbooks even now are not so crucial to know about with the advent of modern computing power. Computers can invert large matrices very quickly, whereas a similar calculation might take days or longer using manual methods. In areas such as quantum chemistry, a calculation that may take up an entire PhD via manual calculations can now be done in seconds or less using modern computing.

The statistician of the first half of the 20th century would be armed with logarithm tables, calculators, slide rules and special types of graph paper, and in many cases would tackle less data-rich problems than nowadays. However, there was a gap between the mathematical literature where quite sophisticated methods could be described, often in intensely theoretical language, and the practical applications of much more limited and in most cases simpler approaches. Many of the more elaborate methods of those early days would not have much widespread practical use, but modern-day multivariate statistics can now take advantage of them. The chemometrician can routinely use methods that on very large spectroscopic or chromatographic datasets that were inconceivable prior to the widespread availability of modern computers.

In the post-war years, chemical manufacturing was of increased importance and multivariate methods were applied by industrial chemical engineers [22]. G.E.P. Box worked with a group in the chemical company ICI in the UK for some years, before moving to the US. His text [23] written together with two co-authors, is considered a classic in modern statistical thinking for applied scientists emphasising experimental design and regression modelling and brings the work of the early 20th century into the modern era.

In the 1970s, mainstream applied statistics started to diverge from chemometrics. In chemometrics, we often come across short fat datasets, where the number of variables may far exceed the number of samples. For example, we may record thousands of mass spectral or NMR or chromatographic data points for each of perhaps 20–100 samples. These sorts of problems were not conceivable to the original statistical pioneers, measurements were expensive, so variables were scarce. Fisher's classic iris data [15] consisted of 150 samples but only four variables. Once sample sizes are less than the number of variables, some classic approaches for multivariate data analysis are no longer directly applicable, an important one is the Mahalanobis distance [24] and the corresponding method of LDA (linear discriminant analysis) [15] which have to be adapted and cannot be directly applied to data whose variable to sample ratio exceeds 1. More recent methods such as PLS (partial least squares) [25, 26] and SIMCA (Soft Independent Modelling of Class Analogy) [27] were advocated in the 1970s and 1980s very much with the needs of chemometricians in mind and coped well with such data. Chemometricians are often very interested in variables, for example, which are most significant out of possibly hundreds of candidates, whereas statisticians emphasise the significance of factors, in many cases using univariate tests. Although both types of thinking may come to similar types of conclusions, some traditional statistical approaches are invalid, as an example if there are more variables to samples, we cannot use LDA to provide us information as to which variables are the most significant, whereas PLS might provide an answer. As such problems were inconceivable before the 1970s, the impetus to providing niche solutions for the chemometrician is of 50-year vintage. Metabolomics is often a rich source of information where the number of variables far exceeds the number of samples so chemometrics is needed for statistical interpretation.

Although there are still a few workers in the chemometrics field who identify themselves as statisticians, their influence became quite limited after the 1980s, judging by attendance at chemometrics meetings, development of texts and courses and so on. In a way, this divergence is similar to those in areas such as quantum mechanics or statistical mechanics, which are also founded on statistical principles but are primarily led by numerate scientists.

However, the strong statistical parentage is an important cornerstone of chemometrics. As many applications are moving away from the original ones of quantitative analytical and physical chemistry into more hypothesis-based science such as metabolomics, the original aims of measuring more precisely or predicting more accurately are gradually being supplemented by more statistical aims to generate and test hypotheses. In the latter case, chemometrics, whilst once routed in the physical sciences, is being adapted to tackle problems from those in biology, geology, psychology and so on, and requires a return to core statistical thinking. It is questioned for the future whether this will attract more applied statisticians back into the field, or whether niche chemometrics experts will return to the mainstream statistical literature and then incorporate more statistical thinking into their publications and software without the need to bring mainstream statisticians directly into their collaborations.

Another parent of chemometrics was quantitative chemistry. Historically, it is important to understand that interdisciplinary research was not very widespread during most of the formative years in the 20th century. Academia was highly compartmentalised, with students opting for specialist courses in, for example, chemistry, or mathematics or biology. Most university teachings except at the basic levels would have been by staff from a single department. Students might be introduced to statistical concepts at an early stage but if by mathematicians in a somewhat abstract manner. If carried forward in courses such as physical and analytical chemistry, statistical methods would have been strongly oriented towards univariate measurement and estimation or in specialist areas such as quantum chemistry.

At a research level, departments would have their own libraries and staff rooms. In many countries, to obtain academic positions staff would have to be very focused.

An analytical chemist might have to pass a committee for tenure or habilitation, which is narrowly defined in part because subsequent teaching duties will also be highly compartmentalised. Libraries, journals, books, staff rooms, even sports teams and social events would often be organised by departments. A chemist, if working in the right subdiscipline, could snatch some snippets of statistical thinking where needed, but it would mainly be developed independently within their own environment. An analytical chemist would be unlikely to visit a mathematics library or read a mainstream statistics journal, but would gather their statistical knowledge from analytical journals and books; the internet was not yet available so the ability to search for papers from a wide selection of journals was mainly restricted to visits to other departments' libraries. Analytical chemists would be introduced to niche statistics, for example, learning about precision and accuracy of measurements, but most would have limited exposure to other statistical texts except at a basic level.

Industrial cross-over with mainstream academics was also rather limited until the fourth quarter of the 20th century. Thus, the work, for example, G.E.P. Box and colleagues were doing in ICI in the 1950s would be unknown to chemists in many prestigious universities. Formal statistical design of experiments, so important for improved industrial productivity, would not be adopted by mainstream synthetic chemists in an academic environment until the last decade of the 20th century.

Hence, the development of statistical thinking within mainstream chemistry, such as multivariate analysis and statistical experimental design, developed in quite unique ways. A few brave workers did, however try to introduce statistical and computational ideas to the chemistry community. Statistical methods such as univariate linear regression, determining accuracy and precision of measurements and so on, have been part of the analytical and physical chemists' toolbox for more than a century. However, many techniques now recognised as part of chemometrics were shown to be recognised within the mainstream chemistry community. In 1949, Mandel draws together a number of statistical techniques, including design of experiments and ANOVA [28], in a paper which has been cited just seven times at time of writing (December 2023), an almost forgotten paper. W.J. Youden, a pioneering statistician, wrote 37 articles for *Industrial and Engineering Chemistry* between 1950 and 1957, many on experimental designs and very relevant to analytical chemists and chemical engineers, which in turn have been cited only 60 times in total. In 1952, he published a review in the journal *Analytical Chemistry* [29] citing 154 references, which has, in turn, only received seven citations. Yet a paper in the journal *Cancer* by the same author published in 1950 [30] has received over 7000 citations at the time of writing. It would be a value judgement as to whether these different publications contained more in-depth or original information, and the difference in reception would primarily relate to the difference in readership.

Hence, although analytical chemists were aware of traditional statistical methods, for example, how to fit a straight line or how to determine the 95% confidence in a mean, they were relatively uninterested at the time in the statistical revolution of Fisher and colleagues, and a traditional course in chemistry would not cover systematic experimental design, ANOVA, multilinear regression, multivariate pattern recognition, etc. It took until the 1970s before the importance of these approaches, well established 50 years ago in the mainstream statistical literature, started to slowly become recognised within the chemistry community. Chemometrics by name was fundamentally born within the chemical sciences, where the first applications of for example chromatography and spectroscopy for the analysis of complex mixtures were developed. As time has progressed, although the first uses of these instruments were within the analytical and physical community, their application to data-rich sciences such as in the biomedical sciences where large mixtures of chemical compounds have a metabolic significance, has led to a much wider applicability of these instrumental techniques, and so the concepts from chemometrics. The original applications of NMR and MS in chemistry were primarily for the structural elucidation of individual molecules and it took several decades before they became established for the studies of mixtures.

It was not until the 1970s that analytical chemists started widely recognising the potential of statistical principles for experimental design. Stan Deming, whose first paper was published in 1967, published a well-cited paper in analytical chemistry on simplex optimisation in 1973 [31] followed up by a more comprehensive statistically based and well-regarded book in 1987 [32].

A parallel development happened within the physical chemistry community. Spectroscopists studied problems involving overlapping peaks of mixtures. They probably did not have much access to the statistical literature at the time, but did read the physical literature for example about eigenanalysis, which is related to PCA. There were several papers in the 1960s of which two are cited [33, 34]. Physical/analytical chemists of the time developed their methods in isolation to statisticians and, due to the limited availability of computers, applied their approaches to what we now regard as quite simple problems.

The prolific pioneer Ed Malinowski put together many of the first approaches to multivariate resolution of mixtures in the 1970s with statistical and algorithmic descriptions, culminating in his classic book published in 1980 [35]. Many of the methods now recognised as multivariate curve resolution that have a high profile in chemometrics, emerge from Malinowski's early work. Unlike most chemometrics methods, for example for classification or exploratory data analysis or regression, factor analysis (as defined by Malinowski) or multivariate curve resolution have a very specific role in chemometrics, so play a unique role; even now many approaches for resolving peaks in coupled chromatography incorporated in elaborate software have their origin in concepts first advocated by Malinowski.

By the 1980s, ideas from quantitative analytical and physical chemistry had started to become formalised both in the area of multivariate analysis and experimental design, and were slowly recognised initially by a rather small group primarily of analytical chemists. In the 2000s, basic chemometric concepts were starting to be introduced in general analytical chemistry courses and books and rapidly developed into software that was essential for burgeoning applications of instrumental analysis such as metabolomics as discussed in Section 1.4.

The third catalyst was scientific computing. Many of the approaches described in the first half of the 20th century remained theoretical without the availability of good computer power, and prior to the 1970s, only really quite simple problems could be solved by chemometric methods.

In the 1950s, very few scientists and engineers had access to computers for daily calculations. Computers were large and expensive institutional machines often developed for defence (as the initiative originally came out of the Second World War). Individual researchers rarely had direct access, programs had to be sent in using punch cards or paper tape to trained operators, who would then feed the instructions to a mainframe and output would usually be in the form of printer paper sent back to the user. If there was a mistake in a program, this was very costly.

Early programs were in assembly language or machine code, which would be very hard for non-experts and take time to develop simple sets of instructions. In the 1950s, IBM developed a high-level language, called Fortran (Formula Translation) [36], which was the basis of most scientific software for the next two decades. The language continues to be developed with regular updates. Most classical scientific software was written using Fortran. The vast majority of quantum chemistry packages are still written in Fortran, and some build on subroutines over 50 years old. The 1960s was a particularly fruitful time for quantum mechanics computing. Prior to this era, it could take a graduate student almost an entire PhD just to calculate a few quantum mechanical integrals, so the use of scientific computers revolutionised this field.

However, access to mainframes capable of running large scientific programs was very limited until the late 1960s and early 1970s. Scientists who were not viewed as hard-core physical scientists had limited possibilities, and it took some years before access to significant scientific computer power was broadened.

In the very late 1960s, a few chemists did get access to significant computing power. Peter Jurs and his co-worker Bruce Kowalski were some of the first writing a series of papers in the analytical literature [37]. The impetus had been from Djerassi and coworkers who had pioneered the use of computers in structural elucidation [38] leading to the field of artificial intelligence. The Arthur program [39], developed for both mainframes and VAX minicomputers was one of the first widespread chemometrics packages but still one had to be very expert to install and use it.

In the 1980s, two important developments happened in computing that would move chemometrics from a specialist discipline to one that had the potential to be more widespread. The first and most important was the development of microcomputers in the late 1970s to early 1980s, the most successful scientific models of the type based on the original IBM PC [40]. Once micros became widespread and relatively user-friendly, chemometrics software could extend to a far wider user base and allow laboratory-based scientists rather than primarily specialists with access to expensive communal mainframes, to access chemometrics software.

Another important development of the time was MATLAB, originally developed by Cleve Moler [41] in 1981. Most chemometricians like to think in terms of matrices, and languages such as Fortran and BASIC were not naturally oriented towards matrix operations at the time. MATLAB, however, allows the programmer to develop code using matrix and vector algebra directly and contains fast algorithms for operations such as inverting large matrices. Operations such as PCA are also built-in. Since its early days, the software has been substantially expanded with GUIs (graphic user interfaces) and extensive graphics. MATLAB had a significant role in the development of chemometrics because workers could quickly develop matrix-oriented algorithms and often swapped code. Although many developing chemometrics methods now prefer R or Python, there still is an important group of MATLAB users, and this environment strongly catalysed the fundamental developments from the 1980s onwards.

The three catalysts, namely applied statistics, analytical/physical chemistry and scientific computing, converged in the 1980s to provide a fertile environment for the establishment of the discipline. Several events were responsible to create a specific launching pad for what is now well regarded as a coherent body of knowledge.

The NATO-sponsored meeting in Cosenza, Italy, in 1983 [2] brought together many of the experts working in this field at the time, not all would have regarded themselves as fundamental chemometricians until then. The journal *Analytical Chemistry* started publishing biennial fundamental reviews under the name 'chemometrics' as from 1980 [7] bringing together the last two years of papers in the field. Several regular workshops were established. The first comprehensive texts were published [5, 6] although some more specialist books had been published covering specific areas, but without the name 'chemometrics' in the title, in the years before. Two journals were established, published by Wiley [3] and Elsevier [4]. Several series of conferences were started at this period. These attracted mainly niche workers, most of whom had a back in computing within a chemistry environment and were not yet attracting biologists. There was a separate and very well-established area of biological statistics, mainly oriented towards univariate data.

The applications at this phase were fairly simple, with NIR (near infrared) calibration and HPLC (high-performance liquid chromatography) deconvolution predominating, and relatively small sample sizes. Some approaches such as PLS [26, 27] and SIMCA (self independent modelling of class analogy) [27] as well as Malinowski's extensive methods which he called factor analysis [35] were very oriented towards and widely reported within the chemometrics community rather than using a general statistics toolbox, although PLS (originating in economics) has found a home more generally since. The emphasis at the time was primarily to be able to measure and estimate accurately, often in areas such as pharmaceutical and food science, where the concentration of an ingredient or of a reactant had to be estimated by spectroscopic or chromatographic means accurately and quickly. Pattern recognition that has a high profile in modern chemometrics and metabolomics was not a very large part of the original literature. Many dedicated packages were developed over this period, most of which are still in existence now, these will be described in the section on software (1.4) below.

Hence, most of the tools now recognised as chemometrics were available in the 1980s. However, the early promise did not materialise at the time, unlike many other data-rich areas such as bioinformatics and QSAR (quantitative structure–activity relationship) which had a similar vintage. As from the mid-1990s, there was a tremendous interest in the application of chemometrics to fundamental analytical chemistry, for example the resolution of overlapping peaks in HPLC, which generated a large number of technically sophisticated but in most cases not particularly widespread papers and a very introspective number of groups and conferences. There was less scope for huge innovation compared to the 1970s when scientific computing was rapidly evolving and multivariate statistical methods were quite new to numerate chemists. In 2008, Paul Geladi and Phil Hopke ask 'Is there a future for chemometrics?' [42]. Many viewed the subject as rather a technical niche area without much general applicability. The number of specialist chemometrics groups in the world had hardly changed over the decades, just with a few new faces replacing those that had retired or left for other fields. It was not a particularly attractive area for researchers, without much funds, and appeared to flatten off.

In contrast, during the last 15 years or so, there has been a substantial renaissance. This is because chemometrics has moved out of its original comfort zone of quantitative analytical and physical chemistry and been embraced in areas such as metabolomics, among others. Big datasets are now readily available using instruments, such as LCMS (liquid chromatography mass spectrometry), NMR (nuclear magnetic resonance), GCMS (gas chromatography mass spectrometry) and so on. The capability of instruments to analyse many samples, each in turn containing large quantities of information, means a new and urgent need to correctly interpret these immense datasets and also to understand how to correctly design the experiments and perform the sampling. Thus, over more than a decade, there has been a renewed urgent need for chemometrics expertise. The sort of problems tackled nowadays often differs from the traditional analytical chemistry problems. The latter often involved measuring more accurately. For example, can a spectroscopic technique estimate the concentration of an analyte in a mixture without chromatography, and if so how well? We might create some reference standards or perform some independent and slower method of analysis, such as HPLC, and use multivariate analysis of a series of mixtures to create a calibration model. However, in many metabolomics problems, we are not certain of the answer in advance. For example, we may obtain the LCMS of donors' serum with and without a disease. How certain there is enough information in the serum to distinguish each group? It may depend on whether the disease was correctly diagnosed and how far it has progressed. There will be confounding factors such as age or diet or genetics – how representative were the samples? And then if we think we can separate groups, which metabolites are most likely to be markers for the disease? How confident are we? In such a situation we do not know the answer in advance and are generating and testing hypotheses. In fact, most of science outside the core physical sciences is primarily based on hypothesis testing. Much of early chemometrics was developed by programmers good at algorithms and matrices, whereas the needs of biological scientists are more hypothesis testing. This text aligns chemometrics methods primarily from the point of view of hypothesis formulation, which communicates more closely with the language of clinicians and biologists.

Chemometrics has had a renaissance because its methods are being applied to many scientific problems outside core quantitative chemistry. Nevertheless, most of the original methods, first pioneered over 50 years ago, are still relevant, and techniques such as molecular spectroscopy or chromatography, once the domain of chemists, are now widely used in many scientific fields. Understanding the fundamental statistical basis of chemometrics is an essential aid to safely and usefully employing these techniques, which have become widely available due to a plethora of software and datasets.

1.2 METABOLOMICS

The central dogma of biology is illustrated in Figure 1.1. In its simplest form DNA makes RNA, which makes proteins which make metabolites. The metabolic profile influences phenotype and so the characteristics of all organisms.

The study of systems biology is to connect these steps.

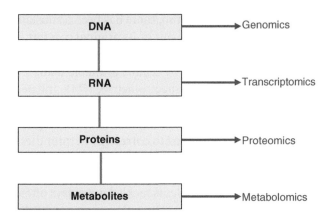

FIGURE 1.1   Central dogma of biology.

The concept of a genotype is the oldest and was first defined early in the 20th century by a number of papers by Johannsen [43]. The concept of a genome was first described by Winkler in 1920 [44]. There is in fact no universally agreed definition of these two terms, despite widespread usage; however, a common distinction is as follows. A genome is an organism's complete DNA, which includes all of its genes (coding/non-coding) and intergenic regions. The genotype refers to the genetic information for a particular trait. In humans, the genome, for example, includes the portion of human DNA that codes for hair colour and decides the genotype for that trait.

Genomics therefore studies the entirety of an organism's DNA. With improvements in high-throughput sequences, the first whole genomes were sequenced in 1980s and 1990s. The first complete genome sequence of a eukaryotic organelle, the human mitochondrion, was reported in 1981 [45] and the first chloroplast genomes followed in 1986 [46]. In 1992, the first eukaryotic chromosome, chromosome III of brewer's yeast *Saccharomyces cerevisiae,* was sequenced [47]. The first free-living organism to be sequenced was that of *Haemophilus influenzae* in 1995 [48]. It was however the human genome project that catalysed this scientific discipline with the complete sequencing announced in 2003 [49], although a small number of sequences still remained.

Whereas the concept of a genome was well developed, the name genomics and the concept of this as a discipline is rumoured to have been proposed in 1986, with a birthplace in the journal *Genomics* in 1987 [50]. This was the first recognised omics discipline and great-grandparent of metabolomics. With genomics arrived a large amount of data and this catalysed the arrival of computationally intense disciplines such as bioinformatics to mine these large datasets, placing computing at the centre of modern biological research.

Next up was transcriptomics, this time concerned with RNA. The earliest known use of the noun transcriptome is in the 1990s. The earliest known use of this term in the scientific literature is from 1997 [51], which was the first key work to report the transcriptome of an organism describing 60,633 transcripts expressed in *S. cerevisiae* using serial analysis of gene expression. With the rise of high-throughput technologies and bioinformatics and the subsequent increased computational power, it became increasingly efficient and easy to characterise and analyse large amounts of data.

Third up, but with an earlier start, is proteomics. The first studies of proteins that might now be regarded as proteomics began in 1975, after the introduction of the two-dimensional gel and mapping of the proteins from the bacterium *Escherichia coli*. However, the first formal mention of the idea of a proteome was in 1995 [52]. The concept of proteomics followed soon after [53]. There was a rapid increase in publications in this area in the late 1990s culminating in the journal *Proteomics*, founded in 2001 [54], which published 156 papers in its first year, an unusual volume for a new journal. Several other journals were founded and named with the word proteomics in the title subsequently.

With the growth in proteomics came a growth in instrumental methods to characterise proteins. Important to readers of this book is the development of mass spectrometric methods which are the most widespread currently used. Most studies now involve some form of LCMS/MS. There are a large number of approaches for the collection and interpretation of such data for proteomics, but the result is usually a table in which the identified proteins and their relative concentrations or intensities are listed. This table can then be subjected to statistical methods, in a very similar way to metabolomics, although the methods for dealing with the raw instrumental data can be quite complex. But there is a lot of similarity in the protocol for interpreting and analysing proteomic data as for metabolomic data, except that the raw datasets tend to be more complex. In proteomics, good instrumental analysis, data analysis, statistical interpretation and computing are integrated and have been from the early days.

Stepping into this stable is metabolomics. The chain DNA to RNA to protein has the main outcome of creating metabolites that result in so-called phenotypic expression, so metabolites are the end product of this sequence of events. For many years, metabolism was the province of biochemistry and bio-organic chemistry and somewhat divorced from mainstream biology. The original omics techniques were developed with nucleic acids and proteins in mind. Chemists focused on small molecules by isolating and synthesising individual metabolites rather than studying entire metabolic profiles.

With the advent of coupled chromatography; however, it now became possible to detect and characterise whole metabolic profiles by GCMS [55] and LCMS [56]. NMR was also demonstrated as a technique to analyse whole metabolic profiles [57] although it took some years to improve sensitivity to develop this as a widespread technique in metabolomic studies. However, with advances in instrumentation, all three techniques are now essential workhorses for metabolomics.

The first published use of the concept of a metabolome was in 1998 [58, 59]. The metabolome refers to the complete set of small molecule (<1.5 kDa) metabolites found within a biological sample. The first recorded use of the derivative term metabolomics in the formal scientific literature is from 2003 [60]. The journal *Metabolomics* was established in 2005 [61]. Many of the early pioneers from the 1990s are still in leading positions within the discipline, unlike genomics or proteomics where the leaders have mainly passed the torch onto future generations. A large number of metabolome databases have since been developed, of which the human metabolome database is the most prominent [62]. These are distinct from metabolite databases, which are much larger, but less specific, some of which are described in Chapter 2. The aim of the databases is primarily the identification and characterisation of metabolites, which then can be used for statistical analysis.

Public domain datasets consisting primarily of analytical LCMS, GCMS and NMR data have been uploaded in various formats to a number of large and well-maintained databases, such as the Metabolomics Workbench [63] and Metabolights [64]. These are quite distinct from the metabolite databases described above that characterise analytical data such as spectra of individual metabolites rather than whole profiles.

It is common the define the metabolomics workflow as illustrated in Figure 1.2. There are variants in the literature, but this term started gaining currency in the late 2000s and is now widely employed to illustrate the various steps in a typical analysis.

Steps 1 and 2 involve sample collection and preparation (usually by extraction). In some descriptions, these are combined into a single stage. The chemometrics expert is sometimes involved in step 1, using experimental design to choose experimental conditions or sampling designs to choose which samples to analyse. In large clinical sampling projects that may take several years; however, the chemometrics expert may have less initial input although he or she may be able to select samples for pattern

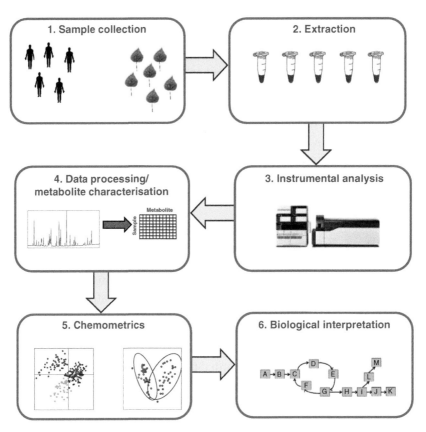

FIGURE 1.2   Metabolomics work flow.

recognition often several years after initial collection, from a larger set of samples as discussed in Chapter 6. In smaller laboratory-based studies, it is usually good practice that the expert who analyses the data also advises on the experimental design where feasible and is therefore involved in planning the work from the beginning, rather than after the data becomes available.

Step 3 is normally the experimental domain, often laboratory based involving analytical chemists. However, chemometrics experts can have a role, for example, in advising on instrumental stability, quality control, quantification, number of runs and so on.

Step 4 is sometimes combined either with step 3 and viewed as an essential part of processing raw spectroscopic or chromatographic data, or as part of step 5, the chemometrics. This book will not cover this stage in detail. As an example, there are an ever-increasing and evolving number of software-based approaches for processing and annotating (or identifying metabolites in) LCMS datasets; these are often very technical and bespoke sometimes with only some steps in the public domain and will change over the period of the lifetime of this text. A book that covered these would be as long as this text and would become rapidly out of date. However, the main principles and, where appropriate, webpages for some of the more widespread packages at time of writing, are described in Chapter 2. Some laboratories develop their own in-house approaches, others use free public domain packages, and others use commercial products often oriented towards specific instruments. The outcome is usually a matrix, often a peak table, which can then be used for subsequent statistical analysis. In some cases, raw spectroscopic data such as NMR frequency domain intensities can alternatively be used without initially identifying metabolites, so stage 4 is rather straightforward. However, most users of both coupled chromatography and NMR would prepare a peak table using a variety of software. In the case of NMR the columns of the data matrix might consist of the concentrations of identified metabolites corresponding to several spectroscopic peaks in different regions of the spectrum.

Step 5 involves chemometrics and statistical data analysis or pattern recognition. This key step is the main focus of this text, although we place it in context in Chapters 1–3. Most metabolomics data is very complex and often cannot be successfully interpreted without statistical and computational approaches and hence chemometrics is integrated as an essential step in almost all metabolomics studies. This book is restricted to describing statistical techniques although there is some trend towards machine learning. Although the latter is very powerful and takes advantage of the ready availability of modern computing, most metabolomics data, whereas it is substantially more complex compared to information obtained from single metabolite studies, its complexity is much lower than encountered in problems that require typical bioinformatics data mining, as example in human genomics. Hence, currently and probably for the foreseeable future, chemometric and statistical techniques are likely to be of the correct sophistication for metabolomic studies. The end product of this step often results in proposed biomarkers, which can be investigated further.

Step 6 involves biological interpretation, usually in the form of pathways, in favourable cases leading to clinical or toxicological explanations. Some metabolomic studies might stop at step 5, for example, just to identify potential biomarkers or classify samples. In this book, we will not extend our discussions to the interpretation of our results, which inevitably must be done on a case-by-case basis according to the individual aim of the experimentation.

Although the most widespread applications of metabolomics involve the use of NMR or coupled chromatography for biomarker exploration, not everyone who would describe their work as metabolomics uses these techniques, and it is anticipated that there will be some readers with other viewpoints. Single wavelength HPLC, Raman and FTIR can be used to study whole metabolic profiles either from extracts or intact. Some studies do not involve biomarker discovery but may involve just classification using metabolic profiles or calibration to a biological property. Whereas the focus of this book will be on biomarker discovery, we will also discuss other applications, which some will view as part of metabolomics. The overriding theme is studying whole metabolic profiles using spectroscopic or chromatographic techniques in order to relate these to biological properties.

The term metabonomics is sometimes used as an alternate [65]; in practice, there is no well-accepted difference with the term metabolomics and we will not distinguish between these in this book. Metabonomics is more commonly used in NMR in a clinical context but there is no exclusive use of this term. Lipidomics, is also another term some readers may have come across and can be considered a subfield of metabolomics [66]. It is the study of the lipidome, that is, the total lipid content within a cell, organ or biological system. Lipids are often simply defined as hydrophobic biological substances generally soluble in organic solvents. We will not distinguish these fields in this text, as similar statistical methods are required.

There have been several general articles about the application of chemometrics to metabolomics of which we cite some early well-known ones [67–71]. Review articles within specific application areas are widespread also. A recent edited book contains a number of articles by a variety of authors mainly related to chemometrics and metabolomics [72].

This book is distinct in that it provides in-depth numerical and graphical descriptions of many of the common chemometrics methods used in metabolomics, illustrated by applications.

1.3   CASE STUDIES

The methods described in this text are illustrated by a mixture of case studies and simulations.

The case studies are all public domain and the datasets are available on the publisher's website, to allow readers to reproduce results using their own favourite software. Sufficient details are available in this text to allow the reproduction of the numbers and graphs in this book, which are not tied to any specific software package.

The experimental case studies are described in greater detail in Chapter 3, so we only summarise in brief below, providing readers with references to descriptions of the source experiments. It is not the purpose of this book to reproduce a full step-by-step analysis of each dataset, as this is already available in excellent published articles, but to use datasets where appropriate to illustrate the application of different methods in appropriate chapters. Some datasets may be more useful for illustrating variable selection, classification performance, calibration or exploratory methods. Some datasets are only used a few times in this book, others are used in several chapters. We do not always employ the same approaches as in the original papers, and it is not the purpose of this book to copy existing literature, and for reasons of brevity, this book only describes some of the most common approaches that in this author's opinion are widespread. However, some readers will be interested in the full chemometric analysis and biological interpretation, and we list source papers below for readers wishing a more comprehensive background to the data.

Case Study 1: Presymptomatic study of humans with rheumatoid arthritis using blood plasma and LCMS was analysed in Umeå, Sweden, from samples obtained from the Biobank of Northern Sweden [73].

Case Study 2: Diagnosis of malaria in human blood plasma of children using GCMS, was analysed in Umeå, Sweden, from samples obtained in Rwanda [74].

Case Study 3: Measurement of Triglycerides in children's blood serum using NMR was obtained from a large study of school children in Norway, and the data were analysed in the University of Bergen, Norway [75].

Case Study 4: Glucose intolerance and diabetes in humans as assessed by blood serum using NMR involved clinical data collected from Beijing, China, and analysed in Wuhan, China [76].

Case Study 5: Metabolic changes in maize due to cold as assessed by NMR involved plant growth experiments performed and analysed in INRAE, France's National Research Institute for Agriculture, Food and Environment [77].

Case Study 6: Effect of nitrates on different parts of wheat leaves as analysed by FTIR was supplied by the University of Manchester, UK [78].

Case Study 7: Rapid discrimination of Enterococcal Bacteria in faecal isolates by Raman Spectroscopy was analysed in the University of Manchester, UK, with samples obtained from Belfast, UK [79].

Case Study 8: Effects of salinity, temperature and hypoxia on *Daphnia magna* metabolism as studied by GCMS, involving environmental toxicology, was obtained and analysed in IDAEA-CSIC, Barcelona, Spain [80].

Case Study 9: Bioactivity in a Chinese herbal medicine studied using HPLC was obtained in Hong Kong, in collaboration with the University of Bergen, Norway [81].

Case Study 10: Diabetes in mice studied by LCMS was obtained from UC Davis, California, USA [82], and is also described in case study ST000075 in the Metabolomics Workbench [63].

1.4   SOFTWARE

Crucial to chemometrics or all forms of statistical data analysis is software.

There are a large number of approaches, and which is chosen will depend on the background and expertise of the user, also how much time they have to develop code or analyse data, how important user friendliness and good graphics are, and of course cost and technical support. Some packages are also well supported via dedicated courses and books. Often, different people have very strong views as to which package or environment is most suited. It may depend on whether you are in a large organisation with limited time to learn the basics of software development, whether you have a background in programming, whether you are intending to develop your own methods or just trying to learn the basics of different approaches before you use them.

This author wrote all programs for the methods described in this book, created all graphs and performed all calculations, using MATLAB, a large suite of software maintained by MathWorks [83]. It was originally invented by Cleve Moler in the 1960s [41] but was first released in a compiled version in the 1980s. Many early chemometricians used this as it is a very matrix-oriented environment. Over time the basic package has been substantially enhanced especially with very sophisticated graphics and very many toolboxes. The company runs many courses and there are a large number of MATLAB-based books available. There are several chemometrics-based software libraries that can be downloaded for free. It became very popular in the early days of the subject as its initial expansion took place at about the same time as the first events and organisation of chemometrics, so was a useful way chemometricians could talk together and exchange code and data, and was a matrix-oriented language with an extensive scripting language. Today there are around four million users, many in engineering. Scripts can be compiled so users without access to the package can use software written in MATLAB.

There are of course many other approaches.

Fortran is the oldest scientific programming language, first developed by IBM in the 1950s [36]. Not many people use Fortran in the chemometrics community currently, but it is considered a much lower level programming environment than MATLAB, in that there is much more flexibility in accessing the computer and it can be compiled into fast code. There are numerous numerical and graphical libraries, and matrix operations are easy. For development of fast code, it is preferred but it is more difficult to program than MATLAB. Since its original release in 1954, Fortran has undergone numerous updates to become a modern-day widespread numerical programming environment and still remains a well established programming language, and it is entirely possible to program in the methods from the text using Fortran. Some higher-level packages have been originally developed using Fortran so is still a choice for developing packages or applications in chemometrics and very good for numerical computations.

BASIC (Beginners' All-purpose Symbolic Instruction Code) was originally developed in the 1960s as an instructional language for college students [84], but was more widely used and refined in the next few decades, especially for use on microcomputers which started to emerge in the 1980s; Fortran was at the time more oriented towards large mainframes. Early versions of the SIMCA package (see below) running on micros were written in BASIC. As time moved on there were an increasing number of modifications for use in the early micros, and VBA (Visual Basic for Applications) [85] was developed by Microsoft with a particularly popular application to produce Add-ins for Excel: VBA was first launched with MS Excel 5.0 in 1993. There are several add-ins to perform common chemometrics methods for chemometrics in Excel [86, 87], which can be used for many basic procedures such as PLS and PCA. These together with

Excel's data analysis tools such as ANOVA or regression and numerous downloadable tools means that most chemometrics calculations can be performed in Excel although graphics are limited. BASIC though is an interpreted language and so slow for larger calculations, but can be sped up using DLLs (Dynamic Link Libraries) for functions often using compiled code in C (see below) and Visual Basic still remains a popular programming language.

C and C++ can be used for chemometrics but do require a good knowledge of programming and how computers work. C was originally developed at Bell Labs 1972 and 1973 to construct utilities running on Unix [88]. The likelihood many readers of this book will use C is quite low unless they are involved in developing a package from scratch, although it could be useful for compiled DLLs to link to VBA as above. C++ is particularly flexible and often used for programming operating systems but requires a good grasp of object-oriented programming.

At a higher level, the R programming language [89] is gaining substantial popularity within the chemometrics and statistics community, and has a comparable chemometrics user base to MATLAB. It is a free software originating from the University of Aukland, New Zealand, based on the precursor statistical language S. It has been widely developed subsequently, with thousands of contributed routines. The Comprehensive R Network (CRAN) [90] at time of writing consists of around 20,000 contributed routines and is the largest depository. The R journal published by the R foundation [91] was founded in 2009. A great deal of chemometrics and multivariate statistical software has now been implemented in this environment and is freely downloadable. Users with a background in bioinformatics are more likely to favour R than MATLAB, although users from an engineering background are more likely to prefer MATLAB. Despite R's substantial popularity in the statistical community, a survey of popularity of programming languages at the time of writing [92] places R lower than some of the other languages described elsewhere (Python, MATLAB, Visual Basic, FORTRAN) but free availability and a large depository of statistical routines makes R an increasingly widespread environment for the type of analyses discussed in this book, and worth learning for newcomers.

Python is another high-level programming environment that is freely available, created first in the late 1980s [93]. Interestingly, at the time of writing, Python, which is freely available, was listed as the most popular general programming language beating even C [92]. It has undergone a large number of releases, with version 3.8 released in October 2019 being the oldest still supported. There is some chemometrics software available in Python, which is open source and free to download [94].

There will be many readers of this book that do not wish to program methods. Some may be satisfied for example with Excel Add-ins or able to use MATLAB or R packages with limited programming knowledge, but others will want all-inclusive packages. In any organisation there may be a variety of users all with different computer skills and experiences. Some of the developers of packaged software produce excellent courses on how to handle data using software that has been developed for many years or decades and is oriented towards the laboratory-based user without much time to learn to program. This of course depends on how a unit is structured, some may have data analysis specialists who can develop their own procedures, but others will have laboratory-based scientists who want to immediately analyse the results of their data as it becomes available without much programming but perhaps after attending one or more hands-on courses. Most of the methods in this book are possible to reproduce using niche packaged software, although aspects such as data pre-processing, variable selection and so on might be under default control of the software, and some software is more flexible than others. The less flexible the software, in general, the more automated but the more the decisions have been made for you in advance. The more flexible, the more you can control the outcome but may have to understand some of the procedures in processing the data. In all the environments described above the numerical results in this book should be obtained if the various details are followed carefully, although graphical output may differ. For the packaged software described below, in some cases especially using default methods, it may not always be possible to obtain the same results as in this book; however, it should be possible to understand why there are differences: in some cases, this can be corrected by changing parameters in the processing steps or by macro languages/scripting extensions to the basic package where available. Being able to compare numerical results between packages is a very important step in the understanding and safe use of statistical approaches for data analysis. In this text sufficient numerical and graphical output is presented to allow users to check against their own software and if necessary develop modifications for themselves where there are differences. If in doubt about a method it is recommended to use more than one package and compare results.

PLS Toolbox [95] was developed within the MATLAB environment. It was first released in 1989 at a time chemometrics was becoming established, with a demand for MATLAB-based scripts. Eigenvector Research, based in Seattle, which distributes the software, and supports this by courses and consultancy, was founded in 1995. A version called Solo which does not use the MATLAB command line but consists of compiled MATLAB code with a GUI is also available. This software is continually evolving and contains a wide range of chemometrics methods; it also allows users to flexibly develop their own approaches without programming from scratch. It would be suitable for both users without much or any programming experience or for users who are more comfortable doing some programming and want flexibility via macro commands. Eigenvector provides extensive training and consultancy based on their software.

The SIMCA package (to be distinguished from the method) was released and first developed at about the same time [96], based on the group in Umeå, Sweden. It was first released by Umetri, and at time of writing is now marketed via Sartorius [97]. It is a well-known software, based very much on the philosophy of Svante Wold and his colleagues in the early days of the subject. Unlike PLS Toolbox, it is very much a packaged software, allowing less flexibility, but very much oriented towards the laboratory-based user without requiring much software expertise. Many of the approaches such as SIMCA, OPLS and Hotelling's T^2 etc are embedded into the software, which will guide the user along a certain pathway of data analysis, and many of the original developers were leaders in the early days who developed protocols for analysis of multivariate analytical data. There are extensive books and courses associated with using this package.

The Arthur program [37], written in Fortran developed for both mainframes and VAX minicomputers, from the University of Washington in Seattle, was one of the first widespread chemometrics packages but still one had to be very expert to install it. From this seed emerged the company Infometrix in 1978 [98] which was probably the first software-based company primarily focussed on chemometrics, who developed the Pirouette package which continues to be updated and available today.

The UNSCRAMBLER software was originally developed in 1986 by Harald Martens from Norway and later by CAMO Software [99]. It was released at about the same time as SIMCA from Sweden. The late 1980s were a pioneering time for chemometrics software, with the change from mainframes and minicomputers to micros as the workhorse, and the first organised literature, meetings and courses in chemometrics, reflecting the need of the time. This software is still available and has been updated over time.

The Sirius package was also developed at around this time [100] from Bergen, Norway. It is distributed from Pattern Recognition Systems (PRS) and has facilities for many multivariate approaches of interest to readers of this book, including PLSDA, SIMCA and many variable selection approaches.

Some users of metabolomics software will have a prior background in applied statistics. Although many will be able to use R, there is a large community that prefers packages originally developed primarily for statisticians.

SAS (Statistical Analysis System) is the largest and most broadly based package oriented towards statisticians. It originated at North Carolina State University from 1966 until 1976, after which SAS Institute was incorporated [101]. SAS is the largest privately owned software company in the world. It has a very extensive suite of statistical and graphical software. SAS software includes a Base SAS component that performs analytical functions and more than 200 other modules that add graphics, spreadsheets or other features. Even some statistical concepts like Type I, Type II and Type III tests in ANOVA were first defined by SAS institute which has had a strong influence on applied statistics for many decades. The software is very comprehensive in areas such as ANOVA and regression. Most methods described in this book can be performed using SAS. There are extensive books, highly informative manuals, dedicated conferences and courses. There is a strong user base and certification [102]. However, the software is regarded as harder to use than packaged software such as SIMCA, PLS Toolbox, UNSCRAMBLER, Pirouette or Sirius, and not oriented towards the hands-on laboratory-based user, but the statistically trained modeller; some general programming skills may be desired, but this is not a low-level programming language. For the applied statistician moving into chemometrics though, most of the tools will be available in SAS, although it is not very commonly used in this area, probably due to the background of most users. SAS has a scripting language that allows users to develop their own approaches in addition to the add-on packages.

SPSS (Statistical Package for the Social Sciences) is one of the earliest general statistical packages, now maintained by IBM [103]. It was first released in 1968 for batch processing on mainframes, but in the 1980s was also implemented for micros. Although originally developed for social sciences, its applicability is wider and there are many multivariate procedures available. Scripts can also be developed using a macro language. Some biologists use SPSS for pattern recognition. It is regarded by many as somewhat easier to use compared to SAS, so would be suited as an entry-level statistical package for those without much statistical training, but is less comprehensive.

Although there are several other statistical packages readily available, the use of niche statistical software is not very common in metabolomics, so we will not produce a comprehensive list of statistical software and list only two of the most common and widespread above. Nevertheless, if using scripts or a macro language these statistical packages are likely to be suitable for most of the methods described in this book, and for readers working in large institutes such as universities it is quite common for there to be institutional licences and often in-house support for general statistical software.

REFERENCES

1. S. Wold (1972). Spline functions, a new tool in data-analysis. *Kemisk Tidskrift*, 3, 34–37.
2. B.R. Kowalski (ed) (1984). *Chemometrics: Mathematics and Statistics in Chemistry*, Reidel, Dordrecht.
3. *Journal of Chemometrics* (1987). Wiley, Chichester.
4. *Chemometrics and Intelligent Laboratory Systems* (1986). Elsevier, Amsterdam.
5. M.A. Sharaf, D.L. Illman and B.R. Kowalski (1986). *Chemometrics*, Wiley, New York.
6. D.L. Massart, B.G.M. Vandeginste, S.N. Deming, Y. Michotte and L. Kaufman (1988). *Chemometrics: A Textbook*, Elsevier, Amsterdam.
7. B.R. Kowalski (1980). Chemometrics. *Analytical Chemistry*, 52, R112–R122.
8. K. Pearson (1900). On the criterion that a given system of deviations from the probable in the case of a correlated system of variables is such that it can be reasonably supposed to have arisen from random sampling. *The London, Edinburgh and Dublin Philosophical Magazine and Journal of Science 5* (50), 157–175.
9. Student (aka W S Gossett) (1908). The probable error of a mean. *Biometrika*, 6, 1–25.
10. J. Arbuthnot (1710). An argument for Divine Providence, taken from the constant regularity observed in the births of both sexes. *Philosophical Transactions of the Royal Society of London*, 27, 186–190.
11. D. Bernoulli (1735). Recherches physiques et astronomiques. *Pieces qui ont Remporte le Prix Double de l'Academie Royale des Sciences en,* 1734, 93–122.
12. P. Laplace (1778). Mémoire sur les probabilités. *Mémoires de l'Académie Royale des Sciences de Paris*, 9, 227–332.
13. R.A. Fisher (1925). *Statistical Methods for Research Workers*, Oliver and Boyd, Edinburgh.
14. R.A. Fisher (1935). *The Design of Experiments*, Oliver and Boyd, New York.
15. R.A. Fisher (1936). The use of multiple measurements in taxonomic problems. *Annals of Eugenics*, 7, 179–188.
16. H. Hotelling (1933). Analysis of a complex of statistical variables into principal components. *Journal of Education & Psychology,* 24 (417–441), 498–520.
17. H. Hotelling (1936). Simplified calculation of principal components. *Psychometrika*, 1, 27–35.
18. J. Neyman and E.S. Pearson (1928). On the use and interpretation of certain test criteria for purposes of statistical inference: part I. *Biometrika*, 20A, 175–240.
19. D. Salsburg (2001). *The Lady Tasting Tea*, W. H. Freeman and Co., New York.

20. R.A. Fisher (1921). Studies in crop variation. I. An examination of the yield of dressed grain from Broadbalk. *The Journal of Agricultural Science,* 11, 107–135.

21. F. Yates (1937). The design and analysis of factorial experiments, Technical Communication of the Commonwealth Bureau of Soils 35 *Commonwealth Agricultural Bureaux,* Farnham Royal.

22. O.L. Davies (ed) (1956). *Statistical Methods in Research and Production,* Longman, London.

23. G.E.P. Box, W.G. Hunter and J.S. Hunter (1978). *Statistics for Experimenters,* Wiley, New York.

24. P.C. Mahalanobis (1936). On the generalised distance in statistics. *Proceedings of the National Institute of Sciences of India,* 2, 49–55.

25. S. Wold, M. Sjöström and L. Eriksson (2001). PLS-regression: a basic tool of chemometrics. *Chemometrics and Intelligent Laboratory Systems,* 58, 109–130.

26. P. Geladi and B.R. Kowalski (1986). Partial least squares: a tutorial. *Analytica Chimica Acta,* 185, 1–19

27. S. Wold (1976). Pattern-recognition by means of disjoint principal components models. *Pattern Recognition,* 8, 127–139.

28. J. Mandel (1949). Statistical methods in analytical chemistry. *Journal of Chemical Education,* 26, 534–539

29. H.J. Hader and W.J. Youden (1952). Experimental statistics. *Analytical Chemistry,* 24, 120–124.

30. W.J. Youden (1950). Index for rating diagnostic tests. *Cancer,* 3, 32–25

31. S.N. Deming and S.L. Morgan (1973). Simplex optimization of variables in analytical-chemistry. *Analytical Chemistry,* 45, A278–279.

32. S.N. Deming and S.L. Morgan (1987) *Experimental Design: A Chemometric Approach,* Elsevier, Amsterdam.

33. R.M. Wallace and S.M. Katz (1964). A method for determination of rank in analysis of absorption spectra of multicomponent systems. *The Journal of Physical Chemistry,* 68, 3890–3892.

34. D. Katakis (1965). Matrix rank analysis of spectral data. *Analytical Chemistry,* 37, 876–878.

35. E.R. Malinowski and D.G. Howery (1980). *Factor Analysis in Chemistry,* Wiley, New York.

36. J.W. Backus, H. Herrick and I. Ziller, (1954). Preliminary Report: Specifications for the IBM Mathematical FORmula TRANSlating System, FORTRAN. Programming Research Group, Applied Science Division, *International Business Machines Corporation.*

37. B.R. Kowalski, P.C. Jurs, T.L. Isenhour and C.N. Reilly (1969). An investigation of combined patterns from diverse analytical data using computerized learning machines. *Analytical Chemistry,* 41, 1949–1953

38. R.K. Lindsay, B.G. Buchanan, E.A. Feigenbaum and J. Lederberg (1980). *Applications of Artificial Intelligence for Organic Chemistry: The DENDRAL Project,* McGraw-Hill, New York.

39. A.M. Harper, D.L. Duewer, B.R. Kowalski and J.L. Fasching (1977). ARTHUR an experimental data analysis: the heuristic use of a polyalgorithm, in B.R. Kowalski

(ed), *Chemometrics: Theory and Applications, ACS Symp. Ser. 52*, American Chemical Society, Washington, DC.

40. A. Pollock (1981). Big IBM's Little Computer. *New York Times*, Section D, page 1.

41. C. Moler (1981). *MATLAB Users' Guide*, University of New Mexico, Albuquerque, New Mexico.

42. P. Geladi and P.K. Hopke (2008). Is there a future for chemometrics? Are we still needed?. *Journal of Chemometrics*, 22, 289–290.

43. W. Johannsen (1911). The genotype conception of heredity. *American Naturalist*, 45, 129–159.

44. H.L. Winkler (1920). *Verbreitung und Ursache der Parthenogenesis im Pflanzen- und Tierreiche*, Verlag Fischer, Jena.

45. S. Anderson, A.T. Bankier, B.G. Barrell, M.H. de Bruijn, A.R. Coulson, J. Drouin, et al. (1981). Sequence and organization of the human mitochondrial genome. *Nature*, 290 (5806), 457–465.

46. K. Ohyama, H. Fukuzawa, T. Kohchi, H. Shirai, T. Sano, S. Sano, et al. (1986). Chloroplast gene organization deduced from complete sequence of liverwort Marchantia polymorpha chloroplast DNA. *Nature*, 322 (6079), 572–574.

47. S.G. Oliver, O.J. van der Aart, M.L. Agostoni-Carbone, M. Aigle, L. Alberghina, D. Alexandraki, et al. (1992). The complete DNA sequence of yeast chromosome III. *Nature*, 357 (6373), 38–46.

48. R.D. Fleischmann, M.D. Adams, O. White, R.A. Clayton, E.F. Kirkness, A.R. Kerlavage, et al. (1995). Whole-genome random sequencing and assembly of Haemophilus influenzae Rd. *Science*, 269 (5223), 496–512.

49. I. Noble (2003). Human genome finally complete, http://news.bbc.co.uk/1/hi/sci/tech/2940601.stm.

50. V.A. McKusick and F.H. Ruddle (1987). A new discipline, a new name, a new journal. *Genomics*, 1, 1–2.

51. V.E. Velculescu, L. Zhang, W. Zhou, J. Vogelstein, M.A. Basrai, D.E. Bassett, et al. (1997). Characterization of the yeast transcriptome. *Cell*, 88, 243–251.

52. V.C. Wasinger, S.J. Cordwell, A. Cerpa-Poljak, J.X. Yan, A.A. Gooley, M.R. Wilkins, et al. (1995). Progress with gene-product mapping of the Mollicutes: *Mycoplasma genitalium*. *Electrophoresis*, 16, 1090–1094.

53. P. James (1997). Protein identification in the post-genome era: the rapid rise of proteomics. *Quarterly Reviews of Biophysics*, 30, 279–331.

54. *Proteomics* (2001). Wiley, Chichester.

55. S.C. Gates and C.C. Sweeley (1978). Quantitative metabolic profiling based on gas chromatography. *Clinical Chemistry*, 24, 1663–1673.

56. B.F. Cravatt, O. Prospero-Garcia, G. Siuzdak, N.B. Gilula, S.J. Henriksen, D.L. Boger, et al. (1995). Chemical characterization of a family of brain lipids that induce sleep. *Science*, 268, 1506–1509.

57. D.I. Hoult, S.J. Busby, D.G. Gadian, G.K. Radda, R.E. Richards and P.J. Seeley (1974). Observation of tissue metabolites using ^{31}P nuclear magnetic resonance. *Nature*, 252, 285–287.

58. S.G. Oliver, M.K. Winson, D.B. Kell and F. Baganz (1998). Systematic functional analysis of the yeast genome. *Trends in Biotechnology*, 16, 373–378.

59. H. Tweeddale, L. Notley-McRobb and T. Ferenci (1998). Effect of slow growth on metabolism of *Escherichia coli*, as revealed by global metabolite pool ("Metabolome") analysis. *Journal of Bacteriology*, 180, 5109–5116.

60. M. Satake, B. Dmochowska, Y. Nishikawa, J. Madaj, J. Xue, Z.W. Guo et al. (2003). Vitamin C metabolomic mapping in the lens with 6-deoxy-6-fluoro-ascorbic acid and high-resolution ^{19}F-NMR spectroscopy. *Investigative Ophthalmology and Visual Science*, 44, 2047–2058

61. *Metabolomics* (2001). Springer Nature, Heidelberg.

62. D.S. Wishart, D. Tzur, C. Knox, R. Eisner, A.C. Guo, N. Young, et al. (2007). HMDB: the Human Metabolome Database. *Nucleic Acids Research*, 35 (Database issue), D521–D526.

63. Metabolomics Workbench, https://www.metabolomicsworkbench.org/.

64. Metabolights, https://www.ebi.ac.uk/metabolights/.

65. J.K. Nicholson, L.C. Lindon and E. Holmes (1999). 'Metabonomics': understanding the metabolic responses of living systems to pathophysiological stimuli via multivariate statistical analysis of biological NMR spectroscopic data. *Xenobiotica*, 29, 1181–1189.

66. M.R. Wenk (2005). The emerging field of lipidomics. *Nature Reviews Drug Discovery*, 4, 594–610.

67. J. van der Greef and A.K Smilde (2005). Symbiosis of chemometrics and metabolomics: past, present, and future. *Journal of Chemometrics*, 19, 376–386.

68. J. Trygg, E. Holmes and T. Lundstedt (2007). Chemometrics in metabonomics. *Journal of Proteome Research*, 6, 469–479.

69. D.S. Waterman, F.W. Bonner and J.C. Lindon (2009). Spectroscopic and statistical methods in metabonomics. *Bioanalysis*, 1, 1559–1578.

70. J. Trygg, J. Gullberg, A.I. Johansson, P. Jonsson and T. Moritz (2006). Chemometrics in metabolomics — an introduction. In: K. Saito, R.A. Dixon, L. Willmitzer (ed) *Plant Metabolomics. Biotechnology in Agriculture and Forestry*, vol. 57, pp 117–128 Springer, Berlin, Heidelberg.

71. R. Madsen, T. Lundstedt and J. Trygg (2010). Chemometrics in metabolomics—a review in human disease diagnosis. *Analytica Chimica Acta*, 659, 23–33.

72. J. Jaumot, C. Bedia and R. Tauler (ed) (2018). *Data analysis for Omic sciences: methods and applications*, Comprehensive Analytical Chemistry vol. 82, Elsevier, Amsterdam.

73. I. Surowiec, L. Ärlestig, S. Rantapää-Dahlqvist and J. Trygg (2016). Metabolite and lipid profiling of biobank plasma samples collected prior to onset of Rheumatoid arthritis. *PLoS One*, 11, e0164196.

74. I. Surowiec, J. Orikiiriza, E. Karlsson, M. Nelson, M. Bonde, P. Kyamanwa, et al. (2015). Metabolic signature profiling as a diagnostic and prognostic tool in pediatric *Plasmodium falciparum* malaria. *Open Forum Infectious Diseases*, 2, 2.

75. E. Aadland, O.M. Kvalheim, S.A. Anderssen, G.K. Resaland and L.B. Andersen (2018). The multivariate physical activity signature associated with metabolic health in children. *International Journal of Behavioral Nutrition and Physical Activity*, 15, 77.

76. X.Y. Zhang, Y.L. Wang, F.H. Hao, X.H. Zhou, X.Y. Han, H.R. Tang, et al. (2009). Human serum metabonomic analysis reveals progression axes for glucose intolerance and insulin resistance statuses. *Journal of Proteome Research*, 8, 5188–5195.

77. M. Urrutia, M. Blein-Nicolas, S. Prigent, S. Bernillon, C. Deborde, T. Balliau, et al. (2021). Maize metabolome and proteome responses to controlled cold stress partly mimic early-sowing effects in the field and differ from those of Arabidopsis. *Plant, Cell and Environment*, 44, 1504–1521.

78. J.W. Allwood, S. Chandra, Y. Xu, W.B. Dunn, E. Correa, L. Hopkins, et al. (2015). Profiling of spatial metabolite distributions in wheat leaves under normal and nitrate limiting conditions. *Phytochemistry*, 115, 99–111.

79. N. AlMasoud, Y. Xu, D.I. Ellis, P. Rooney, J.F. Turton and R. Goodacre (2016). *Analytical Methods*, 8, 7603–7613.

80. E. Garreta-Lara, B. Campos, C. Barata, S. Lacorte and R. Tauler (2018). Combined effects of salinity, temperature and hypoxia on *Daphnia magna* metabolism. *Science of the Total Environment*, 610, 602–612.

81. F.T. Chau, H.Y. Chan, C.Y. Cheung, C.J. Xu, Y. Liang and O.M. Kvalheim (2009). Recipe for uncovering the bioactive components in herbal medicine *Analytical Chemistry*, 81, 7217–7225.

82. J. Fahrmann, D. Grapov, J. Yang, B. Hammock, O. Fiehn, G.I. Bell, et al. (2015). Systemic alterations in the metabolome of diabetic NOD mice delineate increased oxidative stress accompanied by reduced inflammation and hypertriglyceremia. *The American Journal of Physiology – Endocrinology and Metabolism*, 308, E978–E989.

83. Mathworks, www.mathworks.com.

84. J.G. Kemeny, T.E. Kurtz and E. Thomas (1964). *Basic: A Manual for BASIC, the Elementary Algebraic Language Designed for use with the Dartmouth Time Sharing System*, Dartmouth College Computation Center, Hanover, N.H.

85. *Getting Started with VBA in Office,* https://learn.microsoft.com/en-us/office/vba/library-reference/concepts/getting-started-with-vba-in-office.

86. R.G. Brereton (2003). *Chemometrics: Data Analysis for the Laboratory and Chemical Plant*, Wiley, Chichester.

87. A.L. Pomerantsev (2014). *Chemometrics in Excel*, Wiley, Chichester.

88. D.M. Ritchie (1993). The development of the C language. *ACM SIGPLAN Notices*, 28, 201–208.

89. F.M. Giorgi, C. Ceraolo and D. Mercatelli (2022). The R language: an engine for bioinformatics and data science. *Life*, 12, 648.

90. *The Comprehensive R Archive Network*, https://cran.r-project.org/.

91. V. Carey (2009). Special section: the Future of R. *R Journal*, 1, 3.

92. *TIOBE Index*, https://www.tiobe.com/tiobe-index/.

93. *General Python FAQ*, https://docs.python.org/3/faq/general.html#why-was-python-created-in-the-first-place.

94. https://pypi.org/project/chemometrics/.

95. *PLS_Toolbox*, https://eigenvector.com/software/pls-toolbox/.

96. W. Dunn (1987). SIMCA-3B, a pattern-recognition program. *Chemometrics and Intelligent Laboratory Systems,* 2, 126-127.

97. SIMCA https://www.sartorius.com/en/products/process-analytical-technology/data-analytics-software/mvda-software/simca.

98. *History of Infometrix,* https://infometrix.com/company/history-of-infometrix/.

99. V. Tysso, K. Esbensen and H. Martens (1987). UNSCRAMBLER – an interactive program for multivariate calibration and prediction. *Chemometrics and Intelligent Laboratory Systems*, 2, 239–243.

100. O.M. Kvalheim and T.V. Karstang (1997). A general purpose program for multivariate data analysis. *Chemometrics and Intelligent Laboratory Systems*, 2, 235–237.

101. E.S. Nourse, B.G. Greenberg, G.M. Cox, D.D. Mason, J.E. Grizzle, N.L. Johnson, et al. (1978). Statistical training and research: the University of North Carolina System. *International Statistical Review*, 46, 171–206.

102. G. Balfour (1999). Certification program validates SAS users *Computerworld Canada*.

103. N.H. Nie, D.H. Bent and H.C. Hadlai (1970). *SPSS: Statistical Package for the Social Sciences,* McGraw-Hill, New York.

CHAPTER 2

Instrumental Methods

2.1 INTRODUCTION

There are several methods for analysing samples for metabolomics. The most common workhorses are coupled chromatography (LCMS (liquid chromatography mass spectrometry) and GCMS (gas chromatography mass spectrometry)) and NMR (nuclear magnetic resonance). Some classical metabolomics were performed by single-wavelength HPLC (high-performance liquid chromatography) and a few investigators use FTIR (Fourier transform infrared) and Raman spectroscopy.

In this chapter, we introduce the principles behind the methods, and the main steps in data analysis where relevant. In the case of coupled chromatography, there are a huge number of software approaches and databases, primarily aimed at creating peak tables and identifying metabolites, both public domain and commercial. It is impossible in this text to detail all of them, and as the pace of change and evolution is rapid, the main details will change over the lifetime of this book, however, basic descriptions are provided as available to this author at the time of writing where information was forthcoming. NMR data do require established approaches such as Fourier transformation and apodization/baseline correction, which have been part of the spectroscopist's toolkit for several decades. More recent software such as for structure elucidation/deconvolution has a slower evolution than coupled chromatographic software but will still change during the lifetime of this text.

For HPLC, Raman and FTIR, methods are somewhat more straightforward. Most instruments provide relevant software or programmers can fairly easily implement the main approaches using raw data.

Data from instruments and public databases may either be in raw format, for example, LCMS output, which requires sophisticated methods to translate into a meaningful data matrix for pattern recognition, or as already processed data in the form of peak tables according to technique. Of course, the quality of peak tables depends on the

Data Analysis and Chemometrics for Metabolomics, First Edition. Richard G. Brereton.
© 2024 John Wiley & Sons Ltd. Published 2024 by John Wiley & Sons Ltd.
Companion website: www.wiley.com/go/Brereton/ChemometricsforMetabolomics

algorithms used and often the choice by users of relevant tuneable parameters, and the same raw data can result in quite different peak tables according to the methods used.

This text is primarily concerned with the chemometrics (mainly multivariate) analysis of the resultant spectroscopic or chromatographic data rather than the detailed mechanics of the methods of transforming raw chromatographic or spectroscopic data to matrices. In some cases, this transformation is very straightforward, for example, in Raman spectroscopy, whereas in other cases such as GCMS, it can be extremely tricky. This text will be illustrated by a various types of data, for example, we will use both original NMR spectroscopic data in the form of intensity versus frequency spectral data and deconvoluted NMR where individual metabolites have been identified.

In the case of coupled chromatography and NMR, readers are recommended to consult relevant websites for more detailed descriptions of features of individual packages, although the main approaches available to the author with their major features are described below.

2.2 COUPLED CHROMATOGRAPHY MASS SPECTROMETRY

Possibly the most widespread workhorse in metabolomics involves coupling chromatography with mass spectrometry.

There are three important steps:

- Chromatography to separate the compounds from the mixture in an extract.
- Ionisation to obtain charged particles in the gas phase.
- Detection to record these particles.

Subsequent to this, most studies convert the data into a peak table of which there are three main steps:

- Converting the raw data into a matrix where each sample is characterised by a number of variables related to metabolites. The variables correspond to chromatographic peaks and hence the matrix is usually called a peak table.
- Using human expertise to clean up and simplify this matrix, both at this stage and at the end.
- Assigning all or many of the peaks in the data matrix to chemical structures, to allow interpretable metabolomics.

2.2.1 Chromatography

An extract, for example, of plasma or urine or leaf tissue, is introduced onto a chromatography column, using either gas chromatography (GC) where a mixture is volatilised at high temperature and then carried through a capillary column (stationary phase) in a carrier gas such as He where molecules are adsorbed at different efficiencies, or liquid chromatography (LC) where a sample is dissolved in a mobile phase and then injected onto a stationary phase in which molecules have a different partition between the solid and liquid.

- There are a very large number of combinations of these two phases, but common to all analyses, compounds elute (or come off) the column (or stationary phase) at different rates.
- These may depend on the polarity, size, solubility and several other factors. The time a compound elutes, is called the retention time (RT). In GC, this is often converted to a Kovats Retention Index (RI) which is more stable across different analytical systems.
- In the ideal situation, different compounds are distinguished by their unique chromatographic peaks, whose areas in turn relate to their concentrations.
- How a peak's area relates to its concentration depends on the method of detection.
- Because the stationary and mobile phase are different according to analysis, for HPLC, there is no fixed order in which compounds elute, for example, compound A may elute before compound B under some conditions but after using different conditions.

Columns are not always stable.

- The RT of a compound can vary even using the same type of column and the same mobile phase.
- The condition of the column can influence the quality of the chromatography. Column overload is also difficult to control, especially when there are analytes of widely different concentrations.
- Ideally, a new column should be used for each analysis, but this can be very expensive for certain laboratories, and not always possible if many samples have to be analysed in a short time. Some columns can be stable for hundreds of injections. Some analytes progress very slowly through columns and can poison the analysis. Sometimes chromatographers snap the ends of columns to prevent this, and there should always be a balance between changing columns too fast and too rarely.
- The instability of biological matrices, and the problems of their storage, means that there may be unknown degradation occurring if there is a lengthy wait for analysis, which could confound (or interfere) with the actual metabolic effects to be studied. In addition, if a large study, perhaps involving 200 samples, is to be analysed within a couple of days, rapid chromatographic conditions are needed, to resolve hundreds of unique compounds on a column for each analysis.
- Usually, laboratories go to substantial effort to create reproducible and often rapid chromatographic conditions.
- Obtaining adequate numbers of samples for a statistically meaningful analysis often requires large sample sizes, but analytical and chemical factors make reproducible analyses over a significant period difficult, so a compromise is frequently necessary. QC (quality control) samples may be analysed throughout a study and if they fail, the entire dataset might have to be discarded.

Hence, the quality of chromatography is important for meaningful interpretation of LCMS and GCMS data. Unlike spectroscopy where compounds are often characterised by fixed features such as m/z values, chromatography is less reproducible.

More information about single wavelength HPLC chromatography is discussed in Section 2.3.

2.2.2   Ionisation and Detection

There must be a detector to record each compound as it elutes. Classically, these were of a single type of measurement, for example, the UV/Vis (ultraviolet/visible) absorption at a specified wavelength (e.g. 254 nm) for HPLC and FID (flame ionisation detector) for GC. Although these can characterise compounds by RT, the RT depends on the stationary phase (column), the mobile phase (liquid or gas), and also to a lesser extent on the condition and history of the column. On its own the mobile phase would not be sufficiently helpful to allow full characterisation of the hundreds (or sometimes more) of metabolites that can be detected in a typical biological matrix, so in most cases, a second detector is necessary, usually Mass Spectrometric (MS). The MS detector is coupled to the LC or GC detector, giving the hybridised techniques of LCMS and GCMS.

There are several different methods for recording the mass spectrum. First, there is a method of ionisation and then detection.

2.2.2.1   GCMS

For GCMS, a common method of ionisation is EI (electron impact) after which there are several approaches for detection of which single quadrupoles and triple quadrupoles are more common, or TOF (time of flight) for high-resolution mass spectrometry.

- In most metabolomics studies, EI is used which produces a reproducible MS, including fragment ions, this allows the identification of molecules often using well accepted libraries developed over many years.
- The m/z value is the mass (of a charged fragment) over its charge. In EI ions, smaller molecules are singly charged so the fragments can be directly related to molecular structural features, although as larger molecules can be found in many metabolomic studies, some may be multiply charged.
- In EI mass spectra often the molecular ion (representing the molecular mass of the compound) is weaker compared to most fragment ions and compounds are normally identified through their fragmentation patterns, which are quite reproducible.
- Most GCMS instruments also have a mass accuracy of only 1 Da (i.e. 1 m/z unit) and are called low resolution.
- The RT can be corrected to give an RI, which is more stable and characteristic of compounds. This index has been employed as a standard in GC for many years and involves interpolating the RT between the elution times of n-alkanes on a column.

A problem with GCMS is that not all metabolites volatilise easily, so it may miss some, especially high-molecular-weight compounds. Nevertheless, if the same sample preparation and analysis technique is used for each extract in a series, it should provide a consistent picture.

2.2.2.2 LCMS

- For LCMS, it is first necessary to transfer the molecules as they elute from the column to the gas phase. Hence, eluents are not directly introduced in a gaseous form unlike in GCMS. Several methods for ionisation have been developed of which ESI (electrospray ionisation) is one of the most widespread.
- This simultaneously vaporises and ionises the molecules.
- ESI can be either in the positive ion or negative ion mode and usually metabolomics analysis is only one of these datasets; however, both can be compared or analysed separately.
- ESI primarily provides information on adducts of each metabolite, although some fragmentation is also obtained. However, these can be used for structural identification.
- APCI (atmospheric pressure chemical ionisation) is an alternate that can also be in positive or negative ion mode.
- Unlike in GCMS, APCI primarily provides information on the molecular ion and does not provide easily interpretable fragmentation papers in contrast to GCMS.
- Often MS/MS is employed, which allows more structural insight. The primary ions are fragmented in a second step.

After ionisation, there is still the need for detection as per GCMS, of which QTOF (quadrupole time of flight) remains one of the most common approaches. The majority of modern LCMS instruments have a mass accuracy of much greater than 1 Da (high resolution).

Most metabolomic LCMS studies rely on a single molecular ion (or adduct) for characterisation rather than stable libraries, unlike GCMS, and we will discuss this below. Of course, there are always exceptions to the rule, such as low-resolution LCMS.

2.2.3 Data Matrices and Peak Tables

The information from each sample can be obtained as a series of numbers. Usually for each GCMS or LCMS, data can usually be arranged in a data matrix, whose rows represent elution times and whose columns represent m/z values (we will not explore in detail MS/MS but assume that only the primary fragment ion is represented in the LCMS data matrices and use the second dimension only for diagnostics). The numerical values in the matrix relate to the intensity at the detector, for example, the current as volatilised and charged ions are detected as they elute from the column, although their charge is amplified prior to detection. The rows of the data matrix can be represented by a chromatogram and the columns by a mass spectrum, if summed up, these

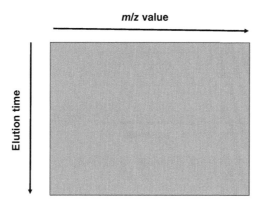

FIGURE 2.1   Data obtained from the LCMS or GCMS of a single extract of a sample.

represent profiles over the entire data, although often people look at selected ions (m/z) or selected elution times. A data matrix obtained from coupled chromatography is illustrated symbolically in Figure 2.1.

This data matrix in itself is often fairly messy.

- There is always noise.
- Sometimes peaks (in the chromatographic direction) are overlapping and may form complicated clusters.
- Peak-shapes, which ideally should be symmetric Gaussians are often distorted, especially in LCMS dependent on the type of column.
- Some peaks are overloaded, that is, their concentrations are very high and detectors and chromatographic columns can only function effectively in certain concentration ranges.
- Some peaks may be below detection limits.
- Baselines may be sloping.
- For high-resolution LCMS, there can be small shifts in the m/z direction.

On the whole, the chromatogram is harder to interpret than the mass spectrum. Ideally, each compound in the analysis results in a single peak in the chromatographic direction, whose mass spectrum is characteristic of its identity, and whose area (over all or one or more selected m/z values) is proportional to its concentration in the original mixture.

A typical metabolomics analysis will consist of several extracts from samples of different and known provenance, many studies consist typically between 20 and 200 of such samples, each of which results in a coupled chromatogram, so a metabolomics study consists of several coupled chromatograms as illustrated in Figure 2.2. However, in many cases, chromatograms are hard to align. In addition, chromatographic peak-shapes and resolution can vary from sample to sample, especially as underlying concentrations or even presence of different metabolites will differ in each extract. Furthermore, for high-resolution mass spectra, there can be small shifts in m/z values. There are so many reasons why it is hard to line up the chromatograms and make

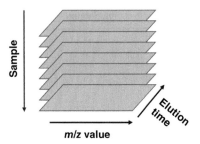

FIGURE 2.2   Typical arrangement of LCMS and GCMS data.

sense. Therefore, treating these huge datasets as one set of information, for example, a tensor (or 3D data box) consisting of I samples, J elution times and K m/z values is not likely to be successful. Only for certain types of data, such as excitation-emission fluorescence are the variables (in this case wavelengths) sufficiently stable that it is possible to treat data as a three-dimensional box and use so-called multiway methods. We will not discuss these approaches in this text.

There are basically three practicable ways of handling such data.

- Sum the RT (or RI in the case of GCMS) information, usually over the entire chromatogram, to give a set of profile mass spectra for each sample. This approach can be used however has two principal problems. First, if all chromatographic information is lost, there is no real point in going to the expense of a chromatographic detector. Second, this loses interpretive value as it is hard to work out which individual metabolites are responsible for given differences between samples.

- Sum the m/z information for each sample, either over all masses or selectively, to give a profile chromatogram for each sample. The problem here is as discussed above, the chromatography may not be stable, although there are ways of aligning chromatograms. The influence of individual metabolites can be obtained, providing there is stability in the chromatography and computational algorithms, sometimes requiring warping, are successful. However, the mass spectral information will be lost, and if there is a complex chromatogram consisting of many hundreds of peaks, it might be hard to align peaks over a series of samples. For more straightforward situations, as may occur in some types of plant metabolomics, one-dimensional UV/Vis spectral chromatograms can be used, and we will discuss this in Section 2.3.

- The commonest is to use a peak-picking algorithm to transform the chromatographic data into a peak table. This involves trying to identify where the underlying peaks are in each chromatogram and using various methods to determine which chromatographic peaks in each sample correspond to peaks of the same origin in other samples (sometimes called features). The result is a new matrix, whose rows now correspond to individual samples or extracts, and whose columns to specific peaks (or features or variables or metabolites). The elements of this matrix are intensity values, which ideally should relate to concentration and are used for subsequent statistical analysis. The matrix is called a peak table as illustrated in Figure 2.3.

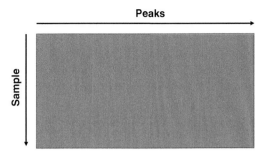

FIGURE 2.3   Principles of a peak table.

The majority of metabolomics studies employing coupled chromatography mass spectrometry involve performing statistical analysis on peak tables. In this text, we will not discuss in depth how peak tables are created, and only a few readers of this text will be writing software for this reason. However, nearly all users of LCMS and GCMS for metabolomics studies will come across peak tables and have to employ software for their creation. We will describe several software packages for the creation of peak tables in Section 2.2.4.

It is therefore essential to understand the general principles of how raw chromatographic data are converted to an interpretable peak table and to be able to choose and develop an approach. There is no universal method, and different groups use different methods. There are of course a lot of downloadable algorithms, which convert raw data, often in the form of files obtained from commercial instruments, to peak tables, but they should all be used with caution, as in all cases there are tuneable parameters which can influence the characteristics of a peak table.

How to create a peak table depends in part on the expertise in an individual laboratory. Some may be able to spend a week or more on a study and will have significant expertise in-house. Others may have less resources. Sometimes this depends on the value of the study, is it a low-funded pilot study, or is it a really important and large project. Below we will describe the main stages in creating a meaningful, statistically and chemically interpretable, peak table. It is essential to understand that if this stage is done wrongly, the resultant biological interpretation may likewise be wrong and the statistics are not very meaningful.

2.2.4   Step 1 of Creating a Peak Table: Transforming Chromatographic Data to an Aligned Matrix of Peaks

The first and most important step is to take the coupled chromatographic data in the form of Figure 2.2 and converting into a peak table in the form of Figure 2.3. There are numerous approaches, some of the main ones are discussed in detail below. This list is not comprehensive.

2.2.4.1   XCMS

XCMS was first developed by the Scripps Institute in 2005. It has widespread popularity for several reasons.

- It is specifically oriented towards metabolomics and has been continually supported for almost 20 years from the time of writing of this text.
- It is open source, so that programmers can modify it or take parts of it to tailor to their own needs. It is written in R. It is also the back end of processing of Workflow 4 metabolomics (a Galaxy-based platform), and an optional peak picker for MZmine (see Section 2.2.4.4).
- There is a separate on-line version with a GUI (graphical user interface) that can be employed by users without programming expertise.
- It can be employed in a high-performance computing environment allowing many samples to be analysed simultaneously and so it is fast with the latest developments increasing speed.

XCMS was originally developed for LCMS data but has been extended to GCMS. Also, the original algorithms were for data with low resolution (1 Da or 1 m/z unit), whereas most of the modern instruments have high mass accuracy, so there has been some evolution in the methods over time.

There are several main steps:

- The first step involves identifying peaks, at each m/z value and at each retention (elution) time for each sample. It takes into account expected peak shape characteristics and noise. These of course can be variable according to instrumental conditions and it is often up to the user to determine the variability they tolerate.
- The next step is to match peaks over the sample set for each RT and m/z value. This takes into account primarily the expected RT range for a specific peak as well as sometimes a smaller m/z variability and of course is also possible to be user defined. Hence, peaks centred around, for example, m/z 423.1–432.2 and RT 579–582 s in several or most of the samples might be grouped as likely to be from the same origin. This is called peak matching.
- The third step involves what is called RT alignment. Let us say the peaks at m/z 423.1–432.2 have RT variability between 579 and 582 s, these are then aligned to the same 'virtual' RT over all samples. The method is rather complicated and primarily tries to identify those peaks that occur in most of the samples at roughly the same RT and (for high-resolution MS) m/z window, called well behaved peak groups. This shifts the chromatographic scales so that they appear aligned according to m/z values. This step depends on there being many compounds detected in the majority of samples, which is usually the case in metabolomic analyses, where samples in any individual study usually originate from the same type of source, for example, mouse pancreas or monkey serum, so they should contain many similar compounds. Once aligned, peaks characterised by different m/z values but at similar RTs can be connected together to give mass spectral profiles or groups where appropriate and so build up a picture as to where metabolites elute and what their spectral features are. Alternatively, another algorithm can be used which aligns all the samples to a median sample.
- The final stage looks at regions where peaks were not detected, for example, a peak may have been detected in 78 out of 100 samples, and the remaining 22

samples will be re-examined to determine the intensities in the missing regions. In certain metabolomic studies, it is expected that not all active compounds will be found in every sample, for example, donors with a specific disease may contain metabolites that are absent in control samples.

The resultant peak table should ideally identify where common peaks are in each extract and what their areas are as illustrated in Figure 2.3, and give MS profiles helping identification if required, for each peak. As in all software described in this section, peak areas involve summing intensities over the RTs of each peak at the selected m/z, sometimes after baseline correction. Errors can creep in at this stage; however, what is important is that for each LCMS (or GCMS) set of data, obtained using a consistent instrument, chromatographic and spectrometric conditions, the areas of individual peaks are proportional to their relative concentrations over a set of samples.

A consequence of this procedure is that there are a lot of so-called tuneable parameters, for example, the tolerance window for RT and in the case of high-resolution MS for m/z values. They can be tuned manually but also automated approaches for tuning have been developed such as autotuner and isotopologue parameters optimization (IPO). Generally speaking, the parameters help to find the global maxima; however, the peak table is always a meaningful representation of the dataset. Parameters optimisation can help the feature table (we are not dealing with individual peaks anymore) be closer to ground truth. These parameters can determine how meaningful the peak table is. In addition, there are often unusual or rare ions that may occur just in a few samples, and are often a consequence of the background or sampling procedure. Sometimes several hundred peaks are identified per chromatogram, but only a few hundred have much significance.

XCMS, in addition to being available from the originators both as R code and on-line, is also embedded in several different software packages.

2.2.4.2   Multivariate Curve Resolution

A completely different approach has been developed within the chemometric community over several decades. MCR (multivariate curve resolution) is a form of ALS (alternating least squares), which is one of the cornerstone techniques developed from work started in the 1970s. ITTFA (iterative target transform factor analysis) is historically one of the forerunners of this approach.

Within coupled chromatography, MCR methods are based on the premise that neighbouring peaks in a mixture, such as recorded in a coupled chromatogram, are often not completely resolved, so their individual profiles and spectra need to be estimated by computational means. As this text is primarily about the later statistical steps in analysing metabolomics data, we will only discuss the procedure in outline.

- The original applications, over 40 years ago, were applied to much simpler problems, for example, in equilibrium studies or the DAD-HPLC of 2 or 3 partially overlapping compounds.

- In matrix terms, the data matrix $X \approx C\,S$ where C represents the concentration profiles of the analytes in the mixture and S their spectra as illustrated in Figure 2.4. A fundamental aim of multivariate signal resolution is to take a cluster of partially overlapping signals, for example, LCMS peaks and deconvolute them into their underlying components as shown in Figure 2.5.

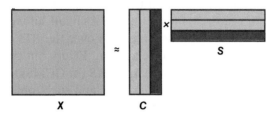

FIGURE 2.4   Matrix decomposition into concentration or elution profiles and spectra.

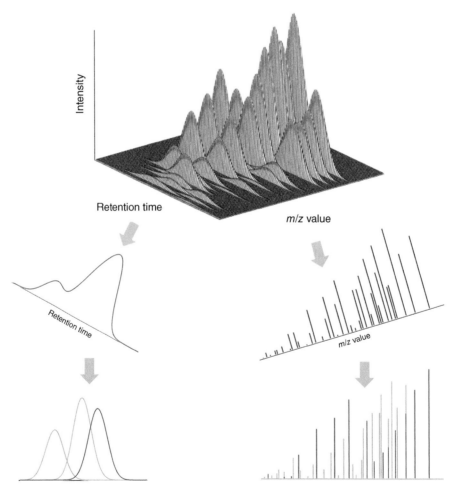

FIGURE 2.5   Deconvoluting a cluster into individual components.

- There is a substantial literature about how to tackle this. ALS (alternating least squares) is an algorithm that is commonly employed. It involves first guessing either the spectra or chromatographic profile of each component by a variety of approaches, then using regression against the observed data X the other part of the profile is estimated.
- The trick is that although there are a large (or in fact infinite) number of solutions; the user imposes certain constraints such as non-negativity and unimodality (each true peak in the chromatographic direction just has one maximum) and a number of others that are possible, and only solutions with these properties are retained.
- The method continues finding solutions alternating between estimating C and S subject to constraints until the algorithm is judged to have converged.
- Of course, it is necessary to decide which constraints are desirable, how many significant components (how many coeluting compounds are in a cluster) and what the convergence criterion is.

If properly used, ALS/MCR is an effective way to resolve out the multivariate signals in a mixture. For some decades applications reported were relatively simple, for example, 2 or 3 overlapping chromatographic peaks, but over the last few years, it has been increasingly used as an aim to pull out the main features of quite complex multivariate datasets obtained in metabolomics in a semi-automated way.

A package for MCR developed over many years from Barcelona is downloadable, although at the time of writing primarily in the MATLAB environment; however, it is ever evolving and will extend to a GUI and an R version. The MCR method is also incorporated as a method in several other packages such as MZmine (Section 2.2.4.4), and hierarchical MCR has been developed by workers in Umeå.

A recent adaptation for large datasets applicable to metabolomics is called ROIMCR, which is based on some of the similar principles of hierarchical MCR.

- It is first necessary to divide each chromatogram into sections called Regions of Interest (ROI) so MCR can be applied to small areas of the data, which can be done automatically. Several parameters need to be optimised to fully characterise these regions and will depend on instrumental parameters and user expertise.
- The ROI contains m/z values above a given threshold intensity, and RTs within the window, and so consists of a small region in both the mass and time dimension which is considered to contain interesting information.
- ROIs in each sample (GCMS or LCMS) are then identified and augmented so that ROIs with similar information are combined to provide augmented ROIs, containing information from all chromatograms in a sample set. Providing the correct m/z tolerance is provided (e.g. minimum intensity, and for high res minimum m/z variability), the ROIs in each chromatogram should line up, even if there are RT shifts.

- MCR is then applied to an augmented data matrix. This combines ROIs from each chromatogram into one long matrix, which should contain similar mass spectra. The trick is that the MS characteristics of components over this large set of data are likely to be the same or very similar. Of course, there would be multiple maxima.
- Then MCR is used to obtain the profiles in each chromatogram and the corresponding MS characteristics.

ROI MCR is a relatively new package and by its origins mainly aimed at MATLAB users, but is rapidly evolving. A version in R is likely to be available during the lifetime of this text. It is primarily used for LCMS data but can be adapted to GCMS.

2.2.4.3   AMDIS

AMDIS (automated mass spectral deconvolution and identification system) was developed and maintained by NIST (National Institute of Standards Technology) based on methods originally developed in the 1970s, primarily for use in GCMS, and has been available for over two decades. It can still be used to obtain peak tables and is free to download.

AMDIS has several steps.

- The first is to identify noise so that it can be distinguished from the signals. This is done in an automated way.
- The m/z values with significant rises and falls are considered diagnostic and are pulled out. If the RIs are similar within a defined deviation, they are combined together to characterise a single compound (or in our case metabolite).
- We will discuss identification of metabolites in Section 2.2.6 but the resolved spectra from AMDIS can be identified using the NIST library, an internal library, or left unidentified.
- Retention data (RI) can be combined with the MS data as well.

In all metabolomics studies, there will be several GCMS traces. AMDIS mainly identifies common metabolites via the MS and occasionally RI data and thus allows the traces to be lined up in a peak table this way. The data can be run in batch mode so that a set of, say 100 GCMSs obtained from a metabolomic study can be lined up to give a peak table, subject to good signal-to-noise ratios.

The method is not as yet suited for high-resolution data, as its origins go back several decades. However, it is useful for creating peak tables for low resolution EI GCMS and is continually maintained and well documented, and very well established. AMDIS can also be applicable in certain cases to LCMS.

2.2.4.4   MZmine

MZmine was originally developed in the mid-2000s, and has been significantly redesigned over the past 20 years; MZmine 3 is the most recent version at the time of writing this text. The improved performance of MZmine 3 over previous MZmine versions now allows the processing of large datasets. It is reported to be the second most used method for peak tables in metabolomics after XCMS according to the cited literature at the time of writing. MZmine is available in the form of open-source code. It was originally developed for LCMS but now supports hybrid datasets from various instrumental setups, including LCMS and GCMS, ion mobility spectrometry (IMS) MS and MS imaging. The method is available for instruments with variable (both high res and low res) mass and RT resolution, peak-shapes and noise characteristics. The software requires Java and for some modules R.

MZmine contains a large number of steps and options but the main stages are as follows:

- Feature detection: This initially finds individual peak maxima at given m/z values and RTs. It then tries to connect these together to produce individual mass spectra. It is called feature detection assuming each mass spectral/chromatograph pair corresponding to an individual metabolite corresponds to a single feature, although is analogous to peak detection.
- Deconvolution: The more recent versions include modules for peak deconvolution, particularly useful for GCMS. There are a variety of approaches, including MCR, whose principles have been described previously (Section 2.2.4.2).
- Feature alignment: This uses spectral and chromatogram features to connect features over multiple chromatograms to the same compounds. There are various alignment algorithms. The Join Aligner aligns detected peaks in different samples through a match score. This score is calculated based on the mass and RT of each peak and ranges of tolerance specified by the user. The RANSAC aligner is an extension of this and corrects any linear or non-linear deviation in the RT of the chromatograms. MZmine takes the premise that the alignment of peaks over multiple chromatograms is an essential step in creating a peak table for metabolomics.
- Feature identification: This attempts to link mass spectra to specific compounds. The prime approach is via the molecular mass and RT, often using a custom--built library (see later sections), and also the isotope ratio patterns. As LCMS does not always often involve recording fragment ions (unlike GCMS), identifying molecular ions is often the best approach for characterisation in such cases. A large number of databases are compatible. This can encompass steps described below in Section 2.2.5 which is integrated into this software, rather than a separate standalone step.

2.2.4.5 Other Approaches

There are a large number of other approaches for obtaining peak tables. There is insufficient room in this test to discuss all in detail, but some are listed below:

- Metalign was developed by the University of Wageningen in the Netherlands and can be used both for GCMS and LCMS in high-res and low-res modes.

It was one of the best-cited metabolomics software for coupled chromatography after XCMS and MZmine in the mid-2010s.

- apLCMS is an R program maintained by Emory University. It can create peak tables either without using any previous knowledge or in a targeted way using previously known metabolites or building up information from the same instruments.
- mzMatch is written in Java. It can integrate its data with XCMS. mzMatch.R is an R version. It is freely downloadable.
- MSDial is a software solution maintained by the Riken Institute in Japan. It focuses on fragmentation data and can benefit from RT index databases. It has a graphical interface and a strong support forum as well as being continually updated and developed.

The websites of these and the approaches described above, at the time of writing, are listed in Table 2.1.

In addition, there are several commercial packages developed by instrument manufacturers, which include as follows:

- Progenesis QI by waters.
- SIEVE and Compound Discoverer by Thermo.
- Mass Profiler Plus by Agilent.
- Metaboscape by Bruker.

The basis of many of the commercial packages is often not described in detail in the public domain and some are quite costly and restricted towards the specific instruments but can be useful for users of these bespoke instruments.

In addition, many organisations who do intense LCMS or GCMS data analysis develop their own in-house approaches. Many are not written up in papers or public domain, sometimes because of secrecy or in other cases because this is not the priority of the developer. Many in-house methods are not constructed as formal packages: the users apply tools as they need them on a case-by-case basis, using, for example,

TABLE 2.1   Main public domain software for LCMS and GCMS peak picking.

Package	Website
AMDIS	https://chemdata.nist.gov/dokuwiki/doku.php?id=chemdata:amdis
apLCMS	https://doi.org/10.1093/bioinformatics/btp291
Metalign	www.metalign.nl
mzMatch	http://mzmatch.sourceforge.net/
MZmine	http://mzmine.github.io/
ROIMCR	http://www.mcrals.info/
XCMS	https://xcmsonline.scripps.edu/
MSDial	http://prime.psc.riken.jp/compms/msdial/main.html

common chemometrics tools such as PCR (principal components regression), MCR, PLS (partial least squares), similarity measures and numerous other computational tools. In addition, time is involved in documenting and maintaining packages, and often there are no additional funds, as most packages are free to download and use. Hence, not every organisation or laboratory that develops software to obtain peak tables in-house has the resources or need to make their methods public domain.

Over time, there are likely to be many more such packages released, some of which will become major tools, others which are primarily for a small user base. Approaches such as XCMS involve a constant need to maintain as new types of data become available, and are supplemented by international courses and videos and tutorial, requiring a significant and permanent staff presence.

2.2.4.6 Which Approach is Most Suitable?

With such a huge number of choices available for constructing peak tables, it can be perplexing to choose the most appropriate approach. There can be no single answer that will suit all laboratories.

Some factors to take into account are as follows:

- Are you primarily using GCMS or LCMS, is your data high-res or low-res MS, and is your MS characterised by a reproducible fragmentation pattern or just a molecular ion? Each package has its niche, and some are better in certain circumstances than others. The more recent ones, or those that are updated more regularly will probably cope better with more sophisticated instrumentation, for example, high-res LCMS, some of the older approaches may be better for low-res GCMS.

- Are you able to program, so can you cope with R (or MATLAB or Java or Python) code, or do you want a GUI? Most public domain packages originate as code but most of the more established ones are also compiled as GUIs.

- What money can you spend? Some are freely available, but others involve quite costly licenses. Sometimes the freeware comes as code, whereas software associated with instruments, with commercial packages and some GUIs, can be quite expensive. However, commercial software is likely to be maintained with technical support and documentation over many years. Some very well established free downloads like XCMS also have good long-term support, with courses and tutorials which obviously cost, but others have been written as one-offs, and may not necessarily be updated in the future.

- How far does the user have to make decisions as to adjustable (or tuneable) parameters, and what is the risk if the decisions are made badly? This depends somewhat on the expertise of the user of course.

Each approach will result in somewhat different peak tables, and of course, it is often a good idea to compare several methods. There are a number of papers that compare how many peaks each method identifies and how many are in common. This to an extent depends on the dataset, as different approaches are optimised to specific types of data, and may also depend on manual choices, for example, noise level, RT (RI) shift tolerance, peak-shapes, etc., dependent on method.

How crucial this is depends on the value of a project, and the time and resources available. The best approach may be to use several different peak picking programs and compare the number and nature of the peaks.

Of course, a key last step is quantification, often by determining the area under a chromatographic peak summed over an elution range and one or more m/z values and this can vary according to software, with methods that involve deconvolution (resolution) often working best, although if coeluting peaks have quite different m/z characteristics this is not always essential.

2.2.5   Step 2 of Creating a Peak Table: Manual Inspection

After creating a peak table by one (or sometimes more) of the methods above, it is advisable to then check its sense using manual inspection. How far this is possible depends in part on the expertise available in-house. A team where the main expertise is in computing may have limited ability to interpret chromatograms or mass spectra. However, many of the more established groups usually have in-house chemical expertise.

Inspecting a series of peaks for a dataset where there are 100 chromatograms and 1000 peaks identified per chromatogram can take a considerable amount of time. However, usually a significant metabolomic study is well funded, and most professional laboratories will cost out this time.

Peak tables could come in one of two forms.

- Only peaks coming from known metabolites are retained, with their mass spectra and RT (or RI if GCMS). In the next section, we will outline ways of identifying metabolites.
- All peaks that pass various criteria of the initial peak-picking method are retained, including those whose identities are not known.

Manual expertise can then be used to see if the peaks extracted look sensible, often by looking at individual chromatograms and mass spectra.

- Are the peaks very rare? They may be due to background artefacts.
- Are the peaks quite sparse, again they may not be very good potential markers? One might want to reject peaks detected just in a specified proportion of samples, however, it may be important to ensure that there are biomarkers within specific groups, so if a group consists of just a small proportion of the samples (sometimes by design) rejecting strong markers just for one group may miss information.
- Unusual peak-shapes are often a sign that something has gone wrong and should be rejected. This can cause a problem if a few samples out of many have unusual peak-shapes in the area a potential marker elutes: should the entire set of measurements be removed from the analysis, or should there be guesses for unresolved values in certain samples?

- Peaks near the detection limit or noise might have to be removed.
- Often there is a problem with overlapping peaks, and so some peak-picking software can make mistakes. Automating peak detection algorithms for the use of chromatograms with several thousand detectable metabolites, often with unresolved compounds, can inevitably lead to mistakes.
- An experienced scientist can often see if there are problems with the appearance of mass spectra.
- QC samples are often introduced to ensure the quantification and resolution of chromatograms remain stable over a run of samples that might take a day or more and should be manually inspected. If the QC samples fail, then an entire batch of data may have to be destroyed.
- For some methods, such as AMDIS, it may be necessary to check mass spectra of key peaks between different samples to ensure they are the same.

This step can be particularly time-consuming, especially when there may be several thousand peaks in each of several hundred chromatograms. For an important project, it is worth spending one or two weeks because the experiments may have taken months or more and a significant effort to acquire data. Many of the best-resourced laboratories have dedicated staff who are expert mass spectroscopists/chromatographers whose main job is checking these huge peak tables.

How much time and resources can be invested in this stage depends on the nature of the problem and staffing levels. Although checking these data may be expensive in terms of paying for people's time, so too can the cost of certain commercial software and instruments but should often be costed into the price of an analysis. Trying to skip this step can lead to significant difficulties later on. The various methods of Section 2.2.4 are not designed to be perfect. There are always unexpected challenges with biological extracts, and it is usually wise to check the peak tables carefully before performing further statistical analysis.

No specific guidance can be given except to say that for a large and complex study it is never wise to rely completely on computerised methods to develop peak tables and the results of statistical data analysis are only as good as the data used.

Step 2 can be done before Step 3, providing the peak picking methods have narrowed down the number of metabolites, but can also be done as well or instead after Step 3. There is no set approach and this will depend on individual laboratories and their expertise.

2.2.6   Step 3 of Creating a Peak Table: Identifying Metabolites and Annotating the Peaks

Metabolomics can either be targeted or untargeted.

- For untargeted analysis, it is not necessary to know the identities of all the metabolites. For exploratory analysis, this allows insight into possible new pathways or products.

- Targeted metabolomics involves only studying metabolites of known structure.
- Assigning a peak (or feature) to a structure is called annotating.

Most metabolomics studies using coupled chromatography are targeted and one of the aims is to see which metabolites are associated with which conditions or treatments. For example, if a group of diabetic mice is treated with different pharmaceuticals, which metabolites are associated with each treatment? To obtain this information, peak tables need to be annotated, hence the majority of data has chemical information attached to each column of the data matrix. Figure 2.6 symbolically illustrates an annotated data matrix, which can be used as input to statistical metabolomics analysis. Each column represents a variable but also can be assigned to a compound of known (or postulated) structure.

Some people do include a few unknowns in the analysis, there is no fundamental difference statistically whether a column (representing a chromatographic peak or feature or variable) corresponds to a known structural identity or not. However, many working in the area of metabolomics prefer only to deal with peaks corresponding to metabolites of known structure, so that they can interpret the patterns in terms of metabolic pathways and also often provide practical advice to bioscientists. Of course, this depends on what questions are being asked. For example, we may primarily be interested in whether a group of donors from one genetic group differ in the reaction to a treatment to a different group, or whether there are any outliers (unusual samples) – under such circumstances we do not need to know the identities of individual metabolites at this

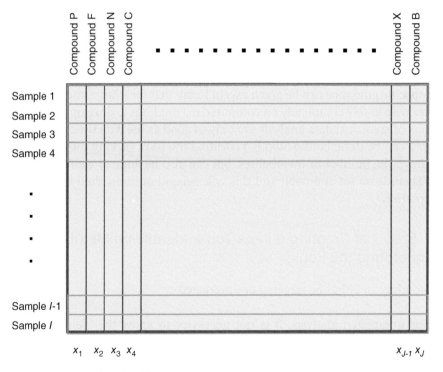

FIGURE 2.6   Annotated peak table.

stage. In addition, even if we are interested in which metabolites cause variation, we may not always be able to assign these to chemical structures a priori, but this information could be investigated later if significant.

However, the majority of analyses using LCMS and GCMS do aim to identify individual metabolites, as chromatography is a very selective technique. Approaches such as vibrational spectroscopy are quicker and cheaper but are not able to provide selective information on individual metabolites. NMR is intermediate and may provide information on groups of compounds in certain cases, and not always be able to quantitatively distinguish, for example, 1000 metabolites in an extract.

In this text, we will mainly illustrate GCMS and LCMS datasets using ready annotated peak tables, as the majority of public-domain metabolomics datasets are in this form. However, all the methods described in later chapters of this text can also be applied where the identities of all or some of the metabolites are unknown.

There are several approaches to establishing the identities of metabolites, some of which can be combined, and used to annotate the peaks (or features or variables) and assign them to metabolites. It is important to recognise that this is a large job, for example, if there are 1000 potentially interesting peaks detected in each of 100 chromatograms, then identifying 100,000 peaks would be an impossible job without assistance. Manual interpretation of 100,000 mass spectra at five minutes each (typical of an expert organic chemist) would take 500,000 minutes. Working at seven hours a day and five days a week for 40 weeks a year means 84,000 minutes work year, or around six years to identify each peak by manual inspection, hence it is not practicable to identify all the detectable peaks in a typical metabolomics experiment without further tools. This does not mean manual interpretation of the data tables, as discussed in Section 2.2.5 should not be done, but that the amount of work can be reduced by using computational tools first.

2.2.6.1 Experimental Libraries

The most reliable is by obtaining sufficient standards in the laboratory for identification, usually by purchase but optionally by synthesis.

- For meaningful metabolomic analysis, one needs a lot of standards, typically around 500.
- The chromatographic conditions should be stable so that the RT (for LCMS) is quite reproducible.
- The MS technique also should be stable and for LCMS it is usual to use a technique such as ESI, which provides a good molecular ion.
- The chromatogram should be well resolved, at least at the m/z ion most characteristic of the metabolite.

For LCMS, where usually a MS technique is employed without a diagnostic fragmentation pattern, but the molecular ion can be identified, having in-house standards is a preferred approach. For larger laboratories where there is a significant throughput of studies, and therefore appropriate resources this is a good option, but for smaller groups who may only occasionally need to analyse such experiments, this is an expensive option. It also depends on having very reproducible chromatography, remembering there may be several hundred compounds in a library, and if conditions are changed, the entire experimental database must be reanalysed. However, the library often contains both mass spectral and RT (or RI) data rather than just mass spectral information and is the most reliable approach if resources are adequate.

Sometimes these libraries can be built up experimentally over time, and used to complement the computational libraries.

2.2.6.2   Computational Mass Spectral and Retention Libraries

These are very common and have been developed often over time.

- The most robust are useful in EI (for GCMS) where there is a reproducible and usually unique fragmentation pattern.
- The NIST EI library has been developed over many decades, and at the time of writing consists of over 350,000 compounds. It is supplemented by an RI library consisting of nearly 140,000; over 100,000 compounds have both RI and EI MS characterised. MS Spectrum interpreter links analytical information to chemical structures.
- For GCMS data, the NIST library and associated software is almost certainly the most comprehensive in the world. It can also be extended to tandem mass spectrometry.
- For LCMS, there are a large number of databases, but these primarily assign structures to accurate mass molecular ions (or $[M+H]^+$ ions as appropriate) often using a specified mass tolerance. Some of the main databases, among many, are as follows:
  - PubChem contains nearly 100 million structures.
  - KEGG (Kyoto encyclopaedia of genes and genomes) contains nearly 20,000 structures.
  - HDMB (human metabolome database) contains over 100,000 structures.
  - ChemSpider contains over 67 million structures.
- Some packages such as MZmine as described in Section 2.2.4 will combine the peak picking and annotating in the same software.

As time goes on, during the lifetime of this text, such databases are going to grow, and new ones will evolve. Larger is not necessarily better as too many possible structural candidates may be predicted if the database is very big.

Coupled tandem MS/MS can provide further structural insights as well. Metlin is primarily an MS/MS database worth using where such data is available.

In this text, we will not compare all the various and endlessly evolving databases in depth as the field will change during the period of this text, but suggest readers refer to the most up-to-date literature.

Many of the databases do not provide conclusive structural results but can be used to narrow down the number of possible compounds—if very high res mass spectra are obtained, this information can be useful. Combined with knowledge of what is reasonable from metabolic pathways or expectations, sometimes a good candidate can be postulated, although often these databases are just used as a first step to more detailed analysis. Some of the larger more general compound structural databases are too broad for conclusive assignments to metabolomics and could easily result in nonsensical candidates.

The more information, for example, fragmentation patterns, accurate mass, RIs and tandem MS the better the chance of obtaining a meaningful result via library searching.

2.2.6.3   Expert Systems

One of the main origins of AI (artificial ntelligence) in the 1960s by Djerassi, Lederberg, Feigenbaum and Buchanan, involved organic structural elucidation via the DENDRAL program, using mass spectral as well as NMR data. Hence, the assistance of expert systems in this application area has a long and continuing vintage. The original DENDRAL was not widely employed at its time but its achievement was to spawn a new field, rather than be actively promoted as a widespread commercial package. Originally, structure elucidation by MS was primarily the domain of synthetic chemists and it was said that organic chemists found this interesting and resisted being taken over by machines.

However, computerised structure elucidation has had a renaissance, because determining structures of hundreds or thousands of metabolites by their MS patterns is too much work for the solitary organic chemist, so expert systems and machine learning can now offer assistance, although should not replace human expertise at the final confirmatory phase.

All sorts of information can be included as input to expert systems, including MS/MS, isotope patterns and fragmentations, often to narrow down candidates from the molecular ions. We will not describe these in detail.

2.3   SINGLE WAVELENGTH HPLC

We will briefly discuss single-wavelength HPLC. The principles of LC have already been introduced elsewhere in this chapter and so we will not repeat them.

Single-wavelength HPLC was used in classical chemometrics for early pattern recognition studies and a large classical literature has developed. Its use is fairly restricted in metabolomics, but some workers in this field do employ this technique, which may be of interest to a few readers of this book, so we will summarise the main principles briefly below. It is primarily useful where the metabolites are common to a series of samples and the main interest is in how their concentrations vary or whether one is present or absent in a key group.

The method is only suitable if metabolites of interest have chromophores in the UV/Vis range (100–800 nm). The technique has for example been used in herbal medicine, where the properties of a medicinal plant extract can be related to its metabolic profile. The method is cheap compared to LCMS, GCMS and NMR. As herbal extracts tend to be fairly consistent in their metabolic profile, it is relatively easy to identify peaks in common to a series of extracts. Single-wavelength HPLC is not very suited for human metabolomics where there is often large variability between extracts and often a significant challenge to identify metabolites in each chromatogram, usually requiring an additional MS dimension as described in Section 2.2.2.2.

- The first step is to choose a wavelength for monitoring metabolites. The usual and default wavelength is 254 nm. If there is no specific reason for choosing a specific wavelength most studies are at this wavelength.
- It is important to choose a column (stationary phase) and solvent system (mobile phase) suitable for a study. Most people use RP (reversed phase) HPLC which has a non-polar stationary phase and an aqueous, moderately polar mobile phase. In the RP methods, the substances are retained in the system the more hydrophobic they are.
- In some cases, it is necessary to optimise the mobile phase to improve the resolution of the peaks of interest. This is a big subject that people have written entire books about and was an important early application of statistical experimental design in analytical chemistry. There is a balance between running the chromatogram rapidly, for example, over 10 minutes, so many chromatograms can be obtained in a specified period of time, often useful if sample stability is of concern, or improving chromatographic resolution, that is ensuring neighbouring peaks are well separated, which may require longer runtimes. Optimising conditions can improve both speed and chromatographic resolution for a given type of separation.
- It is also important to ensure that the column is not overloaded. Peaks with absorbances between around 0.1 and 1.5 AU are within the Beer–Lambert region and hence their intensities can be modelled linearly using common chemometric methods. Peaks outside these regions may exhibit non-linearity and solutions are to change the amount of extract or use a more/less sensitive detector.
- In many cases, the volume of extract can be kept constant, so the relative peak intensities between each sample are proportional to the concentration of each metabolite, but where this is not possible either some form of internal standard is needed or the data will need to be row scaled (as described in Section 4.6.1).
- It is important to establish the digital resolution of a chromatogram. This is the number of intensity measurements recorded per unit time, for example, 1 data point per s equals 60 per min, or $20 \times 60 = 1200$ data points over a 20-minute analysis time.

The next step is to analyse a series of extracts.

- If there are large number (50–100), column stability is often a problem. Changing or flushing columns every so many runs may be necessary, as is the need to run QC samples.

- It is first necessary to eliminate baseline problems. These can be done physically (by improving the chromatography) or computationally using a variety of mathematical algorithms. In the latter case, often regions of the chromatogram where no compounds elute are taken as the baseline, and polynomials or related mathematical functions are fitted to these empty regions, which are then subtracted from the entire chromatogram. Baseline correction is often important as the concentrations of metabolites relate to intensity above the baseline, which should be flat.

- Most chemometric approaches involve lining up data points in individual chromatograms so each data point corresponds chemically to the same eluent (metabolite). The intensity at each elution time is the element of a data matrix X of dimensions $I \times J$ whose rows correspond to samples and whose columns to elution times.

- Sometimes some preprocessing is required at this stage. Examples involve smoothing, commonly with Savitzy–Golay functions, if there is a lot of noise, and binning, involving averaging a window of several neighbouring elution times and replacing with this average. These methods may reduce resolution but to the benefit of reduction in noise. If peaks are very narrow relative to digital resolution, these approaches may degrade the information, and a solution is to increase the digital resolution in the chromatogram when acquiring data where possible.

- The biggest problem is chromatographic alignment. A peak eluting at, for example, 6.78 min in one chromatogram may elute at 6.69 min in the next. Depending on the digital resolution, these could be several data points different, and so the variables in the X matrix will not properly line up. There are a huge number of algorithms but correlation optimized warping (COW) is one of the most popular. A full description of these algorithms could fill an entire chapter, but there are many automated packages available, and this method is also discussed in Section 2.4.3 in the context of NMR.

- Once an X matrix of baseline-corrected aligned chromatographic data has been obtained, it is possible to perform pattern recognition in such data. Sometimes it is useful to remove uninformative regions of the chromatogram as discussed in Section 4.7.3, especially if the data are to be standardised, as these chromatographic regions would primarily correspond to noise.

- In many cases, the objectives are to find regions of the chromatogram or peaks that are most significant or diagnostic of the metabolic effects or provenance of the samples. As there is limited diagnostic information, usually it is then necessary to coelute with known standards to identify the peaks of interest.

These are the main steps of preparing a single-wavelength HPLC chromatogram prior to chemometric pattern recognition. When reading articles or reports it is always important to understand the details of how the data were obtained first and usually requires some considerable expertise in an initial phase.

2.4   NUCLEAR MAGNETIC RESONANCE

NMR along with coupled chromatography, is one of the most widespread techniques for metabolomics. It has the advantage that each sample is normally represented by a single one-dimensional spectrum rather than a two-dimensional chromatogram, so two-dimensional peak picking is not required. It is also a rapid method with individual NMR spectra acquired in s, although often several have to be averaged to improve signal-to-noise ratios. The disadvantage is that all the metabolites are part of a single spectrum, and most are characterised by several peaks (or resonances) that are spread over different regions of the spectrum; however, experienced spectroscopists can identify relevant regions. The majority of metabolomic studies at the time of writing use ^{1}H NMR and 1 D spectra. NMR has the advantage that it can easily be automated and is non-destructive. However, compared to coupled chromatography often many less compounds can be detected, and spectral libraries are somewhat less extensive. In addition, under conditions usually employed (fully relaxed protons) the area under each peak (or cluster where peaks are split by coupling as described below) is proportional to the concentration and number of protons, unlike LCMS or GCMS, within a single spectrum. Most NMR instruments are quite stable quantitatively, although less sensitive and selective compared to coupled chromatography. Structures can often be elucidated much more confidently than via MS-based techniques. Nevertheless, compared to coupled chromatography often many less compounds can be detected, but NMR spectral libraries although somewhat less extensive are more robust than MS ones. NMR is an inherently less sensitive technique compared to MS.

There are many technical issues with the correct acquisition of a series of spectra for metabolomics, which an expert spectroscopist will be able to advise. Pulse sequences, solvent and chemical shift references all must be carefully controlled and chosen.

The chemometrician will be presented with transformed, aligned, data, but should be aware of the earlier steps required to obtain the dataset. In this text, we will restrict our illustrations to 1 D ^{1}H NMR data, which is the most common.

2.4.1   Fourier Transform Techniques

2.4.1.1   FT Principles

All modern NMR spectra are obtained using FT (Fourier transform) methods. We will not discuss the mathematics in detail, but it is important to appreciate the principles.

- NMR spectra are recorded using a time domain FID (free induction decay), which is acquired in a few seconds.
- Most metabolomics uses ^{1}H (proton) NMR although ^{13}C NMR is rare but can be used. In this text, we will be restricted to ^{1}H data.
- The FID is a rapid method of obtaining data, a typical acquisition time is around 5 s. The longer the acquisition time, the better the digital resolution in the resultant frequency domain spectrum as discussed below.

- In order to improve the signal-to-noise ratio (S : N), typically several spectra are acquired from each sample and averaged. The S : N improves by a factor of $\sqrt{M}$ where M is the number of scans. Therefore, acquiring 16 spectra improves it fourfold over taking a single spectrum. There is usually a relaxation delay between successive spectra, this delay is 2 s and each scan takes 5 s, this would require $7 \times 16 = 112$ s or nearly 2 min per sample for 16 scans. Of course, the number of scans required is very dependent on sample.

- The FID, also called the time domain, contains all the signals mixed up simultaneously. In order to make the spectrum comprehensible, it has to be transformed into a frequency domain or spectrum, as illustrated in Figure 2.7 using a FT. It is the spectrum that can be directly interpreted in terms of molecular abundances. The Cooley–Tukey algorithm was developed in the 1960s and is called a Fast Fourier Transform (FFT). In order to perform a FFT, the number of data points must equal a power of 2, which makes computational FFTs feasible within a short period of time.

- In fact, an FT is a little more complicated than illustrated in Figure 2.7. The time domain is usually converted into two spectra, a real spectrum and an imaginary spectrum (using the notation of complex numbers). Usually, as discussed above, the time series is sampled a number of times equal to a power of 2, as an example $2^{14} = 16{,}384$ times over 4 s. There are $2^{13} = 8192$ real and the same number of imaginary frequency domain spectra. This is illustrated in Figure 2.8.

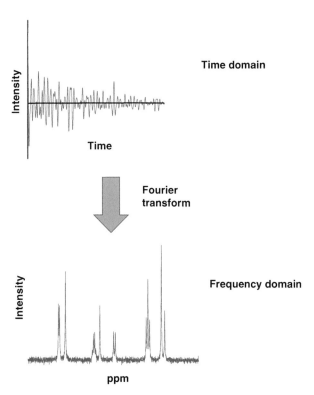

FIGURE 2.7   Fourier transformation.

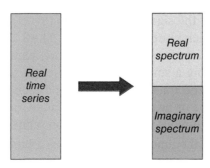

FIGURE 2.8   Converting a time series to real and imaginary spectra.

- The real frequency domain should ideally contain Lorentzian peak-shapes, which appear as unimodal positive peaks. However, these are normally mixed without-of-phase components called dispersion peak shapes. The spectroscopist has to mix the real and imaginary parts of the spectrum using a procedure called phasing to obtain an absorption mode spectrum consisting of unimodal positive Lorentzians, which can be further processed.

- The number of data points N and acquisition time of the FID T in s relate to the observable spectral width W in the frequency domain in Hz by $W = N/(2\,T)$.

- So if 16,384 data points have been acquired over 4 s, the frequency domain is $16,384/(2 \times 4) = 2048\,Hz$.

- NMR spectra are obtained by applying a radio-frequency pulse, usually measured in MHz. Typical metabolomic spectra for clinical purposes are recorded at 600 MHz, although other frequencies are occasionally employed, with lower field strengths still sometimes used in foodomic studies.

- Hence, if spectra are recorded over a range of 8192 Hz, the range of observable frequencies is $(8192/6 \times 10^8) \times 10^6$ ppm (parts per million) = 13.65 ppm as 1 ppm equals 600 Hz at 600 MHz. The ppm scale is often called a chemical shift scale or denoted by the symbol δ. Peaks can resonate outside this range, but they are 'folded over' in other words they are mixed up with the observable frequencies over the spectral width, and cannot be distinguished. Hence, it is best practice to ensure the spectral width in ppm encompasses all possible resonances of interest.

- The chemical shift in ppm is characteristic of a specific proton in a given environment. A typical molecule will have several different environments, so proton signals will be found in different regions of the NMR spectrum. There are many comprehensive books about how to interpret chemical shifts and assign them to specific groups in molecules, but we will not discuss this in depth here.

- In most ^{1}H NMR studies we are interested in spectra over a 10 ppm range. In organic solvents usually a chemical shift standard TMS (Tetramethylsilane) is added with a defined shift of $\delta = 0$ ppm. Most signals are downfield (lower ppm) from TMS, so a signal 2 ppm downfield (lower frequency) has a chemical shift of 2 ppm. TSP (trimethylsilyl)propionic-2,2,3,3-D_4 acid sodium salt) is often used as a chemical shift reference in aqueous solvents. Metabolomic studies normally involve aqueous extracts.

2.4.1.2 Resolution and Signal-to-Noise

The quality of NMR spectra is characterised by their resolution and their signal-to-noise.

Resolution is important for two reasons. First, different signals in the spectrum might overlap, so the sharper they are the better. Second many signals have multiplicity, that is, they do not consist of a single peak, but of several, in ratios such as $1 : 1$, $1 : 2 : 1$, $1 : 3 : 3 : 1$ or even more complicated patterns, called coupling; the coupling constant is in Hz and not ppm, and equals the distance between each of these peaks. Unless peaks are well resolved it may be difficult to determine this pattern, especially where signals are overlapping from other groups and compounds. At higher field strengths, coupling patterns appear better resolved because coupling is in Hz but resonances are separated in ppm; a 0.1 ppm separation will correspond to 60 Hz separation at 600 MHz but only 40 Hz separation at 400 MHz.

Better resolution can result in lower signal-to-noise and so both have to be balanced.

The ability to interpret these complex spectroscopic patterns depends on how good the resolution is, without compromising signal-to-noise, and there are several main ways of improving this.

- By tuning or shimming the instrument: This can be done manually by the operator and reduces the linewidth of individual resonances if done correctly or automated with an automatic tuning and matching probe.
- By apodization: This involves multiplying the FID by a function, for example, $h(t) = g(t)f(t)$ where $f(t)$ is the original time series, and $g(t)$ is an apodization or convolution function. This is performed prior to FT and can be mathematically related to a corresponding moving average filter in the frequency domain via the convolution theorem. There are many choices of apodization functions, but the most effective try to reduce the linewidth of each resonance in the spectrum whilst at the same time moderating the noise. An inappropriate function may increase resolution but also noise so that although the signals are sharper the noise is also amplified. For a meaningful analysis, if a set of several samples are to be analysed, the same method for apodization should be performed on all of them.
- By increasing digital resolution: Whereas the chemical shift in ppm is independent of field strength, coupling is in Hz and coupled protons will appear to be separated at different apparent δ according to field strength. Coupling constants in Hz are characteristic of the neighbouring environments of individual protons. The digital resolution in Hz in the frequency domain is dependent on the acquisition time in s, so one point in the frequency domain is obtained every $1/T$ Hz. Hence, if the acquisition time is 4 s, this results in a data point every 0.25 Hz in the transformed spectrum. To improve digital resolution, the acquisition time can be increased, but this takes time, which might be better spent obtaining more spectra and then apodizing the FID. An alternative is zero-filling as illustrated in Figure 2.9, in which the FID is doubled in length by adding zeroes. This works providing the FID has suitably decayed by the end of acquisition, otherwise it results in truncation of the end of the FID, which can have the consequence of strange peak-shapes in the transformed spectrum.

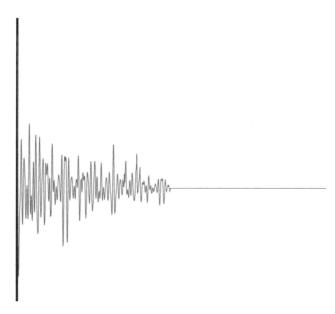

FIGURE 2.9 Zero-filling.

- Changing field strength is also feasible. Most metabolomics NMR data are recorded at 600 MHz, but different field strengths are possible. Increasing field strengths improves the signal-to-noise ratio. The improvement going from B_{old} to B_{new} MHz is $(B_{new}/B_{old})^3/2$, so increasing the field strength from 400 to 600 MHz increases the signal-to-noise ratio by $(6/4)^3/2 = 1.68$. However, this also has a consequence for resolution. Peaks separated by 1 ppm will be separated by 400 Hz at 400 MHz, but 600 Hz at 600 MHz. Coupling constants, though are field independent. This means that for spectra acquired over the same acquisition time at different field strengths, peaks of the same separation in Hz will be characterised by less data points and so if narrow may be harder to define.

An understanding of FT techniques can lead to improvement of the quality of the spectrum and is essential before performing the next steps. In this book, we will deal with transformed NMR spectra but advise hands-on users to be aware of the early steps in preparing the data.

2.4.2 Preparing the Transformed Spectra

After the first step, there will be a set of transformed, phased and ideally, well resolved spectra. A typical metabolomics experiment may result in 100 or so spectra. Several steps are required before the data are useful.

There is often a wavy baseline. As much statistical analysis involves analysing intensities above a baseline, it is always necessary to ensure a straight baseline. Ideally, the areas of noise in the spectrum (where no resonances can be detected) should have a mean of 0. Most baseline correction procedures involve determining which areas of the spectrum do not contain detectable spectral peaks, and then fitting a function, usually a polynomial to these areas. The new function is subtracted from the spectrum. Figure 2.10 illustrates this, with the red line representing the baseline, which is subtracted to give a corrected spectrum.

2.4.3   Preparing the Data Table

The next step is to prepare a data table for chemometric analysis. This transforms the set of spectra into a matrix whose rows represent samples (or spectra) and whose columns represent variables, which may be intensities or spectral areas at different chemical shifts (ppm or δ) or from the deconvoluted profiles of different compounds. In this text, we have datasets that involve intensities at different chemical shifts and also another that involves deconvoluted and identified metabolites. There are two fundamental approaches:

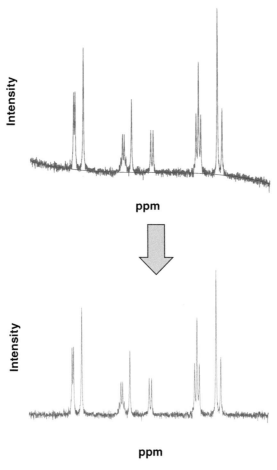

FIGURE 2.10   Baseline correction.

2.4.3.1   Chemometric Approach

In this approach, the net result is a data matrix whose columns consist of the intensities of resonance frequencies, either at single data points or over a range (bin or bucket) of data points.

2.4.3.1.1   Alignment

A very important step is to align spectra from different samples as each column has to correspond to the same type of information.

- Very small changes across samples are due to factors such as pH but also dilution, ionic strength, temperature and so on. It is critically important to align peaks across a dataset, otherwise individual variables may correspond to different compounds or not at all. For example, a proton resonating at 3.45 ppm in one sample may correspond to an equivalent type of proton resonating at 3.48 ppm in another sample.
- At 600 MHz, an 8192 data point frequency domain spectrum over 6000 Hz (10 ppm) would correspond to 8192/6000 = 1.36 data points per Hz or 819.2*0.03 data points for a 0.03 ppm shift at this field strength would equal a 24.57 data point shift.
- At even higher digital resolution (e.g. with zero-filling or longer acquisition time), this digital shift will be even greater.
- So for a meaningful data matrix where the columns for each variable correspond to the equivalent protons, in different spectra, spectral alignment is essential.

There are many methods for alignment. Some require expert programming, for example, in MATLAB or R or Python, and others are more automated. There is no room to discuss all the various approaches in detail, but a few are listed. The basic principle is that regions of spectra are stretched or compressed so that resonances from identical origins align, often called warping.

The methods can be grouped as follows:

- Methods that require peaks to be picked first. The chemical shifts of individual peaks are calculated and spectra are aligned to these. Methods differ in how they find corresponding peaks and their regions, how shifts are computed, and how they are applied. Recursive segment-wise peak alignment (RSPA) with its origins in 2009 as a method for NMR metabolomics is available as R code as the routine align_mQTL.
- Methods that involve aligning to a reference spectrum. This reference either is an experimental reference spectrum or could be computationally generated, for example, by PCA. Dynamic time warping (DTW) is a well known approach.
- Methods aligning segments of spectra. For MATLAB users, at the time of writing, the *icoshift* method is becoming increasingly popular, reported first in 2010. The method divides the spectrum into segments (normally equal) and aligns these separately. For non-MATLAB users, several methods for aligning segments of spectra have been included in NMRProcFlow, a user-friendly

open-source software to process and visualise 1D NMR spectra datasets. COW is the classical method, originally reported in 1998 for chromatographic data, but adapted for NMR also.

- Methods aligning whole spectra. Spectra are warped and the difference between them minimised. Fuzzy Warping (FW) and SpecAlign (originally developed for MS) are representatives of this group of methods.

To fully describe all available methods, of which there are 20 or so available, would require a lengthy chapter in its own right, and the aim of this text is primarily as an introduction. The choice of method depends on laboratories' expertise, some are in MATLAB, others in R, and others packaged. Some work better dependent on the data (is it complicated, are there lots of different peaks in each spectrum or are the spectra fairly similar, etc.). Some are in the form of downloadable freeware that is not maintained, but others are quite well established and maintained. It is not the purpose of this text to make recommendations, and it is likely there will be continual development of new approaches reported during the lifetime of this book.

Not all problems have been solved, one being of cross-over where the relative positions of peaks shift, so alignment cannot always be done perfectly just by warping spectra, but this often depends on how diverse the spectra are.

It is important to inspect spectra visually at this stage to make sure that there is a sensible solution. Where almost all spectra contain similar compounds but just at different relative concentrations, this is quite easy, however, where the composition of samples is very different it can still be challenging.

2.4.3.1.2 Binning or Bucketing

It is not always desirable to retain each single data point in a series of spectra. Reasons may, first, be that alignment, even with the best approach, is not always perfect to within a single data point, and also there may be small cross-overs in more complex cases. Second, with large datasets, it can be very time-consuming to process huge spectra, especially where some regions contain limited information.

Binning (aka bucketing) involves averaging the intensity within specific regions of a spectrum. Traditional binning sums intensities over a specified spectral range, typically 0.01–0.04 ppm. At 600 MHz, 0.04 ppm corresponds to a 24 Hz range. If there are 8192 data points recorded over a 13.65 ppm spectrum, this in fact corresponds also to exactly 24 data points. These bins allow for small shifts and imperfect alignment to be evened out.

An alternative is adaptive binning, where the regions are uneven in size, often using a variety of algorithms. Sometimes the widths of the bins correspond to spectroscopically interesting regions that can be directly interpreted in terms of chemistry. In some cases, small regions can be selected manually as an alternative, but this depends very much on users' knowledge and the complexity of the series of spectra.

It is not always necessary to perform this pre-processing.

2.4.3.1.3 Removal of Uninteresting Regions

Not all the spectrum is useful. Some consist of noise. These can interfere with the chemometrics analysis, particularly if the data are standardised.

2.4.3.2   Deconvolution and Identification

An alternative is for each column in the resultant data matrix to correspond to individual identified compounds. In LCMS and GCMS columns in the data matrix are almost always related to individual metabolites as identified by peak picking algorithms.

The alternative to the computational approaches discussed in Section 2.4.3.1 involves first to identify key compounds in the mixture and then create a data table whose columns consist of the intensity of the resonances of identified compounds or deconvoluted metabolites rather than of spectral frequencies.

This deconvolutes individual spectra separately and then identifies compounds in each of them, lining up identical compounds. Common approaches are described below:

Chenomx NMR Suite involves as follows:

- Using an NMR library appropriate to the matrix and the NMR field of the spectrometer. This database could be obtained experimentally or via any number of public domain libraries. As the pH may differ in mixtures, it should take into account pH variations.
- The acquisition of NMR spectra should be standardised.
- The NMR spectrum is then fitted to a sum of compounds in this database. Compounds not present can be modelled as background.
- Finally, the spectral area of each compound identified in the spectrum is obtained.

BATMAN is freeware.

- This differs from Chenomx in that it does not aim to identify an experimental library of compounds, but creates a template for candidate metabolites, using structural and other information.
- These templates usually are from the HMDB (Human Metabolome Database) but can be edited and extended by the user.

Bayesil is another approach to automatically deconvolute spectra using a known library, although the spectra have to be acquired in a standardised way. Bruker has developed several commercial approaches, AMIX, FoodScreener and JuiceScreener.

There are several other methods available. In all cases, just as for LCMS and GCMS, manual inspection and good expertise in interpreting spectra are important, particularly in tandem with automated methods. Mistakes can be made, for example, due to unusual line shapes, shifts, interferents, etc. A large NMR metabolomics dataset has often taken substantial resources to obtain, and as such it is important to invest time to check the data table obtained is meaningful. Mistakes at this stage can result in errors in the data matrix and so in the resultant statistical analysis, which could take significant time to unravel, resulting in errors incorporated into reports, papers, presentations and misinterpretation of pathways.

The data matrices obtained using this method and that of Section 2.4.3.2 are illustrated symbolically in Figure 2.11.

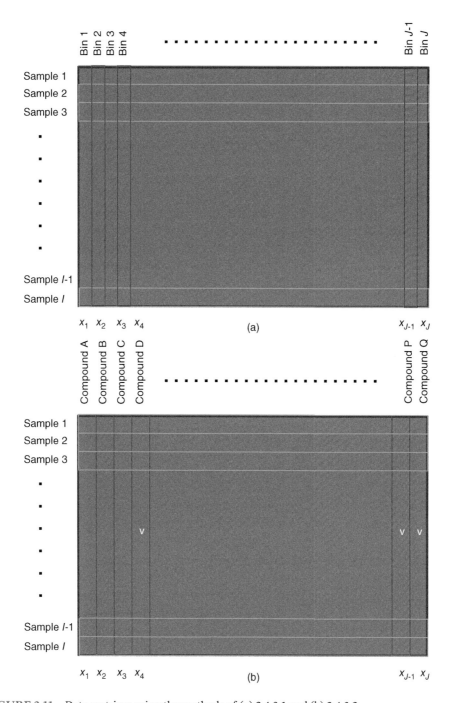

FIGURE 2.11 Data matrices using the methods of (a) 2.4.3.1 and (b) 2.4.3.2.

2.4.4   Identification of Metabolites

This is an important but difficult step. For the method of Section 2.4.3.2, identification is performed at an early stage before the data table is created.

However, using the chemometrics methods of Section 2.4.3.1, it is not essential to identify compounds in the spectra prior to multivariate statistical analysis. Some people might use their knowledge, for example, to choose bins based on certain known compounds they are interested in, or simply to identify specific regions of a spectrum. But to put some meaning to the statistical results, it is important to be able to assign regions of the NMR spectrum to individual metabolites, although this is possible after multivariate analysis.

Structural elucidation requires significant expertise in organic chemistry and is especially challenging in metabolomics where there may be tens or hundreds of compounds mixed up in one spectrum. There is no one accepted approach. To describe all the possible methods would involve writing a text in its own right. For the chemometrics expert, if you do not have sufficient knowledge, always consult an expert. As this is not the focus of this book, we outline just a few of the main tools. Often, several will be used in combination.

- Automated methods can be employed as in Section 2.4.3.2 but after a data table has been obtained rather than before. These depend on having a good spectral library, which takes into account pH variations. There are several spectral libraries that have been built up over the years.
- Special spectroscopic techniques can be employed to characterise certain compounds. 1D selective irradiation (selective excitation of one multiplet in a spectrum) and 2D spectroscopy are important tools. They can be performed on one or two spectra from the series containing representative compounds. So long as all the spectra line up, this will attempt to identify which regions belong to which metabolites. If chemometrics identifies an interesting area of the spectra, can we assign these to compounds of known structure?
- Spiking: Known standards are added to a spectrum and the peaks that increase in intensity are identified as belonging to a candidate metabolite, rather analogous to coelution in LC-based methods. This requires a laboratory to have an experimental library of interesting compounds.
- Using spectroscopic knowledge: In some cases, only one or more specific metabolites are of interest, for example, to calculate the amount of one or more specific compounds in different groups of individuals – a full spectral profile is unnecessary. In such case full spectral profiling may not be needed.

Like coupled chromatographic methods, it is not really possible to successfully interpret NMR profiles without some knowledge of spectroscopy. Chemometricians will often prefer to first perform alignment and where necessary binning, then multivariate methods, and finally once they have found the differences between samples they are interested in, interpret the significant variables at the end, possibly with the cooperation of a spectroscopist. There is no one accepted approach and it depends in part on the expertise of a specific laboratory. For some of the more well established markers, identification may be unambiguous, but for an NMR spectrum consisting of for example 100 possible metabolites, there are bound to be ambiguities and differences of interpretation according to how the spectra have been assigned.

In this text we will take data matrices both obtained by binning and integration and by deconvolution and show how chemometrics can discover trends, but without offering an expert spectroscopist's insight into how these can be interpreted in terms of molecular events.

2.5 VIBRATIONAL SPECTROSCOPY

Raman and Infrared (IR) spectroscopy can also be used for metabolomics and were quite widespread techniques for studying metabolic profiles in the early days, some 20 years before the writing of this book. There are three regions for IR spectroscopy, NIR (Near IR), MIR (Mid IR) and FIR (Far IR); the MIR region is commonly used in metabolomics and will be described below. The NIR region is not commonly used in metabolomics analysis and the FIR region is often used for imaging such as in astronomy.

Their original popularity is that they are cheap and easy techniques and do not require the complexity of data preprocessing compared to LCMS, GCMS and NMR. However, unlike NMR and chromatographic techniques, there is much less diagnostic information obtained from vibrational spectroscopy. Although individual molecules exhibit characteristic patterns, as a typical biological matrix will consist of many detectable molecules, their signals are usually combined to give a unique spectral fingerprint. However, it is a good way to, for example, group samples, look for outliers, validate classification models and so on, and is cheaper and faster than conventional chromatographic or NMR-based methods, whilst less diagnostic.

Both Raman and MIR spectroscopy produce a spectrum over similar wavelength ranges. In the context of structural chemistry, the same functional groups such as carbonyls and alkenes can be observed in addition to the so-called fingerprint region (in total approximately 400–4000 cm^{-1} or 25–2.5 μm over the full mid-IR or Raman spectral range typically used in these studies). Both techniques measure the amount of energy it takes for the molecular bonds to vibrate; however, for a vibration to be Raman active, there must be a change in the bonds' polarizability, whereas for IR it is a change in the dipole moment. The theory is rather complicated and interested readers should reference the specialist literature; however, we will give a simple example of CO_2 as illustrated in Figure 2.12. For non-chemists, the carbon atom has a small

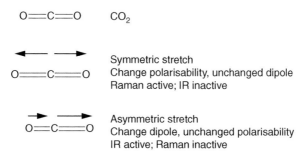

FIGURE 2.12 Symmetric and asymmetric stretches of CO_2.

positive charge and the oxygen atom has a small negative charge. There are specific vibrational modes that can be predicted by quantum mechanics. One is a symmetric stretch. In this case, the oxygen atoms move away from the central carbon atom simultaneously by the same amount. The dipole moment which relates to the average charge in the middle, does not change. However, as the molecule stretches, the polarizability changes because the charge distribution is more spread out as the oxygen atoms move away from the carbon atom. During the asymmetric stretch, the distance between the two oxygen atoms remains the same, hence the polarizability does not change, but the centre of the charge changes so does the dipole. Thus, the symmetric stretch is Raman active and IR inactive and conversely the asymmetric stretch is Raman inactive and IR active.

Of course, this differentiation is in practice much more complicated. Most features of a molecule have some vibrational modes, usually related to their functional groups. A very symmetric but simple compound, such as H_2, has no dipole or polarizability, but that is rare. Most compounds have many atoms and bonds, which in turn have several different vibrational modes, these are not arbitrary but predicted quite precisely by quantum mechanics. Some vibrational modes are Raman active and others IR active, and in some cases, a vibrational mode might be weakly Raman active and strongly IR active or any combination. It is beyond this text to provide a detailed discussion of spectral interpretation, but it is important to understand that these two types of spectra are complementary and emphasize different features. Ideally, both types should be acquired, but other considerations, such as cost, instrumentation available, sample preparation and speed of analysis. are probably more important, especially as we do not tend to interpret the spectra in molecular terms when doing metabolomics.

2.5.1   Raman Spectroscopy

Raman spectroscopy is an increasingly popular technique and is sometimes employed in metabolomics. It does not give detailed molecular insight but can be a fast and cost-effective technique, sometimes allowing hundreds of samples to be analysed quickly.

Raman spectroscopy is a very insensitive method, that is, the signals are very weak, and for many decades was not common because of the difficulty obtaining sufficiently intense spectra. This changed from the 1970s when the use of lasers for Raman spectroscopy was developed. As usual it took some years for the method to move from a somewhat esoteric method used mainly by physicists and physical chemists in specialist laboratories to a widespread technique incorporated into commercial instruments.

- Raman spectroscopy is non-destructive and non-invasive and can be performed directly on tissues, cells, blood, plasma and so on.
- Usually, the sample's vibrational energy is excited at a specific wavelength, often in the NIR region, which reduces fluorescence for biological samples, most commonly 785, 830 or 1064 nm. Note that although we may use the NIR region for excitation this does not mean we are interested in observing spectral transitions in this region.

- The majority of vibrational energy levels return to their original state, so emit radiation at the same wavelength as excitation. This is called Rayleigh scattering.
- A small portion of vibrations return to a higher energy level than previously and emit radiation at a slightly lower energy than previously, this is called Stokes scattering.
- An even smaller portion is already in a higher energy state, and after excitation falls to a lower energy state than previously, called anti-Stokes scattering.

The principles are illustrated in Figure 2.13.

- The top left illustrates an incident source of light, usually a laser.
- As the light interacts with the sample, there is scattering, most (blue arrows) being elastic scattering at the same wavelength of the incident light, but some (green and magenta) at slightly different wavelengths, called inelastic scattering. The small amount of scattered light at wavelengths differing from the incident light can be understood (by those trained in spectroscopy) by the energy diagram symbolised on the top right.

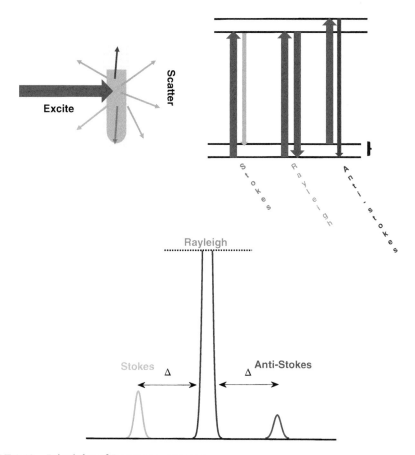

FIGURE 2.13   Principles of Raman spectroscopy.

- These give rise to three lines for each vibrational mode, the most intense Rayleigh bands and the less intense Stokes and anti-Stokes bands, as illustrated on the bottom diagram.
- In Raman spectroscopy, we normally just record the Stokes bands as displaced from the central Rayleigh band.
- The displacement Δ is recorded and is characteristic of different vibration modes, each having its own value, which is generally displayed in a spectrum.
- The range of wavelengths is normally reported in units of cm^{-1} rather than nm to be compatible with the traditional MIR frequencies, and can range from 50 to 5000 cm^{-1}, and is referred to as the Raman Shift, but the interesting region is normally in the centre of this range (ca 400–1800 cm^{-1}) often called the fingerprint region where individual molecules have characteristic patterns.

There are several problems encountered in practical Raman spectroscopy and the experienced user will be aware of these.

- Fluorescence can overwhelm the weak Raman bands if not careful. Biological samples are inherently fluorescent which is why we typically use longer wavelengths of light for excitation. Fluorescence can be overcome also by exposing the sample to extended laser light which 'quenches' or 'bleaches' the fluorophore but the increased laser exposure also increases the potential for sample damage. This can be controlled by using different excitation wavelengths, and also by paying attention to the matrix (solvent).
- Cosmic rays are high-energy particles from out of space that randomly hit the instrument detector and usually result in very sharp peaks. They can often be removed by software or by expert users who can easily identify them and distinguish them from the background. Median smoothing is a popular computational method for removing bands from cosmic rays.

Most datasets suitable for chemometrics processing have been corrected for fluorescence and cosmic ray background problems already but newcomers to the area should be aware of these potential issues in case their datasets show anomalies that are not expected, or make the chemometric processing hard.

2.5.2   Fourier Transform Infrared Spectroscopy

Mid-infrared (MIR) spectroscopy, often referred to as Fourier Transform infrared (FTIR) as the spectra are recorded by FT techniques which are faster and more sensitive than traditional approaches, is an alternative to Raman spectroscopy. FTIR produces spectra in a similar range as Raman spectra, typically 400–4000 cm^{-1}, when using a mid-IR (MIR) source.

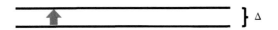

FIGURE 2.14   Principles of infrared spectroscopy.

The principles of IR spectroscopy for a single vibrational transition are illustrated in Figure 2.14. Obviously, like Raman spectroscopy, there will be several such transitions for most molecules, according to their complexity. However, the transitions will be in the same region as Raman, but differ in relative intensity as discussed earlier, according to the symmetry of the vibrational modes.

Compared to Raman spectroscopy:

- FTIR instrumentation is generally cheaper (although this advantage is less significant nowadays compared to a couple of decades ago).
- It is generally more sensitive, having higher signal-to-noise.
- It does not suffer from problems such as fluorescence and cosmic rays.
- However, IR spectra can contain large water peaks that dominate parts of the spectra; in traditional organic chemistry, spectra are recorded with an organic solvent, but this is not usually possible for biological matrices. Water is a very strong absorber of infrared light. Atmospheric water vapour is easily detected by an FTIR spectrometer but can be corrected by software. Water within a biological material, that is, bound water, makes FTIR measurement much more difficult. If examining a protein in a solution, then a specialised sampling accessory with a very small pathlength (~5–6 microns) can be used to minimise the dominating spectral features. In most cases biological cells and tissues are fixed, frozen, or freeze-dried and cross-sectioned to allow for analysis.

There are several ways of recording FTIR spectra, a popular sampling technique is ATR (attenuated total reflectance) which does not require sample preparation but involves pressing a crystal onto the sample of interest on top of the ATR crystal for data collection.

FTIR (aka MIR) and Raman spectroscopy are complementary techniques and partly can be chosen for speed and cost, but also partly to provide different types of spectra from similar samples as appropriate.

CHAPTER 3

Case Studies

3.1 INTRODUCTION

The majority of examples in the text are based on 10 case studies.

These have been chosen to be representative of the different types of data commonly found in metabolomics studies. They also are representative of the geographical spread of potential readers and the types of instruments used in this field.

The main features of the case studies in this text are listed in Table 3.1. From the total case studies, four are from humans, three from plants, two from animals and one from microbes, to represent a range of applications. In fact, different workers in different areas often encounter quite different challenges, and during extensive discussions when developing this text, it was noticeable that workers mainly familiar with clinical human metabolomics or with plant metabolomics often emphasise and use quite different methods. As an example, most studies on plants can be performed under very controlled conditions with only one or two factors varied; the effects of these factors can often be observed in the first principal component. Most clinical studies on humans, in contrast, cannot provide such clear information, as the metabolic profile depends on human behaviour, environment and diet over periods of many years and genetic variability is usually quite large, compared for example, to plants or many controlled laboratory-based animal studies; hence, the main trends are often not obvious in the first principal components or PLS (partial least squares) components. A worker familiar with plant metabolomics may feel that human data is not clear, whereas a worker familiar with human metabolomics might want to use methods that are unnecessarily sophisticated to analyse plant extracts and not be satisfied with more straightforward approaches.

We have also used a variety of instrumental techniques, with 3 NMR (nuclear magnetic resonance), 2 LCMS (liquid chromatography mass spectrometry) and 2 GCMS (gas chromatography mass spectrometry) datasets, representing the major workhorses in metabolomics. There may also be some readers using HPLC, Raman

Data Analysis and Chemometrics for Metabolomics, First Edition. Richard G. Brereton.
© 2024 John Wiley & Sons Ltd. Published 2024 by John Wiley & Sons Ltd.
Companion website: www.wiley.com/go/Brereton/ChemometricsforMetabolomics

TABLE 3.1   Major characteristics of the case studies used in this book.

Case study	Brief description	Origin	Instrument	Regional origin of data
1	Pre-arthritis	Human	LCMS	Sweden (Europe)
2	Malaria	Human	GCMS	Rwanda (Africa)
3	Triglycerides	Human	NMR	Norway (Europe)
4	Diabetes	Human	NMR	China (Asia)
5	Maize	Plant	NMR	France (Europe)
6	Wheat	Plant	FTIR	UK (Europe)
7	Enterococcal Bacteria	Microbes	Raman	UK (Europe)
8	Daphnia	Animal	GCMS	Spain (Europe)
9	Herbal Medicine	Plant	HPLC	Hong Kong (Asia)
10	Mice	Animal	LCMS	US (N America)

and FTIR (Fourier transform infrared spectroscopy), so we also include one case study of each type.

Finally, we hope this book will be useful in a wide variety of geographical regions. Some of the problems that are studied, for example, in Europe, may differ from those from China or Africa. Herbal medicine is important in China, whereas tropical diseases are of significance in Africa. Each culture and region has a different set of problems and sometimes uses different analytical approaches to tackle these problems. As an example, many studies on herbal medicine can be done adequately using HPLC, which technique on its own is probably not adequate for clinical studies. Hence, we exemplify by one dataset where samples originate from North America, two from Asia, one from Africa and six from Europe (UK, France, Spain, Norway and Sweden). Of course, some types of study are universal internationally as well.

In this text it is not the main aim to go through a full analysis of each case study from start to end: there are many excellent papers already describing each of the case studies, all of which are in the public domain and referenced in Chapter 1. The prime aim is to use various case studies to illustrate different methods where appropriate. Some case studies will just be introduced, for example, by a PCplot, whereas others will be analysed in several chapters. Some simulated datasets will also be used if appropriate to illustrate an approach and introduced in the relevant chapter.

3.2   CASE STUDY 1: PRESYMPTOMATIC STUDY OF HUMANS WITH RHEUMATOID ARTHRITIS USING BLOOD PLASMA AND LCMS

This dataset was obtained by the University of Umeå in Sweden and was a study of rheumatoid arthritis in humans. A large cohort of plasma samples was obtained from 479 individuals up to seven years prior to diagnosis of arthritis. Those exhibiting symptoms within seven years were assigned to form one group, and the ones without symptoms were controls.

The aims were as follows:

- To see whether the pre-symptomatic individuals and the controls can be distinguished before the disease is diagnosed and,
- if so, what markers can be identified in the pre-symptomatic group, giving a potential clinical method for predicting the onset of arthritis several years before symptoms appear.

Of the 479 individuals, only females that were non-smokers, with a BMI $< 30 \, \mathrm{kg \, m^{-3}}$ and drug-free, at the time of sampling, were used for the analysis below. Elimination of other factors reduced the dataset to 30 pre-symptomatic individuals who eventually developed arthritis that fitted the criteria; 19 controls were chosen from the remaining samples that likewise fitted the criteria and matched for average age, sex and other criteria with the pre-symptomatic individuals, making a sample size of 49 in total.

Although this sample size may appear somewhat small:

- It is important to minimise variation in the pre-symptomatic individuals, for example, by choosing only one sex and rejecting smokers, otherwise confounding factors may confuse the results.
- Due to the need for instrumental conditions to remain stable, there is usually a limit to the number of samples that can be analysed in one batch. In addition, there will be problems of time and expense if the sample size is very large. Due to the fact that large sample sizes cannot usually be analysed over a period of days or weeks using stable instrumentation, there often is a limited benefit given the additional cost and capacity of a laboratory to increase the number of samples above a certain level, as the additional variability in instrumental performance can become more significant relative to the metabolic signature being studied. It is important that the metabolic markers of interest may be only a very small portion of the metabolic profile, so need quite stable methods to uncover.
- Although there was a much larger number of controls in the original population compared to the pre-symptomatic group, the control group for the chemometric analysis had to be balanced approximately for age, weight, BMI, etc. and so was reduced to 19, compared to 30 that had been identified as developing arthritis.

The precise details are in the published literature and not discussed in depth in this text; however, it is often an important essential step when there is a large sample size to choose which are to be used for subsequent chemometric analysis. In Chapter 6, we discuss sample designs. Given the variability in human behaviour over a period of many years, perfect balance will never be achieved, even through surveys, which can often be unreliable, but it is still possible to remove some confounding factors.

LCMS analysis was employed on the plasma extracts in both positive and negative ion modes. An in-house experimental library of 713 standards was employed to identify the peaks. Metabolites not detected in both positive and negative ion modes and peaks of unusual features were removed, as well as those not able to be identified, leaving 81 identified compounds. For peak intensities, the maximum of the positive and negative ion intensities were entered as elements in the data matrix. For all compounds, the same polarity was chosen in all samples, due to the ionisation process, but some compounds were more intense in the positive and others in the negative mode. Some compounds were detected in only one mode.

Hence, a 49×81 data matrix was available for subsequent chemometric analysis. A few measurements were missing, we discuss the problems of missing data in Section 4.7.1. In this text, for simplicity, we replace missing measurements with 0s, as the proportion missing is quite small. Data can be missing either because the data are below the limit of detection or because of problems such as instrumental noise, overlapping peaks and spectral uncertainty. If the proportion is high, more consideration of the method for replacing these missing measurements might be important.

3.3 CASE STUDY 2: DIAGNOSIS OF MALARIA IN HUMAN BLOOD PLASMA OF CHILDREN USING GCMS

Malaria is a disease primarily found in the tropics, transmitted by mosquitos. Sampling for this study was performed in Rwanda, a landlocked country in central Africa. Plasma was collected from three hospitals in the country, frozen and then transported to Sweden for extraction, preparation and analysis.

In total, 421 blood plasma samples were collected from children between six months and six years old and assigned into three groups, namely

- healthy controls
- mild malaria
- severe malaria

World Health Organisation criteria were used initially for assigning samples to these groups, supplemented by clinical diagnosis.

The samples were further reduced to approximately the same number from each group (19 from the controls, 20 from mild malaria and 21 from severe malaria), plus a total of 12 replicate analyses (4 per group), making 72 samples in total, and the samples were selected using a factorial design from a PC scores plot of clinical data (for a description of factorial designs see Chapter 6). It is often impractical to analyse very large numbers of samples using GCMS, taking into account limited resources and also instrumental stability. If too many samples are analysed, changes in instrumental performance can be a problem and interfere with biological factors. Good stability and QC can be achieved over a smaller sample size. The samples in each group were balanced by sex.

The main aims were as follows:

- To see if GCMS of plasma samples can distinguish between the three groups and therefore is a good diagnostic tool: in contrast to case study 1, this is a 3-group problem.
- To study changes in metabolites as the disease progresses from healthy to mild-to-severe.

Multivariate curve resolution was used to obtain individual spectra and profiles. The NIST library was employed to identify most peaks, with the remaining identified manually. Only peaks that were identified were employed in the subsequent analysis, leaving 84 peaks to give a 72×84 data matrix for further analysis.

The peak areas were scaled in each chromatogram ratioed to nine internal standards. Peak areas of all standards were calculated, in each chromatogram. PCA was performed on the 72×9 matrix of the standards and then the score of the first PC was used to scale the peak areas of each chromatogram as follows. A PCA model (with unit variance scaling without subtraction of the mean) on peak areas of internal standards was calculated and the first PC score for each sample was used to scale the resolved data by dividing the peak areas of each sample with the corresponding score value.

3.4   CASE STUDY 3: MEASUREMENT OF TRIGLYCLERIDES IN CHILDREN'S BLOOD SERUM USING NMR

This dataset comes from a large study of the health of children in Norway, the aim being to see whether their lifestyle, especially as the result of level of physical exercise, can affect their metabolism.

The study was performed on 1202 fifth-grade (ca ten-year-old) children from the Sogn and Fjordane counties in western Norway. The overall aim is to determine the concentrations of several classes of lipoproteins and relate these to physical and other indices of cardio-metabolomic health and anthropometry. There was an approximate 50% balance between males and females.

- For all samples, the concentrations of several classes of serum lipoproteins were analysed using an established HPLC method.
- In this study, the concentrations of
  - TC (total cholesterol)
  - LDL (low-density lipoproteins)
  - HDL (high-density lipoproteins)
  - TG (triglycerides)
  were determined for each sample in $mmol\,L^{-1}$.

In order to reduce the size of the dataset, 200 samples were selected from the overall population.

The procedure for NMR analysis was as follows:

- Blood serum was sampled from each child.
- NMR spectra of extracts were run at 600 MHz over a spectral width of 30 ppm, and an average of 32 spectra was recorded for each sample.
- Each spectrum was processed over 131,072 ($=2^{17}$) frequency domain datapoints to give a resolution of 0.000229 ppm per datapoint ($=0.137\,Hz$).
- The spectra were aligned by a reproducible in-house method.

For the dataset presented in this text, the region 1.5 to −0.104 ppm was selected initially, resulting in 7002 data points where the lipoproteins of interest resonate. By chemical knowledge, a smaller region between 1.305 and 0.602 ppm, representing 3072 data points, can be selected specifically for calibration of TG. Other lipoprotein classes would be better estimated using different regions of the spectrum.

Hence, the full data over the region of interest can be represented by a 200×7002 data matrix, with a subset of a 200×3072 data matrix for the purpose of TG calibration.

There are several aims:

- The main aim is to see if a slow clinical procedure for TG can be replaced by a more rapid NMR method, by calibrating the NMR data to the TG measurements obtained using HPLC.
- The more detailed aim is to determine which variables (or chemical shifts) are most helpful to perform this calibration for TG and so to improve the calibration model by retaining only the most significant variables.
- Also, it is important to see how well the very high-resolution measurements perform, which depends on alignment.

The data in this text is part of a much larger dataset in which 30 lipoprotein measurements were measured. The clinical expense and time to measure all such lipoprotein classes in the laboratory is huge, and would make a large study impractical or very time-consuming, primarily doing routine laboratory analyses. Can a single and rapid NMR spectrum, together with chemometric analysis, replace these laborious tests and make an even larger study feasible?

Unlike the other two NMR case studies in this text, we are not concerned with the identification of individual marker metabolites but with quantifying classes of metabolites that may be clinically interesting.

3.5 CASE STUDY 4: GLUCOSE INTOLERANCE AND DIABETES IN HUMANS AS ASSESSED BY BLOOD SERUM USING NMR

Diabetes is one of the fastest-growing diseases in the world, caused in part by diet and lack of exercise. An understanding of the progression of this disease is important for treatment. Genetics plays a part and so a study, for example, using a European population, may result in different features to that using a Chinese population. This study was restricted to Han Chinese subjects reducing genetic variability.

A total of 231 blood serum samples were collected from residents in the Beijing area, classed into three groups:

- Eighty samples from individuals with NGT (normal glucose tolerance).
- Seventy-seven samples from individuals with IGT (impaired glucose tolerance)
- Seventy-four samples from individuals with T2DM (Type 2 Diabetes Mellitus).

The three groups were age and gender balance matched, with mean and standard deviations of ages 51 ± 9 years (NGT), 51 ± 10 years (IGT) and 53 ± 10 years (T2DM) and gender balance 34/46 M/F (NGT) 33/44 (IGT) and 32/42 (T2DM). Although the age and gender balance are slightly different between groups, it is not easy to have an exact balance. As the ethnicity and living environments were similar for all subjects, some of the major factors that could confound the analyses were reduced using this controlled population.

There were several aims:

- Can the three groups be distinguished by NMR?
- Can we see a disease progression between the three groups?
- What NMR peaks and so metabolites are responsible for the differences between the groups?

NMR spectra were recorded at 600 MHz and zero-filled to 128K. To improve signal-to-noise ratio and eliminate small shifts between spectra, 30 success data points were averaged to give a resultant spectrum with a resolution of 0.003 ppm (=1.8 Hz) and the region $\delta 0.5$–9.5 was divided into 2833 integral segments. The regions at $\delta 4.09$–4.21, $\delta 4.32$–5.17 and $\delta 5.50$–6.50 ppm were discarded to eliminate the effects of imperfect water saturation and the urea signals, with the remaining spectral frequencies between 0.4995 and 8.9985 ppm (2180 data points excluding the removed regions) being used for chemometric analysis, resulting in an 231×2180 data matrix. The data consists of readings of intensity versus ppm (or frequency) rather than deconvoluted spectra.

Several peaks were identified in the spectra using 2D NMR, primarily amino acids. Chemometrics would be used to identify the changes in these metabolites during disease progression.

3.6   CASE STUDY 5: METABOLIC CHANGES IN MAIZE DUE TO COLD AS ASSESSED BY NMR

This work aims to study metabolic changes in maize as the temperature is changed. This has significance as it can provide information about metabolism if seeds are sown earlier in the year when it is colder.

The experiment involves:

- germinating seeds for four days,
- transferring to a growing chamber at a temperature of 20 °C for 21 days,
- changing the temperature to 16 °C for a further two days,
- then to 13 °C for the next two days,
- and finally to 8.5 °C for the next two days.

Leaf samples were extracted at the end of each of the four periods (20, 16, 13 and 8.5 °C) with three replicates at each sampling time.

In order to see whether the genetic makeup of the plants influences the effect of temperature on metabolism, 18 genetically diverse dent maize inbred lines (*Zea mays* ssp. *mays*) were selected according to their diversity, based on pedigree and genotyping while keeping close flowering dates and crossed with the UH007 flint inbred line. In this text, we will not discuss the effect of genetics on maize, although having several maize types means that the conclusions are representative of a variety of genetically diverse maize.

This results in 18 (maize types)×4 (temperatures)×3 (replicates) = 216 samples and is illustrated in Figure 3.1.

NMR spectra were recorded at 700 MHz.

- 34 peaks were found in all samples and integrated.
- Of these, 26 were identified and assigned to known metabolites.
- The concentrations of each of the 26 known metabolites were calculated as $\mu g\,gDW^{-1}$ (gram Dry Weight) using calibration curves from the individual metabolites and for each compound a specified resonance.
- For the 8 unknowns, the concentrations were in arbitrary units, using peak areas per gDW.

In contrast to case studies 3 and 4, the NMR spectra are deconvoluted into individual peaks rather than bins (or buckets).

Hence, a 216×34 matrix was obtained, consisting of samples and NMR peak intensities. Questions that can be answered are as follows.

- Does the metabolic profile differ according to temperature as observed by NMR?
- Which metabolites change with temperature?
- Can groups of metabolites be identified that have similar temperature-based behaviour?
- Is there any effect due to the genetic nature of the wheat? This is not directly studied using the methods in this text, but could be seen if there were discrete clusters, for example, in the PC scores plots.

This work is part of a larger project which involves several different analytical techniques. In this text, we will just illustrate the results using deconvoluted NMR.

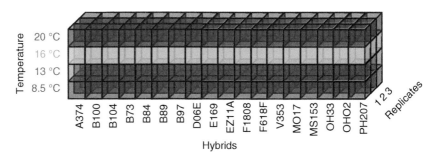

FIGURE 3.1   Experimental design for case study 5 (maize).

3.7   CASE STUDY 6: EFFECT OF NITRATES ON DIFFERENT PARTS OF WHEAT LEAVES AS ANALYSED BY FTIR

What is the effect of limiting the amount of nitrate available to plant growth? Nitrate in the soil is eventually incorporated as amino acids. This study examines how different parts of leaf of wheat are affected according to the nitrate content of the soil.

Plants were harvested after seven or eight days growing in media when the leaf is 12 cm long and switches from fully heterotrophic to fully autotrophic metabolism, either with or without added 10 mM KNO_3 to represent whether added nitrate has an influence on metabolic pathways.

- Six plants were grown with added nitrate and six without.
- Of these six plants, three parts of the leaf, namely
  - base
  - mid
  - tip

  were sampled.
- Each was analysed spectroscopically in triplicate, resulting in 2 (conditions)×6 (plants)×3 (parts of the leaf)×3 (samples) or a total of 108 spectra.

The spectra were analysed by FTIR.

- The region 4000 to 600 cm^{-1} was used for this analysis, at a digital resolution of 1.929 cm^{-1}, resulting in 1764 wavelengths per spectrum.
- To remove baseline effects, all spectra were corrected using SNV (Standard Normal Variates), which is a common approach in IR, described in Section 4.6.1.2. Each spectrum is mean-centred and then divided by its standard deviation

$$x_{ij}^{cor} = \frac{\left(x_{ij} - \bar{x}_j\right)}{s_i}$$ where $\bar{x}$ is the mean of the spectrum and s_i its standard deviation.

This then results in a 108×1764 data matrix for subsequent chemometric analysis. There are several aims of the work:

- Can the effect of added nitrate be observed in the metabolic profile as assessed by FTIR?
- Is there a difference in the effect of nitrate on different parts of the leaf?
- How good is the analytical procedure as assessed using replicates?

FTIR analysis is a quick approach and if successful could be followed using more expensive methods such as LCMS. In this case study we only look at the FTIR data.

3.8   CASE STUDY 7: RAPID DISCRIMINATION OF ENTEROCOCCAL BACTERIA IN FAECAL ISOLATES BY RAMAN SPECTROSCOPY

Enterococcus is a genus of bacteria causing a variety of infections. It is important to be able to distinguish different types. Traditionally, methods such as PCR (polymerase chain reaction) or ELISA (enzyme-linked immunosorbent assay) are employed but these can be time-consuming to carry out, so an alternative is to use rapid spectroscopic assays.

- In this study, 35 faecal isolates of *Enterococcus faecium* from different donors were obtained from a hospital in Belfast, Northern Ireland.
- They were typed using pulsed-field gel electrophoresis, and 12 different strains were detected. Some strains were more common than others; for example, 156UNI was found only in one individual (156), whereas EC13 was found in eight individuals (144, 149, 152, 154, 155, 167, 177, and 185) as some strains are more common than others.
- The isolates were then cultured on plates, each isolate having been divided into four plates.
- In turn, each plate was then sampled at seven different places, because the distribution of biological material is uneven on a plate.

The experimental design is illustrated in Figure 3.2, with the exception of the final level of the seven sampling points per plate. In total, there are 35 (donors/isolates)×4 (plates/cultures)×7 (sampling points) = 980 samples.

For each of the samples, a Raman spectrum was taken between 2492 and 351 cm^{-1} at a digital resolution of 2.169 cm^{-1}, resulting in 988 wavenumbers, using a 785 nm laser. Hence, a matrix of 980×988 is generated. The data are corrected by SNV (see previous case study) like FTIR.

There are several aims:

- The prime aim is to see whether Raman spectra can type the bacteria as an alternative to traditional methods. We will look at this aspect in the next chapter.
- It is also interesting to see which strains are most similar according to their Raman spectra.

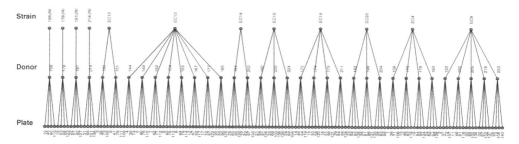

FIGURE 3.2   Experimental design for case study 7 (Enterococcus). Each plate is sampled further seven times at different points, not illustrated above but forming a fourth level.

- Finally, different sources of variability could be studied and compared, for example, between strains, between donors with the same strain (where several samples are available), between plates and within plates.

The last two aspects will not be discussed in detail in this book but can be discovered by readers from the data available.

Although the dataset is large, it is not realistic to perform supervised classification models, because for some strains there is only one isolate, so in practice, we are looking at various types of replication. To obtain a properly validated model, one would need many more donor samples; it is also not easy to control when some strains are rare and so the number of donors containing some strains may be small and this cannot be predicted in advance. However, this dataset is still very rich and is suitable for exploratory analysis as well as quantitative investigation of variability between plates and sampling points.

The dataset in this book is part of a larger study involving several different analytical techniques.

3.9   CASE STUDY 8: EFFECTS OF SALINITY, TEMPERATURE AND HYPOXIA ON *DAPHNIA MAGNA* METABOLISM AS STUDIED BY GCMS

This case study relates to animal metabolomics. *Daphnia* is an important model organism for aquatic environmental and toxicological studies. In this work, *Daphnia* was subjected to three important factors, which may vary due to climate change.

- Cultures of four-day-old *Daphnia* were used for this experiment.
- The cultures were then grown for one day under eight unique conditions arranged as a 3 factor 2-level factorial design (see Chapter 6 for more details) with the following factors:
  - Factor 1: Salinity ($0\,g\,L^{-1}$ (level -1) and $5\,g\,L^{-1}$ (level $+1$))
  - Factor 2: Temperature ($20\,°C$ (level -1) and $25\,°C$ (level $+1$))
  - Factor 3: Hypoxia ($9\,mg\,L^{-1}$ (level -1) and $2\,mg\,L^{-1}$ (level $+1$))
- Level 1 represents normal conditions and level $+1$ stressed conditions, but such that the *Daphnia* do not die.
- Each of the 8 sets of conditions was replicated 5 times, and the 3 most representative chromatograms were chosen (as discussed below using GCMS), to give an overall set for chemometric analysis of 24 samples, as illustrated in Figure 3.3.

After harvesting, samples were extracted, derivatized and GCMS obtained over a mass range of m/z 60–650 and a 49 minutes analysis time and 2606 retention times. MCR methods were employed to resolve out peaks (Section 2.2.4.2), using 12 time windows. From the well resolved peaks, and eliminating some inappropriate peaks, 47 compounds were identified confidently using the NIST standard library.

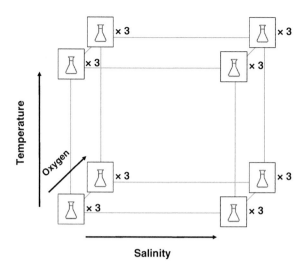

FIGURE 3.3   Experimental design for case study 8 (Daphnia).

Peak areas were calculated from the resolved peaks and ratioed to an internal standard. This resulted in a 24×47 data matrix, which was employed for further analysis. No row scaling was necessary for further analysis.

The main aims were as follows:

- To study the effects of salinity, temperature and hypoxia on *Daphnia* metabolism.
- Evaluation of changes on *Daphnia* metabolome by chemometric methods by determining which compounds change as conditions are altered.

The design used in this study allows interactions between factors to be taken into account. If analysed by a method called ASCA (anova simultaneous component analysis) as described in Chapter 10, it can also take the multivariate nature of the response into account, rather than just the traditional univariate analysis of each individual identified compound in the data.

3.10   CASE STUDY 9: BIOACTIVITY IN A CHINESE HERBAL MEDICINE STUDIES USING HPLC

In China, herbal medicine treatments are very widespread. The scientific basis is that bioactivity is not necessarily a consequence of a single compound, but may be the result of several compounds interacting. Western pharmaceuticals rely primarily on single compounds for pharmaceutical solutions, these are easier for clinical trials and for quality control in manufacturing. However, many biological processes are quite complex, and so the alternative in many Asian countries is to use herbal remedies. The precise composition of a plant will vary with time of year, soil, region, sunlight and temperature and so on, and can be hard to control, so there is considerable interest in studying bioactive herbal preparations as a widespread and often effective alternative to pharmaceuticals.

The Radix *Puerariae lobatae* or Yege is the dried root of *Pueraria lobata* (Wild.) Ohwi. Yege or Gegen has traditionally been used for improving body function, such as promoting circulation and increasing blood flow. In addition, it has been used for the treatment of cardiovascular diseases, such as hypertension, angina pectoris and type 2 diabetes mellitus.

- In this study, 78 plants were collected from different provinces in China.
- The bioactivity of each extract was studied by Ferric Reducing Antioxidant Power (FRAP) assay, to give an antioxidant activity for each extract in terms of a FRAP value. The higher, the more potent the sample.
- Each extract was also analysed by HPLC, and monitored at 254 nm over an elution time range from 9 to 31.4 minutes.
- Five chromatographic standards were run by HPLC on a regular basis to ensure peak areas were stable.
- HPLC intensities were recorded at 1.2 s intervals, resulting in 1120 points per chromatogram.

This resulted in a 78 (sample) $\times$ 1120 (RT) data matrix arising from the HPLC and a 78×1 vector of FRAP values.

The aims of this work were as follows:

- To see how well the bioactivity can be predicted using HPLC.
- To see which regions of the chromatogram, and hence metabolites, are most responsible for bioactivity.

3.11   CASE STUDY 10: DIABETES IN MICE STUDIED BY LCMS

This pilot and feasibility project uses a multi-omics approach to investigate the metabolome (primary metabolites), lipidome (complex lipids) and signalling lipids (oxylipins) in the plasma of Non-obese Diabetic (NOD) mice which progressed to type 1 diabetes mellitus (T1D) and those that did not. Mice ($n = 71$) were assessed as diabetic or non-diabetic based on their fasting (four hours) blood glucose levels at sacrifice, which defined 30 hyperglycaemic (glucose > 250 mg dL^{-1}) and 41 normoglycemic animals. As we will see in the later analysis, four mice diagnosed as non-diabetic were in fact probably diabetic, judged by their metabolomic profile, but this is a common consequence and shows the success of chemometrics combined with metabolomic profiling to identify outliers or misdiagnosed samples; as the proportion is small, these were included in most analyses, although could have been removed first as doubtful. The primary objective of this study was to identify candidate biomarkers associated with beta-cell destruction/survival and T1D progression.

- The dataset consisted of the extracts from blood plasma of 71 mice, 30 of which were diagnosed as diabetic, and 41 of which were diagnosed as non-diabetic (normal glucose tolerance). The mice were numbered 1–30 (diabetic) and 31–71 (diagnosed as not).
- Two datasets were obtained.
- The first is a negative ion UHPLC-QTOF MS/MS ESI lipidomics dataset of dimensions 71×32. The variables consist of the peak heights of 32 identified metabolites.
- The second consists of positive ion UHPLC-QTOF MS/MS ESI data of dimensions 71×146, the variables also being peak heights of 146 identified metabolites.

In order to analyse the data,

- The peak heights for each sample are row-scaled to a constant total of 1.
- The columns are standardised.

These methods of preprocessing are described in more detail in Chapter 4.

The primary aim is to see which metabolites are the most likely markers for distinguishing the two groups.

More details can be obtained from the Metabolomics Workbench website study ST000075 where the analytical protocol and genetics are discussed extensively. In this text, we will mainly use these data for the purpose of illustrating classification and variable selection methods.

Principal Component Analysis

4.1 A SIMPLE EXAMPLE: MATRICES, VECTORS AND SCALARS

In order to illustrate the method, we will discuss a simple example as presented in Table 4.1.

- The dataset consists of three objects (A, B and C) which may for example represent three samples extracted from different organisms, and two variables (1 and 2), which may represent metabolites.
- The six cells within the table represent measurements, for example, the concentrations of the metabolites.
- The objects are usually represented by rows and the variables by columns.
- It is usual to organise the six measurements as a matrix, which has dimensions 3 (rows) × 2 (columns). Note that the number of rows is always cited first when describing the dimensions of a matrix. Each number in the matrix is often referred to as an element of the matrix, and could represent the concentration of a metabolite in an extract.

TABLE 4.1 Example of a 3 × 2 matrix.

		Variables	
		1	**2**
Objects	**A**	7	5
	B	8	4
	C	2	6

Data Analysis and Chemometrics for Metabolomics, First Edition. Richard G. Brereton.
© 2024 John Wiley & Sons Ltd. Published 2024 by John Wiley & Sons Ltd.
Companion website: www.wiley.com/go/Brereton/ChemometricsforMetabolomics

- We will denote this matrix X. Matrices are usually denoted by bold uppercase letters, sometimes italicised. We will use the latter convention in this text.

- The matrix can be represented numerically by $X = \begin{bmatrix} 7 & 5 \\ 8 & 4 \\ 2 & 6 \end{bmatrix}$.

Individual rows and columns are denoted by vectors.

- For example, the row vector [8 4] could represent the concentration of two metabolites in an extract from organism B.

- The column vector $\begin{bmatrix} 7 \\ 8 \\ 2 \end{bmatrix}$ could represent the concentrations of metabolite 1 in each of three extracts.

- Vectors are normally denoted by bold lower case letters, sometimes also italicised, which convention we will employ in this text.

- Often a subscript can be employed to indicate which row or column is being denoted when describing vectors as rows or columns of a matrix, so x_1 could reference either the first row or first column of the matrix. Some authors use the default definition x for a column vector and x' for a row vector, where ' means transpose. However, this could be confused with the transpose of a column vector, so is $x_1' = \begin{bmatrix} 7 \\ 8 \\ 2 \end{bmatrix}$ or is $x_1' = \begin{bmatrix} 7 \\ 5 \end{bmatrix}$? This depends on context so always read an article carefully.

Individual cells or elements of a matrix are called scalars.

- Hence, $x_{32} = 6$, the element of the matrix in the third row and second column.
- When referring to elements of a matrix, always cite the row first and column second.
- Scalars should always be italicised but never bold.

Of course, scalars do not need to correspond to elements of a matrix. Any individual measurement is a scalar, for example, $a = 7.35$ and any set of univariate measurements can be arranged as a vector, for example, $b = [7.2 \; 3.9 \; 5.6]$.

4.2   VISUALISING THE DATA DIRECT

We want to ask several questions from the data, even if the dataset is quite small. The two most obvious questions are as follows:

- Which of the objects (aka samples) are most similar?
- Which variables (e.g. metabolites) are most characteristic of each sample?

One traditional way of helping answer these questions is graphically.

- The objects can be characterised by the values of the variables, and their relationship can be visualised graphically using these values as in Figure 4.1.
- Each axis represents a variable (e.g. the concentration of a metabolite), and we say that the objects are plotted in variable space, as an imaginary graph whose axes represent the size of each of the variables.
- In our case, we see that objects A and B are more similar to each other compared to object C.

For the variables, the graph becomes more complicated in this case as there are three objects and so three axes.

- Figure 4.2(a) is a 3D representation of these two variables plotted against the objects.
- Each represents an object (e.g. a sample), and we can say that the variables are plotted in object space.
- It is not so obvious how the variables relate to the objects from comparing Figure 4.1 and Figure 4.2(a).
- The variable space is two dimensional in our case. However, although the object space is defined by three axes, the variables fall onto a plane, including the origin, in this space, so in practice form a 2D subspace (in technical terms) as illustrated in Figure 4.2(b).

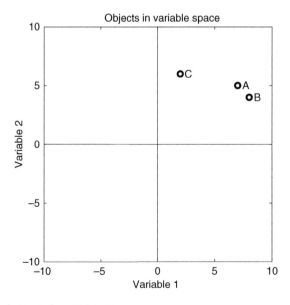

FIGURE 4.1   Plot of objects from Table 4.1.

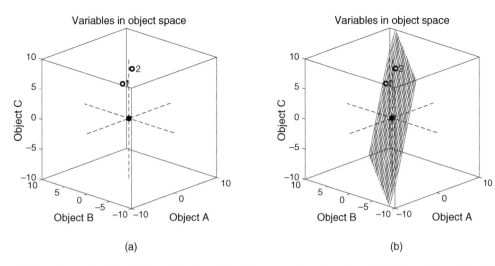

FIGURE 4.2   (a) Variables 1 and 2 plotted in object space (b) plane fitted including objects 1 and 2 and the origin.

- This is inevitable as there are only two variables in our example. Therefore, if we compare a suitable representation of the variables and objects we can find out which variables are most characteristic of which objects. This is analogous to helping us decide which metabolites are most characteristic of which samples.

In practice of course it is much more complicated to view the original data directly using such graphical tools. In traditional investigations, one or two measurements might have been obtained from each sample, with different strains or different diseases just characterised by the presence or by the relative concentration of one or two metabolites. Still, many professionals like to think that a single compound can be found that characterises a group of organisms. However, with modern-day instruments such as LCMS (liquid chromatography tandem mass spectrometry) and NMR (nuclear magnetic resonance) we can record thousands of pieces of information per sample and sometimes identify tens or hundreds of metabolites. Biological processes are complicated and most are unlikely to be the result of the effect of a single metabolite, so it makes sense to try to characterise samples by their metabolic fingerprint.

If we have 50 metabolites that are detected in a set of 100 samples, then

- There will be $(50 \times 49)/2 = 1225$ possible 2D plots in variable space where each axis represents a unique variable,
- and there will be $50!/(47! \times 3!) = 19{,}600$ possible 3D plots in variable space where each axis represents a unique variable.
- Similarly, there will be 4,950 possible 2D plots and 161,700 possible 3D plots in object space.

Clearly, this amount of information is huge and not possible to visualise by eye, so other approaches are required as described below.

4.3   PRINCIPAL COMPONENTS ANALYSIS: SCORES, LOADINGS AND EIGENVALUES

4.3.1   PCA

The ability to make sense of multivariate patterns it aided by PCA (Principal Components Analysis), which is the most widespread technique in chemometrics.

Instead of the original variables, we can calculate the scores and loadings. There are several algorithms for their calculation but a common one is NIPALS (Nonlinear Iterative Partial Least Squares). The experimental data matrix can be decomposed into the product of two matrices plus where appropriate a residual as follows:

$$X = TP + E$$

where

- X is the original data matrix of dimensions $I \times J$ (or 3×2 in our case).
- T is the scores matrix of dimensions I (number of objects) $\times A$ (number of PCs (principal components)) and represents the objects or samples.
- P is the loadings matrix of dimensions A (number of PCs) $\times J$ (number of variables) and represents the variables or metabolites.
- E is an error or residual matrix of the same dimensions as X. If the number of PCs equals the smallest of the number of objects or variables, E is zero (if two variables or objects are exactly correlated this reduces the number still further) as the data are completely modelled.
- In this example, if we calculate $A = 2$ components which fully characterises the data, T has dimensions 3×2 and P has dimensions 2×2 and E is zero.

4.3.2   Scores

For our small simulated example, the scores are presented graphically in Figure 4.3(a) and can be used to represent the relationship between the objects or samples.

- The scores can be visualised as a rotation of the original data represented in Figure 4.1); this rotation is 37.89° clockwise in our example.
- The scores of the two PCs are numerically equal to the distance along the two new axes and are sometimes presented as projections as in Figure 4.3(b).
- The numerical values of the scores can be represented as a 3×2 matrix as in Table 4.2.
- PC scores from different principal components have the mathematical property that they are orthogonal. This means that the sum of the product of the elements of any two PCs is 0, in our case as there are just two PCs, there is only one pair of components, so

$$\sum_{i=1}^{I} t_{ia} t_{ib} = \sum_{i=1}^{3} t_{i1} t_{i2} = \left(\left(8.595 \times -0.354 \right) + \left(8.770 \times -1.757 \right) + \left(5.264 \times 3.506 \right) \right) = 0$$

where t_{ia} is the score of the ith object and ath component.

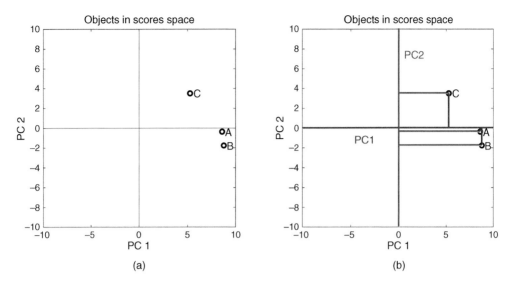

FIGURE 4.3 Representation of the objects (rows) of Table 4.1 in scores space (a) a PC scores plot (b) illustration of projection onto PCs.

TABLE 4.2 Scores of the data from Table 4.1.

		Scores	
		t_1	t_2
Objects	**A**	8.595	−0.354
	B	8.770	−1.757
	C	5.264	3.506

- An alternative is to use vector notation where $t'_a t_b = 0$ and t_a is the scores of component a, in our case a 3×1 vector. The symbol ' is used to denote a transpose in this text; some people us ^{T} instead.
- Because the scores are orthogonal, this means they preserve their relative positions in the original variable space.
- The sum of squares of the scores of all non-zero PCs equals the total sum of squares of the original data, so $7^5 + 5^5 + 8^2 + 4^2 + 2^2 + 6^2 = 8.595^2 + -0.354^2 + 8.770^2 + -1.757^2 + 5.264^2 + 3.506^2 = 194$ in this case.
- PCA can result in two solutions for each component, corresponding to a clockwise and anti-clockwise rotation and it is not possible to fix this, according to algorithm. This means that the scores for individual components can change sign according to package.
- The rotation angle for each component is defined as the angle for which the sum of squares of the scores $\sum_{i=1}^{I} t_{ia}^2$ is a maximum. Different rotation angles will give different sums of squares and will not be the PCA solution.

Scores can allow us to visualise the relationship between the objects, for example, which organisms or extracts have similar metabolic profiles.

4.3.3   Loadings

The loadings can be used to represent the relationship between the variables.

- We can view the loadings in a 3D plot as in Figure 4.4(a), but the third (PC3) axis is 0, as there are only two non-zero components.
- In Figure 4.4(b) it can be seen that the loadings of the two variables and the origin lie in a plane, with the third axis equal to zero.
- The loadings can now be represented in the form of a plane, just like the scores as there are only two non-zero components, as in Figure 4.5.
- The numerical values of the loadings are presented in Table 4.3. Note that we represent the loadings of each PC by rows in this text: some authors transpose the matrix.
- Like the scores, the loadings are also orthogonal in PCA, so

$$\sum_{J=1}^{J} p_{aj} p_{bj} = \sum_{j=1}^{2} p_{1j} p_{2j} = \left(\left(0.789 \times 0.614 \right) + \left(-0.614 \times 0.789 \right) \right) = 0 \text{ where } p_{aj} \text{ is the}$$

  loading of the jth variable and the ath component.

- However, loadings have an additional property, that is, the sum of squares of all the elements of each PC loadings is 1. Hence, for PC1, $0.789^2 + 0.614^2 = 1$. The loadings are said to be normalised, and with the property above are said to be orthonormal. The sum of squares of loadings of all non-zero PCs equals the number of PCs.

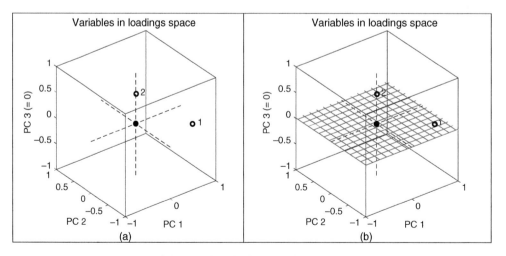

FIGURE 4.4   Representing of the variables (columns) (a) in loadings space (b) with plane at PC3 = 0, including the two variables and the origin.

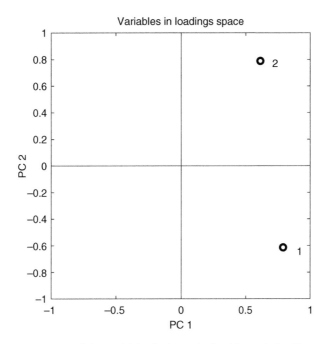

FIGURE 4.5   Representation of the variables (columns) of Table 4.1 in loadings space.

TABLE 4.3   Loadings of the data from Table 4.1.

		Variables	
		1	**2**
Loadings	p_1	0.789	0.614
	p_2	−0.614	0.789

- The loadings also can be represented as a rotation in 3D space, plus a data scaling. Note that the loadings also correspond to rotation angles as discussed below.
- Using the NIPALS algorithm, the magnitude of the PCs is represented in the scores, which correspond to a rotation in the original variable space. Although the loadings could also be represented as a rotation in object space, as they keep the same geometric relationship (such as angles and relative distance) as in the original object space, they are scaled so that their sum of squares equals 1.
- Some PCA algorithms have different ways of scaling the data, notably SVD (Singular Value Decomposition) in which both the scores and loadings are orthonormal. In this text, we will not use the SVD algorithm, but readers should check the packages they use; the difference though is just in scaling.

When there are several variables, the loadings can guide us as to which variables behave in a similar way. If there is a cluster of variables that are close in a loadings plot, they might originate from the same pathway or metabolic process.

4.3.4   Relationship Between Scores and Loadings

It is helpful to be able to compare the scores and loadings, in this case by comparing Figure 4.3(a) and Figure 4.5.

- The position of variable 1 in the loadings plot is in the same direction as objects A and B in the scores plot.
- This suggests variable 1 is at a high value in objects A and B but low in object C.
- There is a similar relationship between variable 2 and object C.

Of course, most datasets are more complicated but by comparing scores and loadings plots one can deduce which metabolites have high concentrations in which samples. We will discuss more complex patterns in later sections.

- The loadings matrix $\begin{bmatrix} 0.789 & 0.614 \\ -0.614 & 0.789 \end{bmatrix}$ can also be considered as a rotation matrix

- as $X = TP$ so $T = XP^{-1}$
- but because P is orthonormal, $P^{-1} = P' = \begin{bmatrix} 0.789 & -0.614 \\ 0.614 & 0.789 \end{bmatrix}$ as can be verified by the reader.

- Since $\cos(37.89°) = 0.789$, $P' = \begin{bmatrix} \cos(37.89°) & -\sin(37.89°) \\ \sin(37.89°) & \cos(37.89°) \end{bmatrix}$ which represents a

  clockwise rotation in variable space.
- So, the loadings can be considered to define a rotation of the original data matrix to produce the scores in variable space.
- T is not a square matrix in this example, because there are more objects than non-zero PCs.
- The number of non-zero PCs is never more than the smaller of the number of objects (rows) or variables (columns).
- Only square matrices have inverses.
- However, for non-square matrices, we can use the pseudoinverse T^+ instead defined so $P = T^+X = (T'T)T'X$. Readers can verify this. The pseudoinverse can be considered as a projection or transformation matrix just as P' can be visualised as a rotation matrix: for simplicity, we will not discuss the algebra in detail here but the reader can consult the more specialist literature on geometric transformations if interested.

Hence, the loadings can be used to transform the raw data onto scores space, and the scores can be used to transform the raw data onto loadings space. For a full rank transformation when all non-zero PCs are used in the model, this is equivalent to a rotation in the original dataspace for the scores, whereas when less than the full number of non-zero PCs are used in the model it is considered to be a projection onto a lower dimensional space (e.g. a projection from 2D space onto 1D space if only one component is used).

4.3.5 Eigenvalues

Another common quantity is the eigenvalue of each PC. This can be defined in a variety of ways but a simple way of calculating this in NIPALS is as the sum of squares of the scores of each component, which we denote by g_a for component a.

- This equals $g_1 = \sum_{i=1}^{I} t_{i1}^2 = \left(8.595^2 + 8.770^2 + 5.264^2\right) = 178.49$ and

$$g_2 = \sum_{i=1}^{I} t_{i2}^2 = \left(\left(-0.354\right)^2 + \left(-1.757\right)^2 + 3.506^2\right) = 15.51.$$

- The sum of eigenvalues for all non-zero components equals the sum of squares for all the elements in $\boldsymbol{X}$ or in this example $g_1 + g_2 = \sum_{i=1}^{3}\sum_{j=1}^{2} x_{ij}^2 = 194.$

- The bigger the eigenvalue the more the component contributes to the variation in the data. Principal components are usually ordered according to their size of eigenvalue, so $g_1 > g_2$ and so on for successive components.

- The size of components can also be expressed as a percentage of the overall variation or sum of squares. In our case PC1 represents 92.01% and PC2 7.99% of the variation.

- Usually, we use the term variance rather than variation, because it is common although not mandatory to column centre data, so most reports use the term 'percentage variance'; we will discuss centring below in Section 4.6.2.

4.3.6 Reducing the Number of PCs

PCA has many roles. Above and in many applications in metabolomics the main job of PCA is visualising the data.

However, consider a 100×50 data matrix. There are up to 50 possible non-zero PCs. We cannot visualise the scores or loadings in a 50-dimensional graph and need to simplify.

Often the later PCs are very small and reflect uninteresting trends, sometimes experimental error or instrumental noise. In metabolomics, although we may be able to detect 50 metabolites, there are likely to be only a few important pathways, so we do not expect all 50 compounds to have independent behaviour.

It is normal to choose only a few PCs, those that are judged most significant. There is a huge literature in the early years of chemometrics about how to make this choice; however, the focus of the early decades of the subject was in analytical and physical chemistry. For example, the number of components in a series of spectra of mixtures might relate to the number of detectable analytes, and getting this right is very important. Even in metabolomics, the current choice of the number of PCs is important, for example, when using MCR to resolve a cluster of peaks in GCMS. However, for modelling and visualising resolved data in which each variable may correspond to the concentration of a metabolite, this choice is much less crucial, often just using the two or three largest PCs is sufficient.

- A full model is one in which $X = TP$ and usually the dimensions of T are $I \times A$ and of P are $A \times J$ where there are I rows and J columns in X and A is the smallest of I and J (there are exceptions if there are correlations between variables in X but we will not elaborate on this). Any components after the Ath are zero.
- However, we can reduce the number of components. An original aim of PCA was data reduction, this made storage and calculations in less powerful micros easier, for example, the NIPALS algorithm calculates components one by one, and so instead of calculating 50 components when there are 50 samples, one could just calculate the first 3 or 4. In modern times this is not such a serious limitation, however still we are often not interested in later components.
- A reduced model, may not exactly reconstruct the original data, so
- $X \approx TP$,
  - or alternatively $X = TP + E$ where E is a residual matrix, sometimes also called an error matrix,
  - or $\widehat{X} = TP$ where the '^' means estimated.
- We can see how well the model estimates the data by the size of E which equals $X - \widehat{X}$ as calculated by its sum of squares. This can be compared to the sum of squares of the original data X.

Consider what happens if we use only one PC to model our small example.

- The estimated values for $\widehat{X}$ are given by multiplying the scores with the loadings for the first PC or $\begin{bmatrix} 8.595 \\ 8.770 \\ 5.264 \end{bmatrix} \begin{bmatrix} 0.789 & 0.614 \end{bmatrix}$ and are given in Table 4.4.
- The residual or error matrix is also presented.
- The sum of squares of the elements of $\widehat{X} + E$ equals the sum of squares of the elements of X as calculated in Table 4.4; they are also equal to the sums of squares of the scores of PC1 and PC2, respectively, when the data are completely modelled by two non-zero PCs, as can be verified.
- The first PC represents 92.01% of the variation in the data in our small example, so we have not lost much by using only one PC and discarding the second one.

We can therefore work out how well the data are represented by the first few PCs of a model and whether we are missing anything very significant by just concentrating on the first PCs, or whether including more PCs is a useful exercise. If we want to keep the first three out of a possible 50 non-zero PCs, just find the sum of the eigenvalues of

TABLE 4.4   (Left) data matrix **X**, (Centre) model using 1 PC, (Right) residual or error matrix.

	X		$\hat{X}$		E	
	7	5	6.783	5.279	0.218	−0.279
	8	4	6.921	5.387	1.080	−1.387
	2	6	4.154	3.233	−2.154	2.767
Sum of squares						
	194		178.492		15.508	

these three PCs and divide by the total sum of squares in order to work out the proportion of variation modelled by retaining just 3 out of 50 PCs.

- In Figure 4.6, we visualise the effect of fitting the data to a one PC model.
- Note that all three points now lie on a straight line including the origin.
- Objects A and B are very close together whereas object C is distant.

When we visualise data in scores space, for example, if we take a dataset of 100 samples and 50 metabolites and visualise the data in a 2D plot of PC2 versus PC1 we are performing a similar type of simplification but from 50-dimensional space (which we cannot visualise) to two-dimensional space, rather than from two dimensions to one dimensions (a line from the origin) as in Figure 4.6.

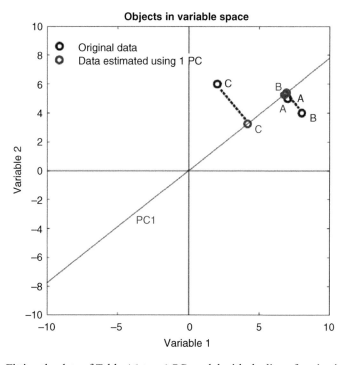

FIGURE 4.6   Fitting the data of Table 4.1 to a 1 PC model with the line of projection illustrated.

4.4   EXPLORATION BY PCA OF CASE STUDY 5 IN DETAIL: NMR STUDY OF THE EFFECT OF TEMPERATURE ON MAIZE

We extend these principles to a real study, that of the effect of temperature on maize. Details of the background are discussed in Section 3.6 and only a brief reminder is provided here.

From the total, 216 extracts are obtained at four different harvesting temperatures, 20, 16, 13 and 8.5 °C, each consisting of 18 different maize types replicated three times. NMR is used to characterise 34 metabolites of which 26 are identified and 8 are unknown, giving a 216 × 34 matrix of NMR intensities: for the known metabolites these are quantified as $\mu g\,gDW^{-1}$, whereas for the others these are reported as peak areas per dry weight.

The first step is to standardise the data matrix as the relative intensities of the metabolites are quite different. This step is discussed in more detail in Section 4.6.3.

4.4.1   Variable Plots

As there are 34 variables (or metabolites) we could produce a very large number of plots of the intensity of one variable against the other. In Figure 4.7 and Figure 4.8, we present the 2D and 3D plots of three of the variables, glucose, sucrose and fructose against each other, with the harvesting temperatures indicated in colour. We can see that whereas there appears to be some trend due to temperature, this is not very obvious. There are a few possible outliers, one at 16 °C and two (if judged from the scores using two PCs, with a possible third one if PC 3 is visualised) at 13 °C.

However, these variable plots are of limited help. They can only investigate a maximum of three variables at a time, so there would need to be a large number of possible graphs, making the job impracticable.

4.4.2   Scores and Loadings Plots of the Whole Standardised Data

Instead, we can perform PCA on the standardised data.

- Figure 4.9 presents the plot of the scores of the first two PCs.

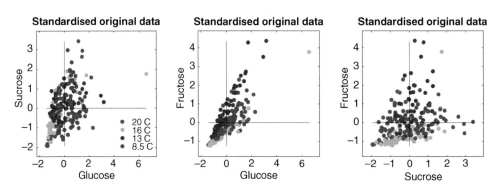

FIGURE 4.7   2D plots of standardised original data for the maize, for three of the metabolites.

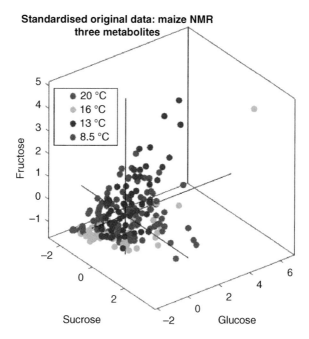

FIGURE 4.8   3D plot of the standardised concentrations of three metabolites for the maize.

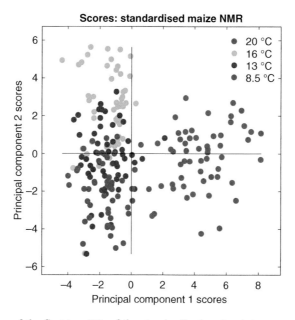

FIGURE 4.9   Scores of the first two PCs of the standardised maize data.

- The plot differentiates the extracts obtained at 20 °C from the lower temperatures, along PC1.
- Lower temperature extracts show some separation along PC2. However, this PC does not discriminate the higher temperature extracts.
- The eigenvalues of PC1 and PC2 are 1833 and 1151; the total sum of squares for the PCs equals 7334 (which is the product 216×34), so PC1 represents 25.0% and PC2 15.7% of the overall variance, or 40.7% in total.
- Figure 4.10 presents the loadings of these PCs, and can help gain metabolic insight. For example, we see that glutamate, malate and citrate are primarily associated (at a high level) with high-temperature harvesting, whereas fructose, glucose and sucrose at low temperatures. Many other deductions can be made by comparing Figure 4.9 with Figure 4.10; however, comparing scores and loadings plots can provide insight into which metabolites are most associated with which samples.

We can go further and plot the first three PCs.

- The third PC has an eigenvalue of 701, so the first three PCs correspond to 50.2% of the overall variance.
- The 3D scores and loadings plots are presented in Figure 4.11 and Figure 4.12.
- Including PC3 allows a more obvious visualisation of the change from high to low temperature and can be interpreted for more metabolic insights, which we will not discuss for brevity.

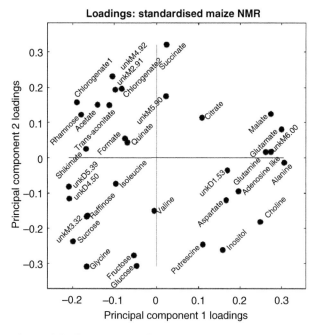

FIGURE 4.10 Loadings of the first two PCs of the standardised maize data.

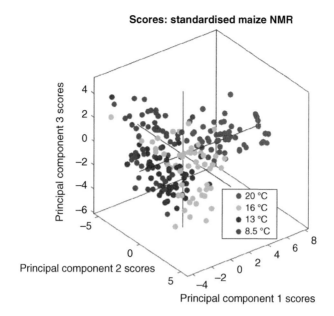

FIGURE 4.11   3D scores plot of the standardised maize data.

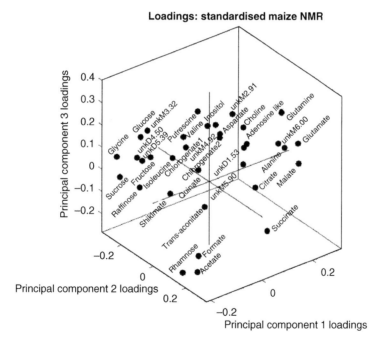

FIGURE 4.12   3D loadings plot of the standardised maize data.

4.4.3   Scores and Loadings of the Low-temperature Data

The 2D PC2×PC1 scores plots suggest that a different process may occur at lower temperatures, so it is possible to reduce the data to only those at the three lowest temperatures, to produce a 162×34 matrix. This needs to be restandardised as the mean and standard deviations of the metabolites of these 162 samples differ from 216 original metabolites.

- In Figure 4.13, we illustrate the scores and loadings of the first two PCs; obviously, many more graphs can be obtained.
- The eigenvalue for PC1 is 1145 and PC2 is 699, the total sum of squares is reduced from 7334 to 5236 (=162×34), so the first two PCs represent 35.2% of the variance.
- Unlike the full dataset (Figure 4.9), in Figure 4.13 we find higher-temperature extracts (16 °C) at the left and lower-temperature extracts at the right. In fact, the sign of the axes cannot be controlled during PCA and we often find there are vertical and horizontal flips on axes even with the same data but different software packages. However, we can still compare the graphs.
- Although we need more detailed analysis quite a few trends can be obtained from comparing the two loadings graphs.
  - A metabolite on the right of Figure 4.9 but left of Figure 4.13(b) is characteristic of high temperature, and its concentration changes as temperature is lowered, for example, malate and glutamate.
  - Metabolites such as sucrose and glycine show the opposite effects, being more characteristic of low temperatures.
  - Some metabolites cluster together in both graphs, for example, sucrose, fructose, glucose and raffinose; having structural similarities suggest that the metabolic pathways are regulated in the same way according to temperature, and probably have a common metabolic origin.
  - A few more metabolites are clustered around the centre in Figure 4.13(b) compared to Figure 4.10. This suggests that they do not change much at lower temperatures, whereas metabolites that are farther from the centre in Figure 4.13(b) continue to change in content as the temperature is lowered below 16 °C.

Naturally, further insights can be obtained by looking at later PCs; however, the more significant PCs give us a useful picture to explore the main trends.

A great deal of information can be explored by performing PCA on subsets of the data, looking at replicates, looking at whether different lines have different temperature profiles and so which are most influenced by lower temperatures and which could be grown earlier in the calendar year, and so on.

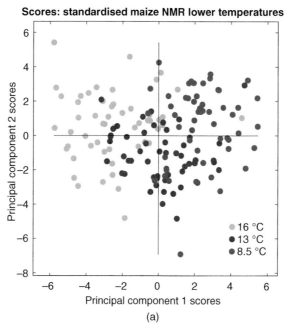

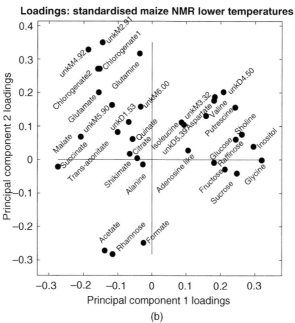

FIGURE 4.13   Scores (a) top and loadings (b) bottom of maize data at the three lowest temperatures.

4.5   PCA OF DIFFERENT CASE STUDIES

In Section 4.4, we looked in detail, how scores and loadings can be used to explore the information from case study 5, concerning the harvesting temperature of maize. PC scores and loadings plots are not always so informative, and below we explore several of the other case studies and discuss what information we can obtain primarily from the first two PCs. Obviously, the data can be explored further via PCA, but for brevity, we restrict the discussion. We will see that PC plots can provide quite different types of information in different situations. Below, we will look at some selective case studies and see how PC scores and loadings provide a variety of interpretations according to circumstances and type of study. We will not attempt a comprehensive discussion of each case.

4.5.1   Case Study 1: LCMS Studies of Pre-arthritis

Human metabolomics often involve very different challenges to interpretation compared to plants. In case study 5 (maize), we could grow plants under controlled environments and genetics and so most of the variability will be in the factors we decide to study. However, in many human studies, this level of control and homogeneity is hard to obtain. They will be more genetically diverse than in a carefully bred and selected line of a given type of plant. Their diet, environment, social habits and medicines will differ. Even if one makes sure their behaviour is fairly uniform in the hours and days running up to a test, and one selects, for example, by gender or BMI, there will still be heterogeneity over a lifetime. Naturally, certain sampling designs as discussed in Section 6.3 do try to reduce this variability, but still cannot eliminate a lifetime's variability. In the pre-arthritis case study, donors will have different genetics, different personal habits and live in different environments. Hence, the trends we are interested in studying are unlikely to be the most obvious ones in the PC plots.

To analyse these data, we first standardise the columns. We discuss this procedure in Section 4.6.3. There was some missing data in the raw dataset, which has been replaced by 0. Some strategies for dealing with missing data are described in Section 4.7.1; there would be several alternative approaches, but the proportion is quite small.

The scores and loadings of the first two PCs are presented in Figure 4.14.

We can see from the scores plot, that unlike case study 5 (maize), there is no noticeable distinction between the groups we are interested in. When studying humans, we cannot control them as well as plants. An additional consideration for LCMS is that it is more sensitive compared to NMR and if the data are standardised will often pick up quite small peaks, NMR usually picks up mainly larger more prominent peaks. Hence, the LCMS study of arthritis picked up 81 peaks, compared to 34 (of which 26 were identified) for the NMR of maize. Many of these peaks may be irrelevant to the main study.

Thus PC scores plots are often not always informative for human data and it is necessary to use other techniques such as PLS (Chapters 7 and 8) to find which compounds are significant markers.

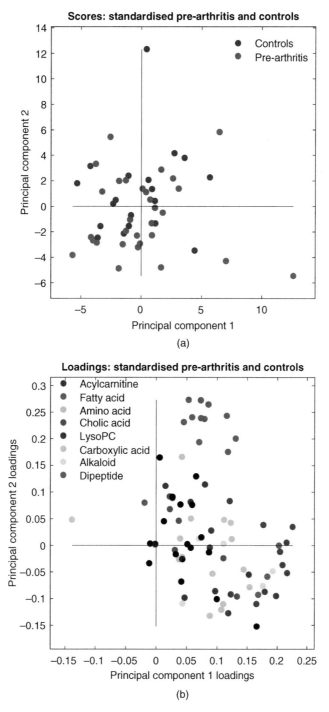

FIGURE 4.14 Scores (a) and loadings (b) of standardised pre-arthritis data; for the loadings only those variable classes with at least three members are coloured, the rest are in black.

Unlike case study 5, where there are 34 compounds to show in a loadings plot, 81 have been identified making the plot too crowded for labelling. However, groups of compounds in given chemical classes can be displayed in a loadings plot as in Figure 4.14(b), where compound groups occurring more than two times are coloured. Some compound types do cluster together, suggesting they arise from similar metabolic pathways. These pathways, however, are unlikely to be related to the cause of arthritis, as they show no strong relationship to the scores. Methods such as PLS as discussed in Chapter 7 would be needed to probe the data further.

This does not mean PCA has failed but suggests more supervised approaches, such as PLS, are necessary in this case. Whether simpler approaches are adequate to investigate a problem or whether more complicated methods are needed depends on the nature of the problem. In the investigation of pre-arthritis, the metabolic signature of interest is likely to be very small and as such harder to tease out than in most applications in plant metabolomics.

4.5.2   Case Study 4: NMR of Human Diabetes

Unlike case study 5, we take the intensities of the spectra at individual frequencies in the NMR spectrum, rather than identifying individual resonances, to obtain a 231×2180 dataset, with three groups according to glucose tolerance.

The data are first row scaled to a constant total of 1. In this case, we will discuss row scaling in Section 4.6.1.1. The data are then standardised down the columns.

The scores and loadings are presented in Figure 4.15, after first standardising the data. In contrast to the pre-arthritis case study, we do see separation between the groups in the scores. There is no overriding rule, however, diabetes measurement is here and now and pre-arthritis is many years before the disease developed. From the PC plot, some information can be obtained although PLS will definitely provide further insight.

The loadings plot of Figure 4.15(b) is much less clear than Figure 4.10, or other loadings plots obtained from the maize. Part is the problem that the spectral frequencies are not assigned to individual metabolites, and most parts of the spectrum consist mainly of noise. Standardisation amplifies noise. In addition, there are 2180 data points spanning the spectral width making the loadings very crowded. The data can be clarified by binning, that is summing the intensities (or taking the average) over several data points. In Figure 4.16 we show the effect of binning over four successive data points, reducing the number of variables to 545. There is very little change in the scores plot, suggesting the main trends are retained, but the loadings significantly simplify. It still is not sufficient to get a very clear indication of which parts of the spectrum contain markers for diabetes; however, the resonances in the top left quadrant and top of the top right may represent some markers. It is possible to look closer at the loadings (we will not in this text for brevity as the emphasis is on techniques), to give a preliminary clue for which parts of the spectrum are most diagnostic. However, there are a number of approaches for variable selection to be discussed in Chapter 9, and because the NMR data are standardised, removing the less informative regions of the spectrum can make a significant difference. We will see that multilevel PLSDA has a

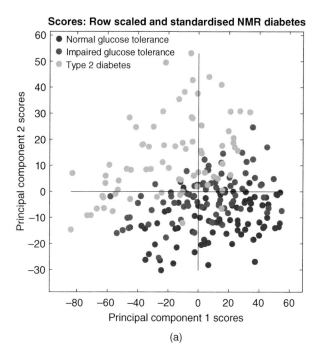

(a)

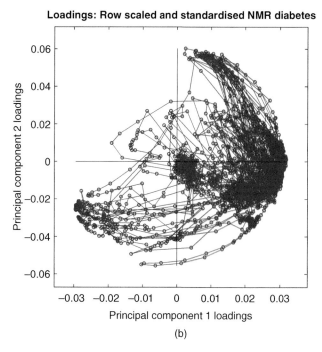

(b)

FIGURE 4.15 Scores (a) top and loadings (b) bottom of standardised human diabetes data; the loadings are joined according to their sequence in the NMR spectrum.

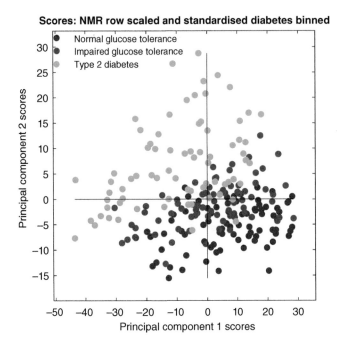

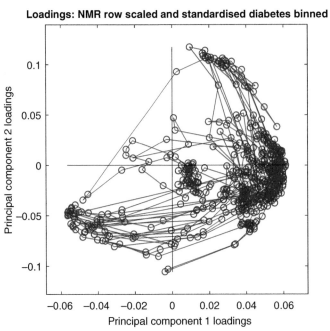

FIGURE 4.16   Binning the loadings to once every four readings for the data of Figure 4.15.

major effect on simplifying the patterns between the NMR spectra. In this case study, there already has been some reduction using domain knowledge, but mathematical (chemometric) knowledge can further simplify the data which we will discuss later.

This case study is intermediate in complexity to case study 5 (NMR of maize) and case study 1 (LCMS of pre-arthritis) with some patterns evident in the PC plots of the samples but not a very clear trend in the loadings.

4.5.3   Case Study 10: LCMS of Diabetes in Mice

In case study 10, we also study diabetes but in mice. Laboratory-based model experiments in mice can be controlled far more carefully compared to human sampling, as one can use genetically identical haplotypes, and control their environment and diet over their entire lifetime. Hence, there is much less sample-to-sample variation compared to humans.

We first look at the positive ion ESI data. To analyse these data, after identifying 146 metabolites, we need to row scale each sample as the instrumental variability needs correction. As discussed in Section 4.6.1.1, row scaling (often called normalisation but this term is ambiguous) can be done in various ways, but below we use a simple approach of summing the intensities of known metabolites in each sample to 1. This does depend on how many metabolites have been identified of course.

The data are then standardised and the PC scores and loadings plots are presented in Figure 4.17.

The scores plot shows a quite good separation between the diabetic and non-diabetic mice. A small number of mice diagnosed as non-diabetic (cut-off blood glucose at $250\,\mathrm{mg\,dL^{-1}}$) might have been at the borderline and could be investigated according to whether they were diagnosed correctly. The loadings plot is harder to interpret, partly because of the large number of metabolites identified in the positive ion ESI.

There are only 32 identified metabolites in the negative ion ESI data, and we can perform the same operations, to give the plots of Figure 4.18. The scores show similar separation, with four of the mice diagnosed as non-diabetic, apparently falling into the diabetic class; the same mice, numbers 26, 33, 54 and 66 appear to fall into the diabetic group in both positive and negative ion datasets. In many situations there may sometimes be misdiagnosed samples, and exploratory methods such as PCA can pick them up as a first visual step.

However, the biggest difference is in the loadings, which show a much clearer pattern, probably because as there are less variables, noisy or non-relevant metabolites have been removed. Further analysis could be used to identify these metabolites by their positions in the loadings plots. As in this case the sign of the scores in PC1 acts as a good indication as to whether the mice are diabetic or not, we can also plot the scores and loadings of PC1 as bar graphs as in Figure 4.18(c and d), with positive scores primarily corresponding to diabetic mice. The positive loadings therefore likewise suggest metabolites that are in higher concentration in mice with diabetes. Of course, we could also use methods such as PLS discussed in Chapters 7 and 9 to get quantitative measures of which metabolites are most characteristic of the two groups, but if we are lucky (and this is not always case for PCA) we can obtain a preliminary view using PC loadings plots.

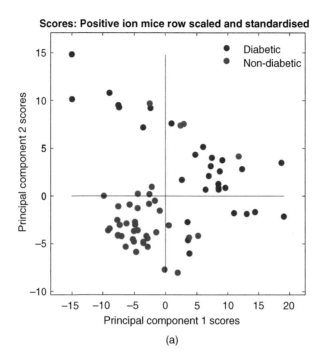

(a)

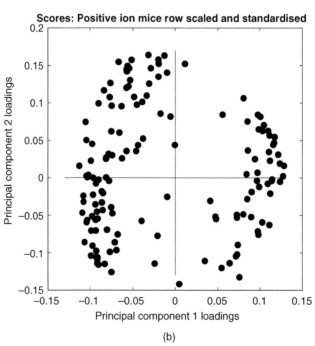

(b)

FIGURE 4.17   Scores (a) top and loadings (b) bottom of standardised and row-scaled mouse LCMS positive ion data.

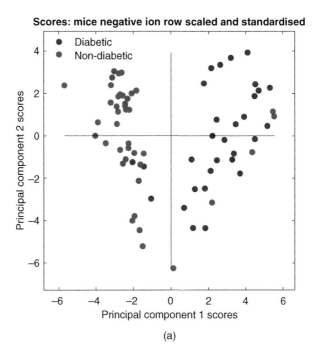

(a)

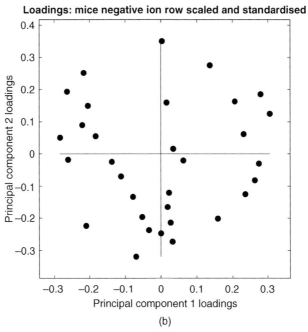

(b)

FIGURE 4.18 Scores (a) top and loadings (b) top middle of standardised and row-scaled mouse LCMS negative ion ESI diabetes data of the first two PCs and (c) scores bottom middle and (d) loadings bottom of PC1 with metabolites indicated.

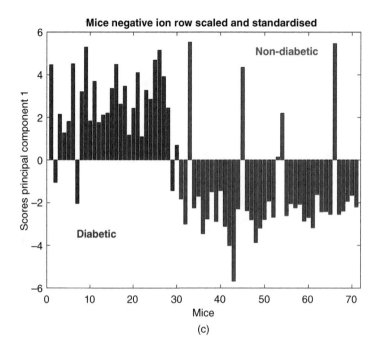

(c)

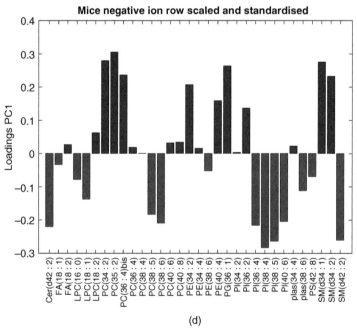

(d)

FIGURE 4.18   (Continued)

The mice show clearer trends in PCA than the human case studies because they can be studied under controlled conditions, so the fact of interest (whether they are diagnosed diabetic) is dominant.

4.5.4   Case Study 6: FTIR of Effect of Nitrates on Wheat

In the case of FTIR, we do not get as much molecular insight as in techniques such as NMR, LCMS or GCMS; however, we can often see patterns quite clearly.

We first scale the data. SNV (standard normal variates) encompasses baseline correction, centring and scaling and so reduces a lot of the variability due to instrument and sample preparation. It is a common approach both for Raman and FTIR.

Figure 4.19(a) shows the scores. The samples with high (or normal) nitrate treatment are mainly separate from those with low nitrate. In particular, the samples from the base (which will be sitting in the soil) are quite distinct in the bottom right of the diagram. Further interpretation could be possible including replicates analysis, which we will not discuss here.

The loadings plots in Figure 4.19(b) can give some clues as to which spectral features are most correlated to the effect of nitrate, for example, 1402 and $1651 \, cm^{-1}$ appear most characteristic of high levels, whereas $1039–1056 \, cm^{-1}$ of low levels. Of course, these would be reinforced by chemical knowledge. They are marked in the average SNV spectrum of Figure 4.19(c).

For spectra and in some cases chromatographic profiles, loadings can provide clues which features are important by comparing them to the scores, providing there is a clear and fairly simple pattern in the scores.

4.5.5   Case Study 3: NMR for Triglycerides in Serum

The next case study we will illustrate in this section relates to the use of NMR to measure the concentration of TG (Triglycerides) in blood serum. The aim is one of calibration rather than pattern recognition, and in this example we are using chemometrics to obtain a quantitative value rather than place the sample into a group or find out which metabolites are most characteristic.

Figure 4.20(a) shows the scores plot, together with the corresponding loadings in Figure 4.20(b). The data have been acquired in a careful quantitative way and so do not require further row scaling. As most of the resonances in the spectral region studied are of similar magnitude, nor is standardisation required; however, data centring down the columns (Section 4.6.2) is a useful technique that we will use prior to PCA. Careful comparison of both plots can determine which individuals have highest resonance intensities in different parts of the spectrum and so have higher relative concentrations of metabolites resonating in these regions, we will leave the reader to compare these, but the loadings plot does show very characteristic features.

What though, is more important, in this study, is to see whether the concentration of triglycerides can be estimated through a model. In Figure 4.20(c), we provide a

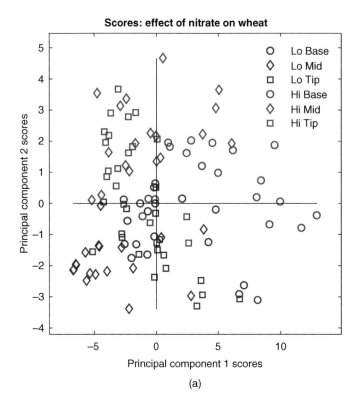

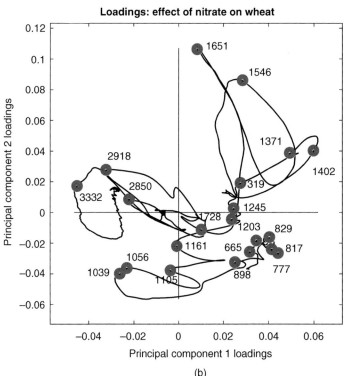

FIGURE 4.19   Scores (a) top of the wheat, (b) middle loadings with spectral features labelled as wavelengths in cm^{-1} and (c) bottom average SNV spectrum with local maxima indicated.

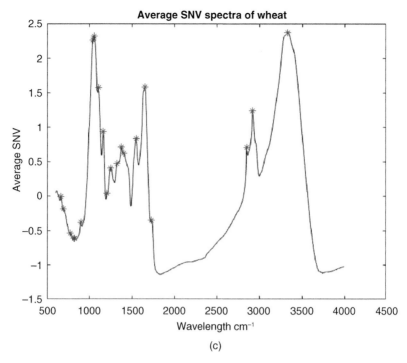

FIGURE 4.19 (Continued)

graph of the scores of PC1 against the concentration of triglycerides, and find there is a good correlation. This shows promise that the NMR model can be used to estimate triglyceride concentration. In contrast, PC2 shows a poor correlation (Figure 4.20(d)).

As the scores of PC1 are well correlated with triglyceride concentration, we can also plot the loadings against ppm in the NMR spectrum, as in Figure 4.20(e), and see that there are regions where the loadings are very positive: these are likely to represent regions of the spectrum correlated to high triglycerides. This technique does not always work and depends on there being a good correlation between one of the PCs and the property of interest, but in this case, we can relate spectral features to triglyceride concentrations.

Most metabolomic studies use PCA in an exploratory manner, in order to visualise groupings or main trends, using other approaches such as PLS (discussed in Chapters 7 and 8) for prediction. However, this case study shows the power of PCA also for quantitative estimation in this situation.

4.5.6   Case Study 7: Raman of Bacterial Faecal Isolates

This is a large dataset of size 980×988, representing 12 strains of bacteria, and PCA initially provides little information, as there are a large number of spectra and a lot of variability.

A full analysis of the data is beyond the scope of this text; however, we will explore how preliminary analysis by PCA can start to show trends.

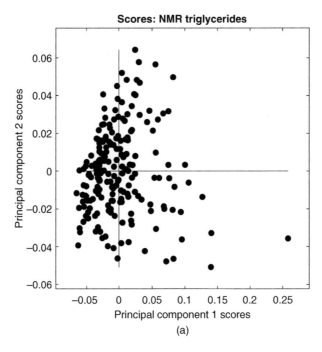

(a)

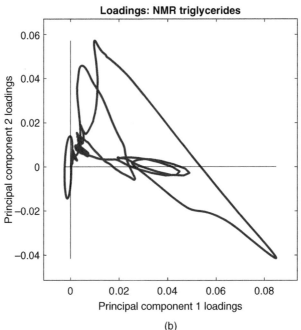

(b)

FIGURE 4.20   From top to bottom (a) top scores of case study 3, (b) top middle loadings of case study 3, (c) middle graph of PC1 scores of case study 3 versus TG concentration, (d) bottom middle graph of PC2 scores of case study 3 versus TG concentration and (e) bottom loadings of PC1 versus ppm.

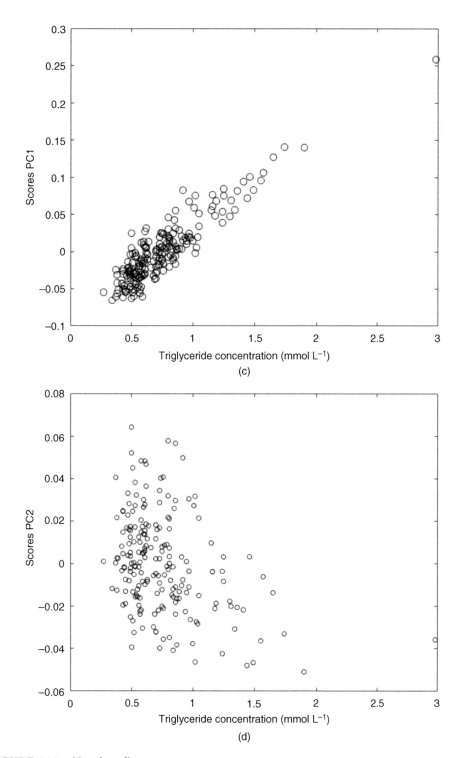

FIGURE 4.20   (Continued)

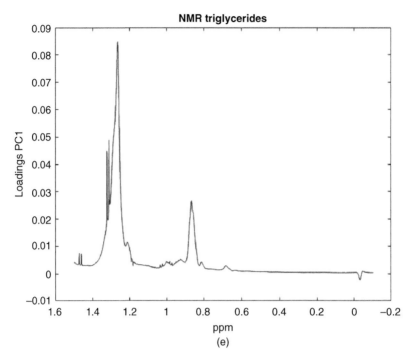

FIGURE 4.20   (Continued)

The scores plot of the first two PCs using the centred Raman data after SNV correction appear somewhat discouraging, as illustrated in Figure 4.21(a), largely because there are 980 data points, which makes interpretation difficult.

However, we can simplify, for example, in Figure 4.21(b) we present the scores of only two strains (EC4 and EC9) and see there appears a distinction between Raman spectra of both strains. Further investigation could be performed, for example, by PLSDA (Chapter 7) which could also determine which parts of the Raman spectrum are best at distinguishing these groups.

We can also average the scores for each of the 12 strains as in Figure 4.21(c) which gives us a good visual representation as to which strains appear most similar, for example, EC9, EC14 and 156UNI appear to form a group, as do EC10 and 178UNI. Often other approaches, for example, unsupervised pattern recognition such as hierarchical cluster analysis are suitable in this case, but we will restrict this text to the most common methods in metabolomics. Classical taxonomy involves many additional approaches outside the scope of this book. Further investigation could be performed to look at how similar replicates both of the plates and samples and human donors are, for brevity, we omit this in this text.

Often different techniques, for example, FTIR or a mass spectrometric approach can provide complementary information to provide a more comprehensive picture. However, in this introductory chapter, our main discussion relates to the use of PC plots to unlock trends in metabolomic datasets, primarily using visualisation.

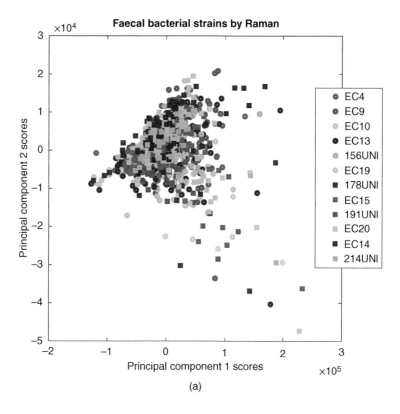

(a)

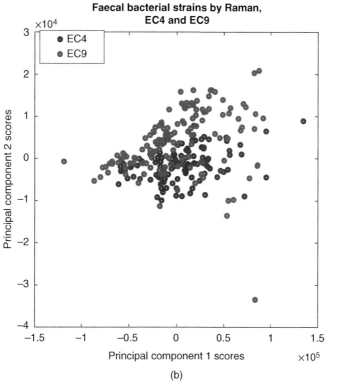

(b)

FIGURE 4.21 PCA analysis of Raman extracts of faecal bacterial strains (a) top scores of first two components for full dataset, (b) middle scores just for EC4 and EC9 and (c) bottom mean scores for all 12 strains.

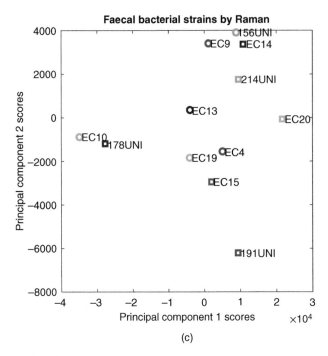

FIGURE 4.21   (Continued)

4.6   TRANSFORMING THE DATA

The first step after acquiring a dataset is to transform or scale the data. This is done before PCA and most multivariate methods and we have referred to the use of several pre-processing methods above. In this section, we provide insights into the basis of different common approaches. There is no hard and fast rule as to what method is best, but this should take into account how the data have been obtained and also what characteristics are of interest on a case-by-case basis.

4.6.1   Row Scaling

This first decision is to whether to row scale the data. This is sometimes called normalisation, but column scaling by standardisation, discussed in Section 4.6.3, is also sometimes called normalisation, so to avoid this confusion we will refer to it as row scaling.

Whether and how to scale the rows depends on the experimental procedure, so cannot be done without an understanding of how the data were obtained.

As an example:

- In case study 5 (the maize), the dry weight of the leaves is measured prior to the analysis.

- For all identified metabolites, peak areas are calibrated against concentration.
- So concentration per dry weight can be calculated for all metabolites, and for unknowns still relative peak areas can be calculated.

As the amount of extract is known and the instrumental conditions are stable, there is no need for further scaling of the data for each sample.

However, for other techniques row scaling is essential. Consider, for example, case study 3, the NMR of blood serum extracts to study human diabetes. In Figure 4.15, the rows have been scaled to a constant total of 1 (see Section 4.6.1.1). The PC scores and loadings of the same data (standardised down the columns) are presented without first row scaling to a constant total in Figure 4.22.

- There is poorer separation in the scores between classes compared to the row-scaled data.
- The loadings now show no easily interpretable pattern.

Hence, considering how the rows are scaled is important for the resultant interpretation of the PC plots and indeed most chemometric techniques. We will describe several of the common methods below. This step should be taken prior to column centring or scaling. It can have a significant influence on the interpretation of the results of PCA and whether it is done depends on how samples are obtained and to a lesser extent on instrumental stability.

4.6.1.1   Row Scaling to Constant Total

This is one of the simplest methods, used classically in many applications of multivariate analysis.

To do this, each row, usually representing a sample, for example, a chromatogram or a spectrum, is summed to a constant total usually 1, so $x_{ij(rs)} = x_{ij} / \sum_{j=1}^{J} x_{ij}$.

The row-scaled data of Table 4.1, are presented numerically in Table 4.5 with their scores and loadings in Table 4.6.

- The distribution of the samples changes when row scaled and now they lie on a straight line as in Figure 4.23(b).
- This changes the appearance of the scores plot as presented in Figure 4.23(c) compared to Figure 4.3(a).
- In fact all the objects have the same score of PC1 (=0.7071) and can only be distinguished by their score along PC2.
- For the raw (not-row scaled) data, we can discriminate using PC1.

The reason why the data fall onto a straight line is because there are only two variables, so the distance along the line of PC2 is monotonically related to the ratio between the value of the two variables.

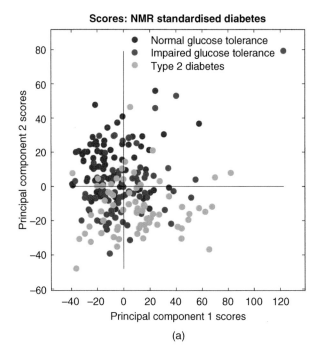

(a)

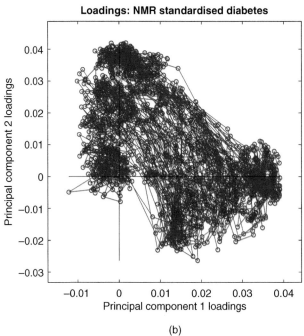

(b)

FIGURE 4.22   Scores (a) top and loadings (b) bottom of standardised human diabetes data without row scaling to constant total; the loadings are joined according to their sequence in the NMR spectrum.

TABLE 4.5   Row scaling the data of Table 4.1 to a constant total of 1.

		Variables	
		1	**2**
Objects	A	0.58	0.42
	B	0.67	0.33
	C	0.25	0.75

TABLE 4.6   (a) Scores and (b) loadings for the row-scaled matrix of Table 4.5.

(a)

		Scores	
		t_1	t_2
Objects	A	0.707	−0.118
	B	0.707	−0.236
	C	0.707	0.354

(b)

		Variables	
		1	**2**
Loadings	p_1	0.707	0.707
	p_2	−0.707	0.707

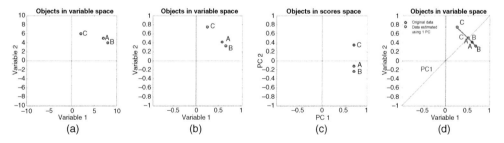

FIGURE 4.23   Effect of row scaling the data of Table 4.1. (a) Raw data, (b) row-scaled data, (c) row-scaled scores, (d) projection onto PC1.

Although the data are now one-dimensional, it is still necessary to use two PCs to reconstruct the data. This is because the data have not been mean centred, prior to PCA. As we can see from Figure 4.23(b), it is impossible to rotate the data so that they are distributed around 0. We will discuss the key step of mean centring in Section 4.6.2 but if we took this extra step, there would only be one non-zero PC. Table 4.7 presents the results of centring after row scaling. We see now that the scores of PC2 are 0, and those of PC1 are identical to the second PC for the uncentred data. If we centre data before PCA after row scaling to a constant total the data loses one dimension or as discussed in Chapters 5 and 10 in more detail, one degree of freedom.

More complicated examples are presented in Tables 4.8 and 4.9. For brevity, we restrict discussion to uncentred data.

- The scores plots of the first two PCs for the 6×2 matrix of Table 4.8 are illustrated in Figure 4.24.
- The objects now lie on a straight line after row scaling to a constant total with equal scores in PC1.
- The number of variables determine the dimensionality of the resultant pattern, for example, if there were three variables, the scores would lie on a plane.
- This does not (for uncentred data) mean the data loses dimensionality, two PCs are necessary to reconstruct the data.
- The scores plots of the first two PCs for the 6×6 matrix of Table 4.9 are illustrated in Figure 4.25.
- As there are more than two variables, the objects no longer lie on a straight line.
- The relationship between the objects changes after row scaling, but only very substantially for object A; B and D cluster close together, F is the extreme sample; C and E are somewhat closer in the row-scaled data but between F and B&D.

Row scaling to a constant total is a common but simple approach in many circumstances, for example, if the amount of extract cannot be controlled, or if there are significant instrumental instabilities. It looks at ratios, for example, how the proportion of metabolites varies. However, if there are a few intense metabolite signals in a sample, these can distort the pattern, as the difference between samples would be dominated by the major constituents, which may not be very interesting, as minor components might be more characteristic of specific pathways.

There is no general rule as to whether to scale rows to a constant total, and it depends on the aim of the experiment and the data analysis. What is important though is to be aware that this step can significantly change the pattern and always to report this initial step. There are several alternatives.

Row scaling should always be done prior to column transformation, and is only meaningful if all values are positive or at least only a few are very close to zero relative to the regions of interest.

Row scaling to a constant total introduces a property called closure, which can in certain cases result in distortions in interpretation. However, for visualisation where sample volumes or quantities cannot easily be controlled it can simplify interpretation and is a relatively common procedure in chemometric processing of metabolomic data.

TABLE 4.7   Centring row-scaled data of Table 4.1 together with PCA.

(a)

		Variables	
		1	2
Objects	A	0.083	−0.083
	B	0.167	−0.167
	C	−0.25	0.25

(b)

		Scores	
		t_1	t_2
Objects	A	−0.118	0
	B	−0.236	0
	C	0.354	0

(c)

		Variables	
		1	2
Loadings	p_1	−0.707	0.707
	p_2	0.707	0.707

TABLE 4.8   A dataset with six samples and two variables.

		Variables	
		1	2
Objects	A	7	5
	B	8	4
	C	2	6
	D	1	9
	E	2	0
	F	8	8

TABLE 4.9   A dataset with six samples and six variables.

		Variables					
		1	2	3	4	5	6
Objects	A	4	20	24	30	28	11
	B	8	4	1	5	12	4
	C	2	6	5	6	3	4
	D	1	5	6	9	7	2
	E	6	16	14	18	9	12
	F	8	8	8	1	0	9

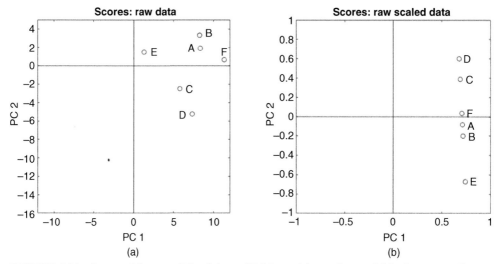

FIGURE 4.24   Scores of first two PCs of data of Table 4.8 (a) raw data and (b) after row scaling to a constant total.

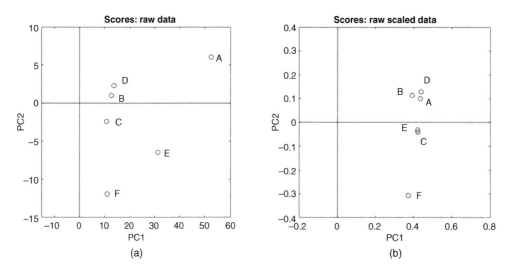

FIGURE 4.25   Scores of first two PCs of data of Table 4.9 (a) raw data and (b) after row scaling to a constant total.

4.6.1.2 Standard Normal Variates

This approach is common in vibrational spectroscopy where there are baseline as well as quantitative challenges.

The row transformation is as follows $x_{ij(snv)} = \left(x_{ij} - \bar{x}_i \right)/s_i$ where $\bar{x}_i$ is the mean of row i, and s_i its standard deviation. In practice, it involves standardising each row and serves a different purpose to standardising the columns as discussed in Section 4.6.3.

For our simple example of Table 4.1, performing SNV on the rows makes very little sense, as there are only two variables. However, we can apply it to case study 6, the FTIR spectra of wheat. Figure 4.19(a) illustrates the scores of the first two PCs and there is a separation between the groups. PCA followed by mean centring has been performed on the raw data without first SNV in Figure 4.26 and the pattern is no longer clear.

Figure 4.27 illustrates the effect of SNV row scaling on the first three spectra. For the raw data, there are considerable baseline problems and PCA will mainly be picking up this instrumental variability. For the SNV-corrected data, we see mainly spectral variation between samples. As well as correcting for the baseline, SNV also corrects if spectra are different in magnitude.

There are several other approaches for correcting instrumental and sampling problems in vibrational spectroscopy which have spawned a huge literature in the early days of chemometrics; however, we leave the reader interested in applications of FTIR and Raman to specialist articles and texts. SNV is possibly the most widespread approach and is recommended to be the first method of choice unless there is a specific reason to solve a tricky problem.

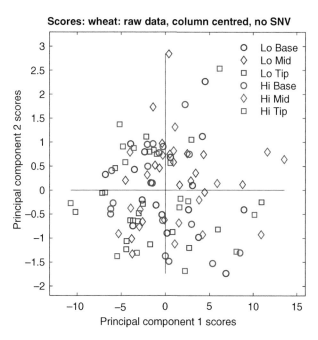

FIGURE 4.26 Scores of the first two PCs, centring the columns of the raw data from the FTIR spectra of wheat, case study 6, compare to Figure 4.19 without first performing SNV.

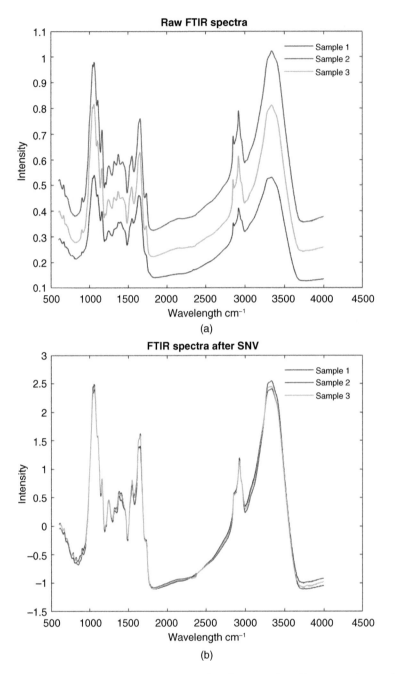

FIGURE 4.27   FTIR spectra of the first three samples of case study 6 (a) top raw data and (b) bottom after SNV.

4.6.1.3   Scaling to Reference Standards

A common alternative approach is to scale to one or more reference standards. There are a number of reasons.

- The first is to correct for instrumental variability.
- The second is if the amount extracted varies, by using one or more internal standards we can make corrections for this variation.

There are several techniques and often different in-house approaches are developed according to laboratory and software packages, so we will only mention a few below.

- Adding an internal standard to the extraction of known concentration. In the case of plants a measured dry weight can be extracted, controlling the amount of material.
- More reliably and more commonly several internal standards can be added to an extract. Peaks can be ratioed to the mean of the areas of the internal standards. In case study 2 (malaria as assessed using GCMS of plasma), nine internal standards are added. The peak areas of each of these standards are calculated for each sample. The PCA of uncentred of the data, scaled by their standard deviations, is performed. The score of the first PC is considered an average and used to scale each sample. A small number of missing values (e.g. peaks in the standard that could not be resolved) can be replaced prior to PCA, for example, by the mean value of the particular variable; if too many values are missing for a given internal standard, the standard is discarded. The first two PCs of the raw and row-scaled data from case study 2, the GCMS of extracts of blood plasma to detect malaria, are illustrated in Figure 4.28. Once each row is scaled by ratioing to the scores of the reference standard, there is a better separation between the groups. The quality of the raw data will depend in part on instrumental variability and the amount of extract.
- External QC (quality control) samples (see also Section 4.7.2) also can be used for row scaling, as is useful in cases where it is hard to control the quantity of tissue extracted. QC samples can be of varying types but a pooled sample will contain metabolites characteristic of the biological extracts being studied. Probabilistic Quotient Normalisation (PQN), involves calculating the ratio between each of the characterised metabolites from each sample and those from the average QC sample. The data are then row scaled to the median of these ratios (quotients). The idea is that the majority of metabolites do not vary significantly, for example, by the treatment. Therefore, their ratio to the QC samples will only change with the amount of the extract for these metabolites. For metabolites that appear to have responded to the treatment the ratio will be different, so are excluded from the procedure. The median of the ratios is therefore a good guess at what the ratio should be if the variation in response is due to variation in amounts of the extracts, and should be used to correct for this.

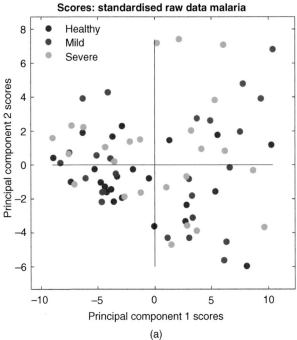

(a)

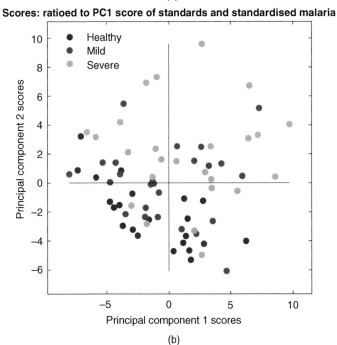

(b)

FIGURE 4.28   Scores of first two PCs of case study 2 of GCMS of extracts of plasma to detect malaria (a) top raw data (b) bottom data row scaled via ratioing to nine internal standards, each followed by standardising columns.

There are several other approaches in the specialist literature, but we only outline some of the most widespread above.

4.6.2   Column Centring

This transformation is very common. PCA, as conventionally applied, looks at variation around an average. If, for example, we examine a set of old coins, their weight, thickness, diameter and so on, might vary, largely due to wear, but in some cases due to manufacturing imperfections. In order to determine whether a coin comes from the same origin (e.g. a historic mint), we might want to determine the characteristics of an 'average' coin, this could have an average weight, thickness and so on, and then take a batch known to come from the same origin, and see how they vary. We are not really interested in variation above zero but around the multivariate mean. Hence, most classical applications of PCA first involve mean column centring the data prior to PCA. This is not mandatory in all cases, for example, if we are interested in variation above a baseline, for example, in chromatography, however, in most cases in metabolomics we are interested in variation around a mean. In the majority of examples in this text we will column centre; note that column standardising also includes a default step of centring.

To illustrate we will return to the example of Table 4.1.

- The column-centred data given by $x_{ij(cen)} = x_{ij} - \bar{x}_j$ are presented in Table 4.10.

- The scores and loadings are presented in Table 4.11. Note that the scores are column centred as are the data of Table 4.10. We could adjust the scores by adding back the column mean of the data [5.667 5] but this is rarely done, although we will illustrate this graphically below to demonstrate the relationship between the scores of the raw (uncentred) data and the scores of the column centred data.

- We illustrate the effect on the scores in Figure 4.29.

- Figure 4.29(a) illustrates how the original uncentred data (represented by blue circles) is rotated to give the scores (represented by red circles) around the circumference of a circle centred on the origin. The red circles represent the scores in the new space whereas the blue circles the original (unrotated) data. The green lines are the arcs of the circle along which the data are rotated.

- Figure 4.29(b) shows what happens when the data are centred about the origin, representing the scores of the column-centred data.

TABLE 4.10   Column-centred data of Table 4.1.

		Variables	
		1	**2**
Objects	A	1.33	0.00
	B	2.33	−1.00
	C	−3.67	1.00

TABLE 4.11   (a) Scores and (b) loadings for the column-centred matrix of Table 4.10.

(a)

		Scores	
		t_1	t_2
Objects	A	1.279	0.376
	B	2.521	−0.302
	C	−3.800	−0.074

(b)

		Variables	
		1	2
Loadings	p_1	0.960	−0.282
	p_2	0.282	0.960

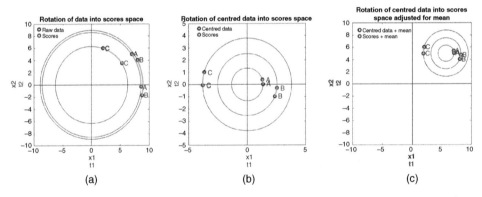

FIGURE 4.29   Rotation of data of Table 4.1 to obtain scores (a) raw data (b) column centred data and (c) column centred data adjusted for the original mean green arcs represent rotation from observed data points to scores.

- Figure 4.29(c) shifts the origin of the centred scores to the mean of the raw data; we very rarely represent scores in this way but is a useful comparison to the raw data.
- We see that the relationship (or distance) between the three points is unchanged, however, their absolute position changes when centred.
- There are however several other differences.

  - The eigenvalues of the PCs change according to whether the data are centred or not. For the centred data, the eigenvalue of PC1 is 22.43 and of PC2 is 0.24. For the uncentred data, these are 178.49 and 15.51, respectively.

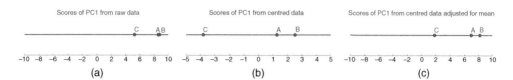

FIGURE 4.30 Scores of the first PC of data of Table 4.1 to obtain scores (a) raw data (b) column centred data and (c) column centred data adjusted for the original mean, corresponding to the plots of Figure 4.29.

- This means that for the centred data, PC1 represents 98.95% of the total sum of squares, whereas for the uncentred data, it represents 92.01%. Hence, different conclusions might be made about the relative importance of each PC or even how many PCs are required for an adequate model according to whether the data are column centred.
- The most important difference comes when we reduce the number of PCs in the model. Although the relationship between the objects is the same if we retain all non-zero PCs, it changes if we keep only one PC to reduce the size of the model.
- Figure 4.30 represents the scores of PC1 both for the uncentred (raw) data and the centred data. By comparing Figure 4.30(a) and Figure 4.30(c) we see that the relative position of the three objects changes according to whether the samples are centred or not.

In general, centring will change the relative position of samples in reduced PC space, so for example, if there are 100 variables the appearance of the plot of the scores of the first two PCs may be different according to whether the data are centred or not. Only if the data could be visualised with all non-zero PCs would the relative position of the samples remain unchanged.

We illustrate the effect of centring on the scores using case study 3, the NMR of serum used for the estimates of triglycerides. Figure 4.20(a) shows the scores of the first 2 PCs in the centred data, and Figure 4.20(c) is a graph of the scores of PC1 of the centred data versus the concentration of triglyceride. The equivalent graphs are presented in Figure 4.31 for comparison. We see that the pattern formed by the scores of the first two PCs differs quite considerably without centring and the scores of PC1 no longer have such a good correlation with triglyceride concentration.

We next look at the effect of centring on the loadings. For our case study of Table 4.1, we find as follows:

- The relative position of the loadings does not change so long as all non-zero PCs are included (2 in our case).
- However, unlike the scores, the loadings are not centred around the origin even though the scores are, but centring the data in variable space and then calculating the scores, can be represented also as a rotation of the loadings of the uncentred (raw) data to the loadings of the column centred data in object space.
- The loadings plot of the two non-zero PCs is presented in Figure 4.32 and should be compared to Figure 4.5.

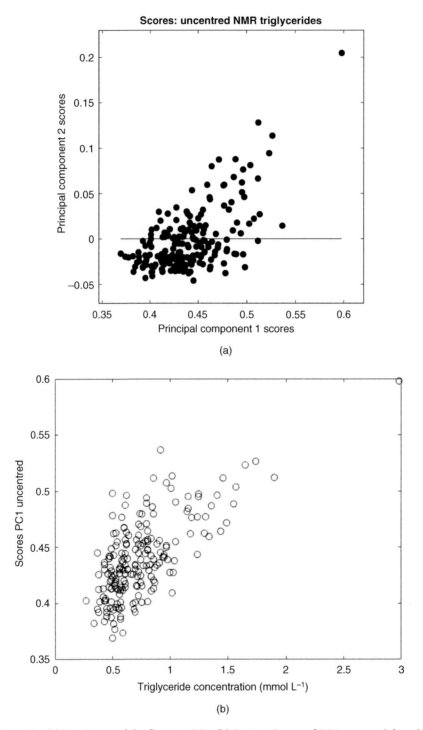

FIGURE 4.31   (a) Top Scores of the first two PCs. (b) Bottom Scores of PC1 versus triglyceride concentration for case study 3 – NMR of serum for measurement of triglycerides without column centring, to be compared to Figure 4.20.

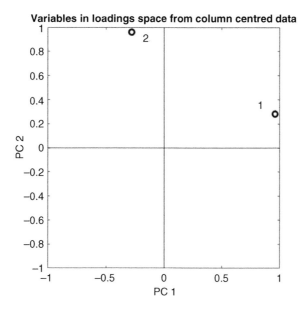

FIGURE 4.32 Representation of the variables (columns) of Table 4.1 in loadings space after first centring column centring; compare to Figure 4.5.

- The loadings of PC1 are quite different according to whether the data are column centred, as illustrated in Figure 4.33. Remember, the values of loadings are between +1 and −1.

Because of this, the loadings plot in reduced PC space can appear quite different according to whether the data are centred. To illustrate this, we show the loadings of the first two PCs of case study 3 (NMR for measuring triglycerides) in Figure 4.34 which should be compared to Figure 4.20(b) which shows a different pattern, just as the scores do.

4.6.3   Column Standardisation

Column standardisation goes under various names, including auto-scaling and variance scaling. It can also be called normalisation confusingly as row scaling to a constant total is also commonly called normalisation in many chemometric articles.

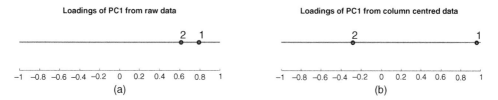

FIGURE 4.33 Loadings of the PC 1 of data of Table 4.1 to obtain scores (a) raw data (b) column centred data.

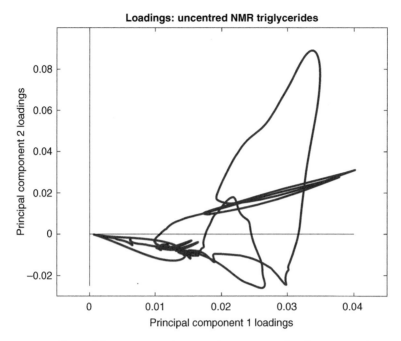

FIGURE 4.34 Loadings of first two PCs of case study 3 uncentred, to be compared to Figure 4.20(b).

This is one of the most common transformations and involves mean centring and then dividing by the column standard deviation so $x_{ij(stand)} = \left(x_{ij} - \bar{x}_j\right)/s_j$ where $\bar{x}_j$ is the mean of column j, and s_j its standard deviation. Strictly speaking, the population

standard deviation $s_j = \sqrt{\sum_{i=1}^{I}\left(x_{ij} - \bar{x}_j\right)^2/I}$ should be used rather than the sample standard

deviation, which would involve dividing by $(I-1)$ rather than I. This is because we are scaling the data and not estimating a parameter. However, some people use the sample standard deviation instead. We will use the population standard deviation for this purpose throughout this text. When there are more than about 10 samples, there is no practical difference in whichever definition is employed.

We will return to our example of Table 4.1

- The standard deviations of the columns of the data are [2.625 0.816] and their means are [5.667 5.000].
- The standardised data matrix is presented in Table 4.12.
- Note that the sum of squares of the standardised data is equal to the number of elements in the matrix or $I \times J = 3 \times 2 = 6$ in this case.
- The standardised data matrix is illustrated in both variable space and object space in Figure 4.35.
- The distribution of objects in variable space is quite different from that of Figure 4.1 (raw data) and Figure 4.23(b) (row-scaled data). Objects A and B are

TABLE 4.12   Column standardised data of Table 4.1.

		Variables	
		1	**2**
Objects	**A**	0.51	0.00
	B	0.89	−1.22
	C	−1.40	1.22

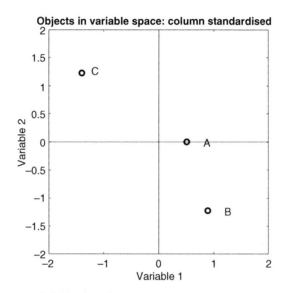

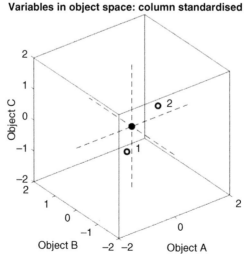

FIGURE 4.35   Representation of the standardised data matrix of Table 4.12.

no longer close together and the distance between A and B is now similar to the distance between A and C. This is because C is separated from objects A and B in the raw data primarily due to variable 1, which has a standard deviation around three times that of variable 2. Once the standard deviation of each variable is the same, variable 1 is no longer so influential.

The scores and loadings of the standardised data are tabulated in Table 4.13 and illustrated in Figure 4.36.

- The scores are centred, since the columns are also centred via standardisation.
- The sum of squares of the scores of PC1 is 5.80 and of PC2 is 0.20. Hence, PC1 represents $100 \times (5.8/6) = 96.7\%$ of the overall variance.
- The relationship between the three objects when projected onto PC1 is quite different from the other forms of data transformation (see Figure 4.30).
- For all forms of data transformation (none, row scaled, centred and standardised) PC1 always represents more than 90% of the variance (if standardised) or variability (if not) in this case and so can be considered a good representation of how similar different objects are. By comparing the different approaches, we might, though, come to quite different conclusions about which objects are more similar to each other.
- Standardisation (or column centring) can also be combined as a second step after row scaling.

TABLE 4.13   (a) Scores and (b) loadings for the standardised matrix of Table 4.12.

(a)

		Scores	
		t_1	t_2
Objects	A	0.359	0.359
	B	1.495	−0.237
	C	−1.854	−0.122

(b)

		Variables	
		1	2
Loadings	p_1	0.707	−0.707
	p_2	0.707	0.707

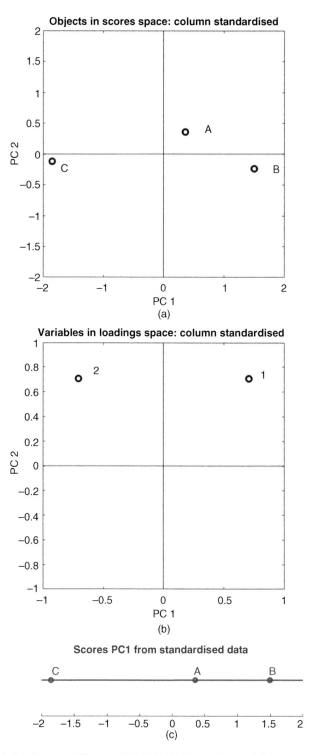

FIGURE 4.36 (a) Top Scores of first two PCs (b) Middle Loadings of first two PCs and (c) Bottom Scores of PC1 for standardised data of Table 4.12.

TABLE 4.14   Row-scaled and column-standardised data of Table 4.1.

		Variables	
		1	2
Objects	A	0.46	−0.46
	B	0.93	−0.93
	C	−1.39	1.39

- The row-scaled data matrix of Table 4.1, which is subsequently standardised is presented in Table 4.14. Note that this matrix is both row and column centred. If there are more than two columns, the row-scaled and standardised matrix will no longer necessarily be row centred, but this property holds for all matrices with two columns.
- The representations in variable and object space are presented in Figure 4.37. Compare the variable space to Figure 4.23(b). In both cases, the three objects fall onto a line, but in Figure 4.37(a) they are centred, and the scales on each axes are different, as a consequence of centring.
- The scores and loadings are presented in Table 4.15.
- We see that the scores all lie along PC1 in Figure 4.38(a) as PC2 is now zero.
- The loadings are presented in Figure 4.38(b). There is some ambiguity in the interpretation of PC2, but these are elements of a rotation matrix. Note that 0.707 is $1/\sqrt{2}$.
- When there are two variables, after transformation by row scaling followed by standardisation it can be shown that $x_2 = -x_1$ so the scores for PC1 for sample i are simply $\pm 1/\sqrt{2}$ and $x_i = 0.707x_1 - 0.707x_2$. For brevity, we will not derive this result in detail.

We shall return to our case studies. Figure 4.39 illustrates the scores and loadings of the raw data of case study 5, NMR of maize. We can compare with Figure 4.9 and Figure 4.10.

- We see that the groups are much better distinguished in the standardised data.
- For the loadings of the raw data, we see that only six metabolites appear to have influential loadings, and glucose and fructose do not appear to be closely associated with the scores of any group.

Many of the interesting metabolites are present in relatively low concentrations and so are low intensity in the raw data matrix, so for this case study, using raw data does not provide a very helpful picture compared to standardisation. In order to illustrate this, we present graphs of the distance of loadings of PC 1 from the centre (point [0 0]) and the mean intensity of the 34 NMR peaks for both the raw and

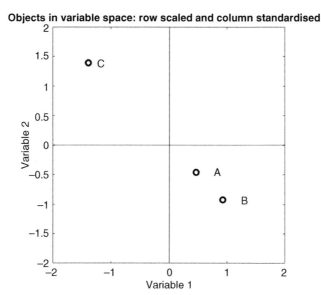

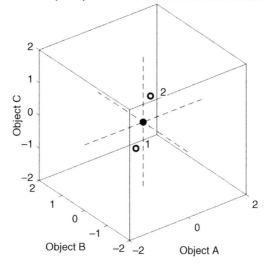

FIGURE 4.37 Representation of the row scaled and standardised data matrix of Table 4.14.

standardised data in Figure 4.40. For the raw data the loadings mainly correlate to the intensity of each metabolite and so are primarily measuring how intense this variable is in the original dataset, while for the standardised data, there is no clear relationship and the loadings represent a better view of the cause of variability at different temperatures.

Standardisation, however, is not always appropriate and Figure 4.41 illustrates the scores. Loadings and relationship between TG concentration for case study 3 (NMR for estimation of triglyceride concentration) and there is no meaningful pattern compared to Figure 4.20.

TABLE 4.15   (a) Scores and (b) loadings for the row scaled and standardised matrix of Table 4.12.

(a)

		Scores	
		t_1	t_2
Objects	A	−0.655	0
	B	−1.309	0
	C	1.964	0

(b)

		Variables	
		1	2
Loadings	p_1	−0.707	0.707
	p_2	0.707	0.707

Hence, although in certain cases standardisation can improve interpretation, in other cases it will destroy important quantitative information, and so must be chosen as a step in analysis on a case-by-case basis.

4.6.4   Logarithmic Transformation

A common alternative is to transform the data logarithmically, so replacing x_{ij} by $\log(x_{ij})$ usually using logs to the base 10.

This also reduces the influence of large peaks.

- In Figure 4.42, we illustrate the result of performing centred PCA after log scaling (using logs to the base 10) the data of case study 5 (NMR of maize) and plotting the scores of the first two PCs.
- We can see that the maize harvested at 20 °C is clearly separated from the rest, and the maize harvested at 16 °C is reasonably distinct also.
- This should be contrasted to Figure 4.39(a) where there is a limited separation between groups (raw data) and Figure 4.9 (standardised data). The effect of standardisation on the appearance of the scores plot is quite similar to using logarithms.
- To understand the effect of log scaling, we present the average intensity of the NMR peaks of the raw data (Figure 4.43(a)) and log-scaled data (Figure 4.43(b)). For the raw data, sucrose and trans-aconitate dominate, whereas once log scaled there is no large difference.
- This can be further understood by presenting the corresponding standard deviations of the NMR intensities of the metabolites in Figure 4.44. We can see that although sucrose has a relatively large standard deviation for the raw data, when log scaled it appears not to be so important as a marker.

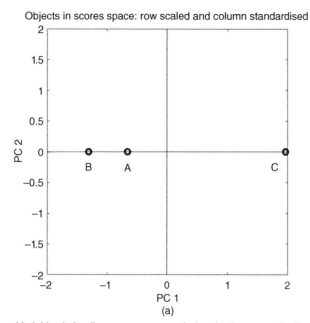

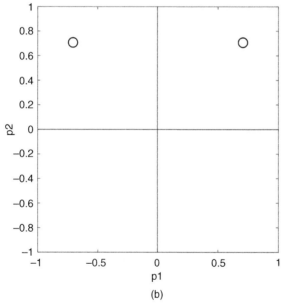

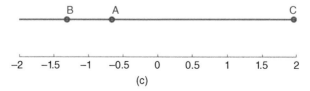

FIGURE 4.38   (a) Top Scores of first two PCs (b) Middle Loadings of first two PCs (c) Bottom Scores of PC1 for row scaled and standardised data of Table 4.14.

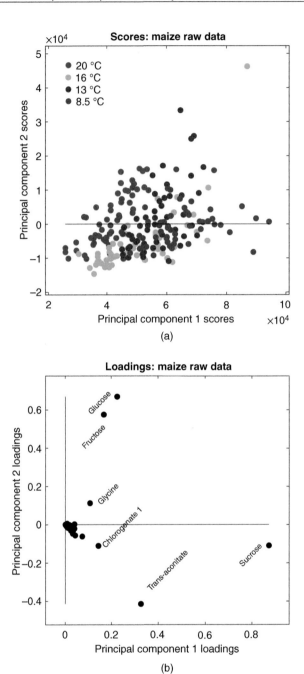

FIGURE 4.39 (a) Top Scores and (b) Bottom Loadings of the first two PCs of the maize data with some of the metabolites labelled.

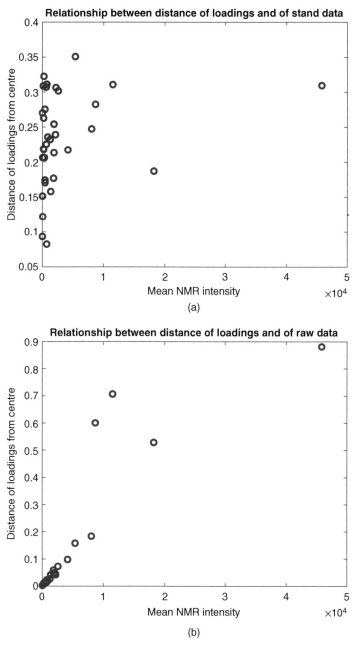

FIGURE 4.40 Distance of loadings of the first PC from the centre for (a) top standardised and (b) bottom raw maize NMR data.

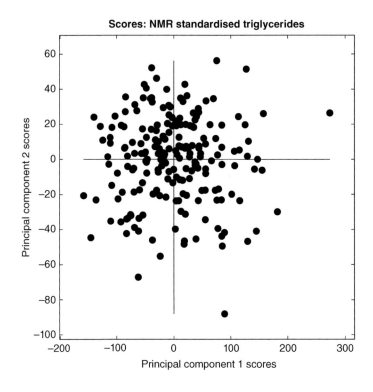

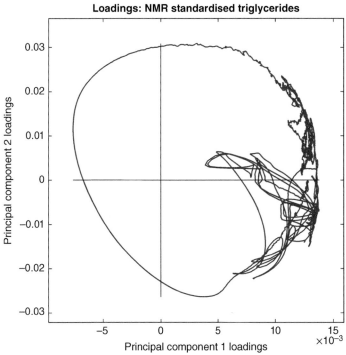

FIGURE 4.41   Standardised data of case study 3, to be compared with Figure 4.20(a), (b) and (c).

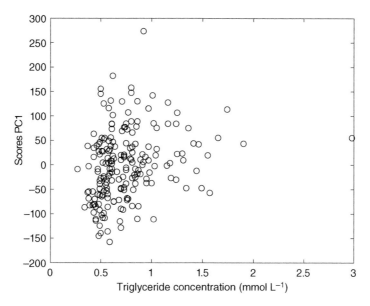

FIGURE 4.41 (Continued)

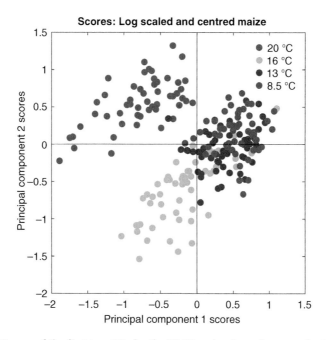

FIGURE 4.42 Scores of the first two PCs for the NMR maize data of case study 5 after first log scaling to the base 10 (no standardisation).

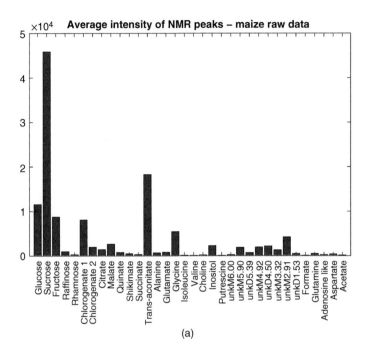

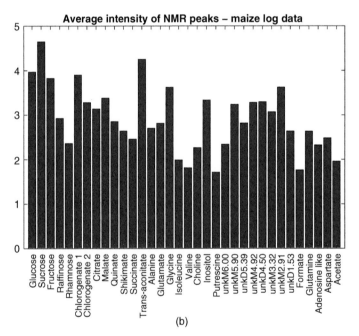

FIGURE 4.43   Average intensity of NMR peaks of maize (a) top raw data (b) bottom log-scaled data.

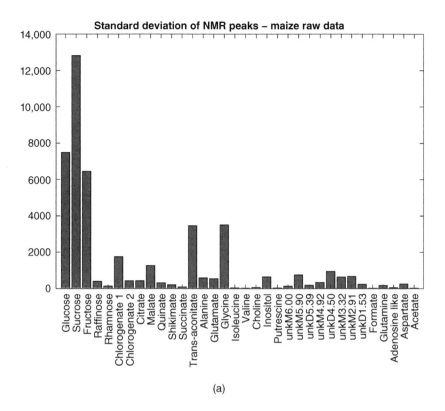

(a)

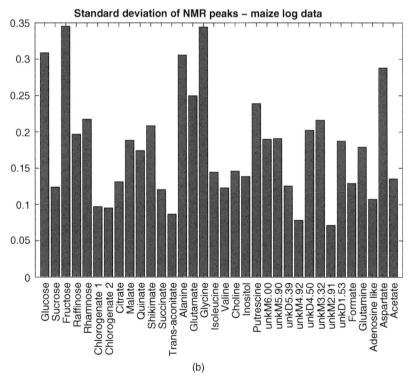

(b)

FIGURE 4.44   Standard deviation of NMR peaks of maize (a) top raw data (b) bottom log-scaled data.

Log scaling often has a similar effect to standardisation.

- It reduces the influence of large peaks or intense variables.
- In addition, sometimes intensities are better modelled using a log-normal distribution as opposed to a normal distribution. In Chapter 5, this is illustrated in Figure 5.6 for the distribution of intensities by LCMS of a metabolite found in mice. As will be discussed in the next chapter, this can result in a better interpretation of p values, as most statistics used in chemometrics assume an approximate underlying (multi)normal distribution.

Although log scaling is appropriate in many cases, it suffers from a major disadvantage in that there is no definition of logarithms for zero or negative numbers. Sometimes in chromatography, individual metabolites may be below the limit of detection and so recorded as 0. In spectroscopy, such as NMR, there may be a noisy baseline, and negative noise is often recorded. Finally, data close to the limit of detection (relatively small positive values) may have an undue influence over the analysis if using logarithms.

If the proportion of unsuitable data is low, there are several approaches for replacing it, for example, by a small number such as half the smallest positive value recorded in the data. In Section 4.7.1, we will look in more detail how to tackle missing data and can treat unsuitable data as missing. There are several choices if there are several unsuitable (zero, negative or relatively small) measurements.

- Discard variables with a defined high proportion of unsuitable measurements.
- Replace these unsuitable measurements using a defined method as described below.
- Do not use logarithms, replace them with an alternative approach such as standardisation.

4.7   COMMON ISSUES

4.7.1   Missing Data

In order to perform PCA or any other multivariate method (such as PLS as described in Chapters 7 and 8) all elements in a data matrix must contain numerical values.

In some forms of data acquisition, such as LCMS or GCMS, it is not always possible to obtain numerical information for every element of the data matrix. There may be a variety of reasons. A common one is that a peak is of low intensity (below detection limit) and so indistinguishable from background noise. A completely different reason is that although the metabolite corresponding to a particular chromatographic feature is present in detectable amounts, it is not identified, for example, due to chromatographic reasons or noisy/distorted spectra. Sometimes manual inspection can sort this out, but if there are hundreds of peaks in hundreds of chromatograms, this can be a time-consuming and tedious procedure, and due to distortions often there can be substantial uncertainty in quantification.

There are a large number of approaches for dealing with such situations resulting in endless papers and theses and conference presentations. In this section, we will just discuss some of the more common methods.

- In order to illustrate this, we will use the 10×5 $\boldsymbol{X}$ matrix as presented in Table 4.16, with five objects belonging to a blue group, five to a red group, and five variables.
- The elements $\boldsymbol{x}_{4,2}$ and $\boldsymbol{x}_{8,2}$ have been removed and are missing.
- This simulation was from an original full dataset where $\boldsymbol{x}_{4,2} = 28$ and $\boldsymbol{x}_{8,2} = 30$, and which was created to show discrimination between both of the groups of objects in the first two PCs.
- The scores of the first two PCs (column centred) for the full original dataset including the two missing values are illustrated in Figure 4.45 with objects 4 and 8 marked, together with the scores plots when various methods have been employed to handle the missing data.
- If a variable is sparse, that is, it can only be identified and quantified in a small proportion of objects, usually the best approach would be to remove it. The exception might be if it is detected in one group but not another, in which case it could be a potential marker. As 2 out of 10 of the measurements for variable 2 are missing, we illustrate the scores when column 2 has been removed, in Figure 4.45. We see that objects 4 and 8 are still in similar positions to the full dataset but some of the objects for example from the red group have changed their relative positions a little.
- In some cases, replacing missing values by 0 is acceptable, if it is suspected that in most cases the reason why they are missing is that they are below the detection limit. Even if this is not so, if the proportion of data that is missing is small, it probably will not make a major difference. If such a variable is identified at a later stage as a potential marker, it could be investigated in more depth later. We have used this approach for case study 2 (GCMS to study malaria). However, as

TABLE 4.16 Dataset for illustrating how to replace missing variables.

Object	Variable				
	1	2	3	4	5
1	16	29	12	14	20
2	27	51	22	24	34
3	36	57	20	30	44
4	20	n/a	8	16	24
5	11	22	10	10	14
6	11	55	38	16	18
7	8	31	20	10	12
8	12	n/a	16	12	16
9	5	40	30	10	10
10	22	77	48	26	32

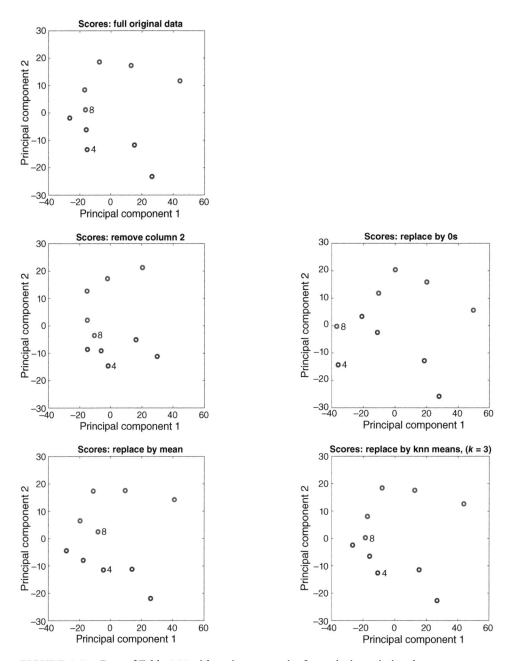

FIGURE 4.45   Data of Table 4.16 with various strategies for replacing missing data.

we see from Figure 4.45 if we set $x_{4,2} = 0$ and $x_{8,2} = 0$, this changes the position of samples 4 and 8 quite dramatically; however, the orientation of the remaining eight samples is very similar as in the full dataset.

- However, if it is desired to log scale the data as discussed in Section 4.6.4 since the logarithm of 0 is undefined, instead of replacing missing values by 0, they can be replaced by a small number, for example, half the smallest detectable value either in the entire data matrix or for the specific variable. In our case, we could set $x_{4,2} = 14.5$ and $x_{8,2} = 14.5$ (half the minimum of column 2) or $x_{4,2} = 2.5$ and $x_{8,2} = 2.5$ (half the minimum of the entire dataset). We will not illustrate this calculation, as it is primarily useful as a preparation for log scaling, and assumes missing data are below detection limits, rather than present but unable to be quantified often due to chromatographic issues.

- Another approach is to replace the missing data with the column average of the existing data. For our simulation, $x_{4,2} = 45.25$ and $x_{8,2} = 45.25$. In our example, we see from Figure 4.45 that this is a reasonable solution but moves both objects to slightly higher PC1 scores; however, they fall well into their respective groups.

- A final and common approach is called kNN (k Nearest Neighbour). There can be very lengthy procedures according to how many and the distribution of missing values, but we illustrate the principles numerically using a simple example for the case study of Table 4.16, using $k = 3$. The detailed calculations are presented in Table 4.17. For our example, following the procedure illustrated numerically in the table, we now find $x_{4,2} = 34$ and $x_{8,2} = 27.33$. The resultant PC scores plot in Figure 4.45 is very similar to the scores plot for the full dataset, and in this case most faithful representation. The reason why kNN works best in this case compared to using an overall mean or replacing using 0s, is that the assumption is that the more similar the objects are (using the remaining fully characterised variables), the more similar they should be using the missing variable. As an example, metabolic profiles may be similar in donors with similar characteristics often due to related metabolic pathways and it is just an analytical problem that a few measurements cannot be obtained often for chromatographic or spectroscopic reasons.

Our conclusion for this case study is that kNN is the most appropriate approach for replacing missing variables. This is not always so and depends on why specific variables (often metabolites) are not detected, and one cannot generalise.

However, one can conclude that the method for calculating missing variables may sometimes have a significant influence on the resultant pattern, and a good approach may be to compare several different approaches. If however the proportion of missing data is small, and there are a large number of variables, it is not likely that the method for imputing missing data has a huge influence on the resultant conclusions. Chemometrics as used in metabolomics is often exploratory and will point the way towards more interesting experiments or detailed investigation at a later stage, and providing the data are of reasonably good quality, and not too sparse, a main aim is simply to obtain a complete data matrix to perform multivariate analysis subsequently.

4.7.2   Quality Control Samples

In many types of instrumental analysis, it is necessary to check instrumental stability regularly over time. This is common in LCMS where there can be significant instrumental drift.

TABLE 4.17   Missing value calculation.

Step 1
Remove column 2 initially.

Step 2
Calculate the distances between the two rows with missing data (4 and 8) and all other

rows, using the remaining columns, 1, 3, 4 and 5, that is, $x_{Mi} = \sqrt{\sum_{j=1,3,4,5} \left(x_{Mj} - x_{ij}\right)^2}$ for $M = 4$
and 8,
These distances are below

	Row									
	1	2	3	4	5	6	7	8	9	10
Row 4	7.21	20.22	31.56	0.00	14.87	31.89	21.63	14.42	30.68	42.05
Row 8	7.21	27.00	41.23	14.42	6.71	22.47	7.21	0.00	16.88	39.70

Step 3
Order these distances from smallest to largest

	Smallest									Largest
Row 4	(4)	1	8	5	2	7	9	3	6	10
Row 8	(8)	5	1	7	4	9	6	2	10	3

Step 4
Choose the k (=3 in this example) smallest distances, excluding itself and excluding any rows where the value is missing.
For row 8, these are rows 5, 1 and 7.
For row 4, the choice is slightly complicated, as row 8 contains a missing value, so we choose rows 1, 5 and 2.

Step 5
Average the values in these k (=3 in this case) nearest objects to replace the missing values 34 and 27.33.

Usually, a sample is analysed regularly every few analyses to check that the instrumental conditions are stable, typically every 5–10 samples. More usually in current practice, this is a pooled sample consisting of a combination of extracts from the analyses. It should cluster in the middle of the data; however, it is alternatively possible to purchase an external QC sample.

Often, the QC samples can be included in the PCA. If there is substantial instrumental drift, the scores of the QC samples will not cluster closely and there could be some problems with the analytical data. Under such circumstances it may be necessary to discard data where the QC sample is deviating, or to discard or repeat analyses if feasible dependent on cost and time, or finally, in certain very specific cases it may be possible to use the QC sample to correct the data. However, the ideal and most common situation is that there is limited change in the nature of the QC sample throughout the analytical process.

4.7.3 Variable Reduction

Sometimes not all variables that have been recorded are appropriate for further analysis using chemometric methods.

Methods for variable selection prior to multivariate analysis are very varied and depend on instrumental method and the aim of the analysis. We cannot comprehensively describe all these approaches, but will briefly describe a few of the most widespread below.

- Removing less discriminatory or informative variables. For example, if we want to discriminate between two groups of samples, many of the metabolites may have no role and their concentrations may be influenced by unrelated effects we are not directly interested in. Often these are retained during the preliminary steps of the analysis, but sometimes can be removed. These can be done using a variety of methods for variable selection as discussed in Chapters 9 and 10 as a preliminary step. The focus of these later chapters is to identify which variables are most significant but they could equally be used to reject less significant variables. Of course, this has a risk of over-fitting, that is exaggerating trends. Over-fitting is discussed in Chapter 7, Section 7.7.2. There is a balance between removing uninformative variables to improve quality of models and over-fitting data. If data are divided into test and training sets, variable reduction can be done on the training set and then the same variables removed from the independent test set.

- Targeted metabolomics retains only those metabolites (usually as identified by LCMS or GCMS) whose structures are known, often via synthetic internal standards, sometimes called annotated metabolites. Untargeted metabolomics may retain some or all of the metabolites, even if their structures are unknown. In this text, we use a mixture of targeted and untargeted data. The chemometrics principles are the same. Targeted metabolomics may involve removing any LCMS or GCMS peaks whose identity cannot be confirmed so is a form of variable reduction.

- Removing variables of low intensity. This can be important prior to standardisation especially when whole datasets such as NMR spectra or HPLC chromatograms are to be analysed. Standardised full data will mean that regions of the spectra or profiles that primarily correspond to noise will have equal influence to those that correspond to detectable metabolites. If these regions are large, they could overwhelm the analysis if not rejected.

- Removing variables that contain missing data. We have discussed ways of handling missing data in Section 4.7.1. One approach of course is to estimate the missing information. If, for example, there are 100 samples and for a specified variable only two of the measurements are unknown, it may lose important information to remove this variable completely. However, if 20 of the 100 measurements are missing, it may be misleading to include this variable and so it may be best to remove it altogether from the analysis. Of course, sometimes individual chromatograms or spectra are of poor quality, and if the missing data are contained primarily within specific samples, removing the samples is an alternative unless this is crucial for a balanced design as discussed in Chapters 6 and 10. The ultimate decision, if the amount of missing data is judged to be large, is to re-run the analysis but this may be expensive and it may be difficult to obtain fresh samples for a specific study.

As can be seen there are a large variety of approaches for variable reduction prior to PCA or other methods of multivariate analysis, but these are very technique- and even laboratory-dependent, and there is no general recommendation, except to consider whether this step is necessary in the analysis of a dataset.

CHAPTER 5

Statistical Basics

5.1 USE OF P VALUES AND HYPOTHESIS TESTING

Classical chemometrics was primarily focused on algorithms, such as PCA (principal components analysis) and PLS (partial least squares) and factor analysis. The early pioneers were primarily chemists good at programming. Many of the early implementations were developed without incorporating statistical concepts such as p values. For example, PLS calibration in its original form aimed primarily to predict concentrations and determine how good these predictions were. MCR (multivariate curve resolution) aimed to resolve overlapping signals and obtain estimations of individual signal intensities or spectra.

Most mainstream statistics, however, is concerned with hypothesis testing and often use p values to provide an indication of the significance for example of an effect on a process. Much of conventional modern-day statistical methods developed from the work of Ronald Fisher and colleagues in the 1920s and 1930s. When applied to areas such as biology or medicine, these traditional approaches are widely taught and used today. Most experiments in medicine, for example, end up with the calculation of p values to determine whether a given underlying hypothesis is significant or not. In most papers in the area of biomedical science, a p value is required for publication.

When metabolomics developed, chemometrics experts, not often used to using p values, and biomedical scientists, whose research often had a mandatory requirement for such indicators, came into contact with each other, and many chemometrics experts were asked to provide p values from their calculations.

Hence, an understanding of p values is important, especially so that the chemometrics expert can talk to the clinical and biological scientist. In parallel, it is also necessary to understand distributions and hypothesis testing.

Data Analysis and Chemometrics for Metabolomics, First Edition. Richard G. Brereton.
© 2024 John Wiley & Sons Ltd. Published 2024 by John Wiley & Sons Ltd.
Companion website: www.wiley.com/go/Brereton/ChemometricsforMetabolomics

In this chapter, we will introduce these principles as simple applications, which we will build on where necessary in later chapters.

There are three main reasons why a chemometrician may be interested in p values.

- The principal use, in metabolomics, is to determine which variable is a possible marker for an effect or a disease. In this case we are computing values for individual markers to predict which are significant in discriminating between groups of samples, often out of tens or hundreds of potential biomarkers.
- It is also important to be able to determine whether a specific sample is a member of a predetermined group. This is a classification or pattern recognition problem. In traditional chemometrics, this is a predominant use of p values, for example, to identify an unknown sample from its analytical signal and predict its provenance. Although less a widespread application in metabolomics, as it is often not the prime aim to identify the provenance of unknown samples, it still has value, for example, to determine whether predetermined groups are homogenous, or to detect outliers.
- Finally, some studies aim to determine the significance of effects on metabolic responses. Much of traditional statistics is based around these applications often using ANOVA, in metabolomics, we could be interested in treatments or whether environmental or biological factors have a significant influence on metabolic signal of a person, animal or plant . In fact, whether a factor or variable is significant often are the answers to the same question but posed differently. If we for example ask whether a metabolite is a significant marker for temperature, we could equally ask whether temperature has a significant effect on the metabolite. Traditional statistics was primarily univariate, and so was concerned with single variables but sometimes several factors; traditional chemometrics was multivariate but often studied single factors. We will see in Chapter 10 how both questions can result in identical p values. Chemometrics does not determine cause and effect.

5.2　DISTRIBUTIONS AND SIGNIFICANCE

5.2.1　Simulated Case Study

We will initially introduce the concepts via a small simulated case study, as shown in Table 5.1.

- The data can be arranged in the form of a 20×5 matrix.
- The first 10 samples correspond to class A (blue) and the second 10 to class B (red).
- Questions we might ask are as follows.
  - Does a sample belong to its proposed class?
  - Which variables are best to discriminate between two classes?

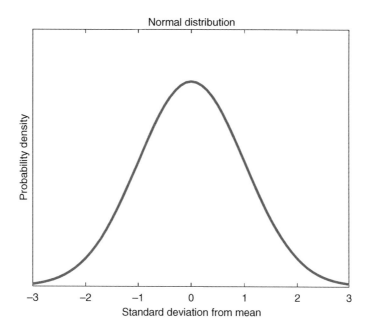

FIGURE 5.1   Standard normal distribution.

5.2.2   The Normal (z) Distribution and p Values

A good way of understanding p values is via the normal distribution. Not all distributions are well approximated this way, but it is common approximation of symmetric distributions and will provide reasonable answers in most cases. In this chapter, we will introduce p values within this context, but there is no inherent requirement a distribution is normal. In some cases, non-parametric tests are used, such as the Mann–Whitney U test as discussed in Section 9.2. These still use the same underlying definition of p values, for example, a p value of less than 0.05 corresponds to an occurrence that is expected to be obtained less than one time in 20.

We will exemplify initially using the first 10 samples and first variable of our simulated case study, forming a 10×1 column vector representing a univariate measurement on samples from class A. For example, we may measure the concentration of a metabolite in 10 samples obtained from a group of donors having a specific characteristic (e.g. a disease marker).

The default statistical assumption which many tests are based on is that the underlying population should be normally distributed as illustrated in Figure 5.1, standardised so that the mean is 0 and standard deviation is 1. This type of curve is often called a probability density function (pdf). Although this of course is by no means always an appropriate model, most common hypothesis tests in metabolomics are based on this assumption.

- If we know the mean and standard deviation of a normally distributed series of measurements, we can calculate what proportion of measurements are expected to lie a certain distance from the mean.

- For example, using a normal distribution, 0.159 (15.9%) of the measurements are expected to be more than 1 standard deviation below the mean. This is called the one-tailed area of the normal distribution 1 standard deviation away from the mean.
- The two-tailed area is $2 \times 0.159 = 0.317$ (taking rounding into account), that is, slightly over 3 in 10 of the measurements are expected to be at least 1 standard deviation either side of the mean. These areas are illustrated in Figure 5.2.
- The p value is an estimate of how frequently a measurement is equal to or exceeds a given value. So, for example, the two-tailed p value of a measurement with at least 1 standard deviation either side of the mean is 0.317 if the underlying measurements are normally distributed.
- Some statisticians speak about the null hypothesis, a concept developed in the 1920s and 1930s originally advocated by Ronald Fisher. In our case the null hypothesis is that there is no significant difference between a measurement of 1 standard deviation from the mean and the population mean, so the measurement could very well be part of the underlying population.
- In more formal terms, the p value that reading at least 1 standard deviation from the mean is 0.317, which is interpreted as the probability that the null hypothesis is correct. More accurately, it is the chance that a measurement of 1 standard deviation or more from the mean is obtained if the measurement is part of the underlying population. So sometimes (almost a third of the time) if a sample is part of a normal distribution, we expect to find a measurement at least 1 standard deviation away from the mean.
- Hence, a very low p value, for example, 0.01, implies that such an extreme reading is obtained very rarely, 1 time in 100. Hence, low p values are often said to guide the investigator to reject the null hypothesis and suggest that the sample may not be part of the underlying population.
- However, it is important to recognise that extreme readings will sometimes be obtained even if a sample is part of the underlying population especially if enough samples are measured. If a samples is not part of the parent population it is likely to correspond to a low p value; however, a few samples from the parent

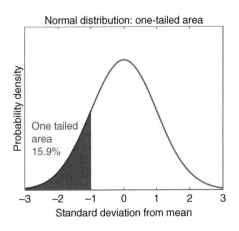
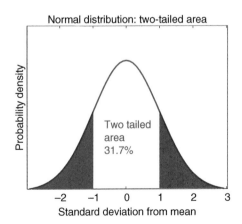

FIGURE 5.2   One and two-tailed areas and 1 standard deviation from the centre using a normal. distribution.

population will nevertheless have low p values but this will not be very common. If a reading has a high p value it is likely to be part of the parent population.

In our case, consider the 10×1 measurements of samples A to J (class A) and variable 1 from the data of Table 5.1.

- The mean is $\bar{x}_{A1} = \sum_{i \in A} x_{i1}/10 = 28.02$.

- The standard deviation (using the population formula) is $s_{A1} = \sqrt{\sum_{i \ni A} \left(x_i - \bar{x}_{Ai}\right)^2 /10} = 10.65$. Note that many authors would use the sample standard deviation for estimation and t statistics, but the population standard deviation for scaling data and z statistics. However, in order to compare different approaches we will stick to the population standard deviation for simplicity and to compare calculations.

- We can now standardise the data. For example, sample J is 2.12 standard deviations from the mean which value we can call z.

TABLE 5.1   Simulated case study.

	1	2	3	4	5
A	22.838	33.317	38.771	8.263	31.397
B	19.294	18.248	46.413	10.113	25.675
C	36.062	32.793	10.644	7.666	25.229
D	37.241	25.657	50.240	7.560	24.917
E	31.542	28.836	26.261	6.677	36.727
F	21.327	26.443	28.444	7.689	6.258
G	13.777	24.666	29.454	8.199	36.699
H	29.825	33.436	27.498	8.178	17.610
I	17.667	35.079	50.505	10.650	9.017
J	50.638	24.623	37.856	10.659	7.364
K	20.790	29.082	34.900	48.864	22.179
L	9.523	26.885	31.592	36.232	29.674
M	12.129	23.520	12.658	23.557	16.290
N	14.257	25.426	13.891	36.172	18.623
O	10.182	27.603	7.093	40.953	18.379
P	19.827	26.093	28.355	40.055	39.586
Q	5.209	29.808	22.410	47.703	16.715
R	18.030	25.516	23.607	27.302	25.807
S	14.331	25.180	32.563	59.576	11.261
T	17.395	25.496	15.350	46.353	26.694

- The p values can be obtained using the normal cdf (cumulative distribution function), which calculates the proportion of the area remaining. This is doubled if we want to do a two-tailed test. P values are equal to $1 - $ cdf. This is illustrated in Figure 5.3.
- For sample J, the p value is 0.034 as illustrated in Figure 5.4. Note that we usually employ the two-tailed p value, as we ask whether how frequently a sample 2.12 standard deviations (in this case) or more on either side of the mean would be expected.
- The p values for all samples in class A using a normal (z) distribution are given in Table 5.2, together with the values using a t-distribution as discussed later in Section 5.2.3. A p value 0.05 implies that a value equal to or greater than the observed measurements (in terms of distance from the mean) is expected to be obtained one time in 20.
- Often people use set criteria to decide whether a sample is likely to be a member of a population, for example, $p < 0.01$ or $p < 0.05$. If we use $p < 0.05$ as our cut-off, we reject sample J as being part of the parent distribution. Whether this is meaningful or not depends on how the numbers are interpreted, and really just suggests that reading this far from the mean is rare.
- Of course, obtaining very low p values such as 0.001 builds up good evidence that a sample is not part of the parent distribution, as such a value or greater would be obtained only 1 time in 1000. If there are just 100 samples, this is evidence that the sample is either an extreme value (part of the parent population but very unusual) or an outlier (not part of this population).

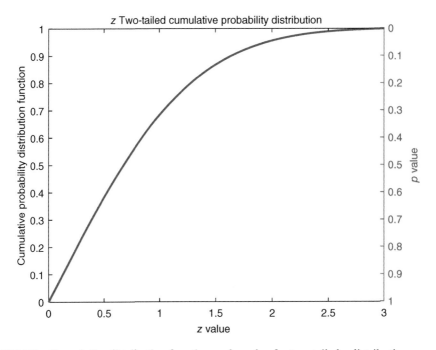

FIGURE 5.3 Cumulative distribution function and p value for two-tailed z-distribution.

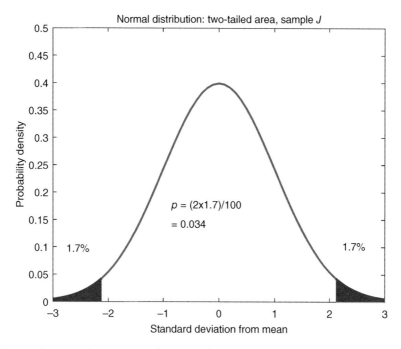

FIGURE 5.4 How sample J corresponds to a *p* value of 0.044 assuming the samples are normally distributed using a two-tailed distribution.

TABLE 5.2 Data for class A, variable 1, standardised and with calculated two-tailed *p* values assuming a normal (*z*-)distribution and a *t*-distribution with 9 degrees of freedom.

	A	B	C	D	E	F	G	H	I	J
Values	22.84	19.29	36.06	37.24	31.54	21.33	13.78	29.82	17.67	50.64
Standardised	−0.49	−0.82	0.75	0.87	0.33	−0.63	−1.34	0.17	−0.97	2.12
p Value (two-tailed) (z)	0.627	0.413	0.450	0.387	0.741	0.530	0.181	0.866	0.331	0.034
p Value (two-tailed) (t 9 df)	0.638	0.434	0.470	0.409	0.749	0.545	0.214	0.869	0.356	0.063

We cannot of course provide a definitive answer just from a single *p* value alone, but could build up a picture, for example, whether the *p* value is low using several independent tests, and low *p* values give us a clue something strange is going on.

- In our case, we have just one sample that has a *p* value of less than 0.05. As there are 10 samples, we may not expect this to be too unlikely but might want to perform some more measurements on this sample or investigate its provenance further.

We will return to experimental data and investigate case study 10 (diabetic mice) and for simplicity discuss the negative ion LCMS data to exemplify the use of p values as a univariate measure.

Consider just the 30 mice diagnosed as diabetic, row scale the intensities of the 32 variables to 1, and then examine the distribution of measured intensities for metabolite 1 (Cer(d42 : 2)), which can be represented by a 30×1 vector. Note that it is not necessary to standardise each variable at this stage as this is done when calculating p values automatically.

- The row scaled intensities of variable 1 (over all 32 variables) together with the associated p value for all 30 mice using the normal or z-distribution are presented in Table 5.3. This is compared to the p value obtained using the t-distribution as discussed in Section 5.2.3 below.
- The p values can be visualised as in Figure 5.5(a) with a critical limit of $p = 0.05$ illustrated.
- This can be compared to the corresponding p values for a normal distribution as in Figure 5.5(b).
- Mouse 14 is clearly below the critical $p = 0.05$ limit. There could be a variety of reasons. For example, it may have been misdiagnosed and is not part of the parent diabetic group. Or it may have some uniquely different metabolism. Or there may have been an analytical problem such as caused by overlapping peaks or misalignments that underestimated the intensity. Although recorded in the table as 0.000 it in practice is computed as 7×10^{-6} although the actual number has little physical meaning.
- Another feature of the p value distribution in this case is that it appears a little asymmetric. This suggests that the data might be best to transform the data.
- Performing a log transformation (see Section 4.6.4) on the data results in a more symmetric p value distribution as in Figure 5.6. This suggests that the data are better approximated by a log-normal rather than normal distribution. More formal statistical tests could be used to distinguish these models, but usually a visual comparison is adequate.
- The problem with logarithmic transforms is that some measurements are below detection limit and so recorded as 0, which cannot be converted into logarithms, so there are various approaches for replacing these values as discussed in Chapter 4. Which is appropriate would depend on why the values were not observed, for example if they were below detection limits, in which case a very low number could be substituted or if they not observed at all in which case an average could be substituted.

Although we could use either the row-scaled data or its logarithm, in practice both come to the same conclusion that the measurement for mouse 14 is not part of the parent group. The actual numerical p value has very little physical meaning as it depends on lots of assumptions, for example, that of normality, which are unlikely to be true. However, because it is so low, it is a flag and can be reported legitimately to allow potentially unusual or extreme samples to be detected.

TABLE 5.3 Case study 10, diabetic mice, variable 1, p values calculated using both a z- and t-distribution for the distribution of the row scaled intensity in all 30 mice diagnosed as diabetic.

Mouse	1	2	3	4	5	6	7	8	9	10
Intensity	4849.384	11,573.54	4330.199	9123.898	9025.85	3825.817	20908.35	11485.58	3224.599	5716.535
Row scaled	0.001587	0.004696	0.002436	0.002004	0.002438	0.00136	0.004173	0.005348	0.001877	0.003354
p Value (z)	0.508	0.535	0.755	0.624	0.756	0.450	0.685	0.373	0.587	0.947
p Value (t(29))	0.513	0.539	0.757	0.628	0.758	0.456	0.688	0.381	0.592	0.947

Mouse	11	12	13	14	15	16	17	18	19	20
Intensity	6184.769	17,470.77	5369.195	31,125.18	4311.571	5433.365	8497.756	4424.711	2523.485	11,341.44
Row scaled	0.002497	0.006169	0.001836	0.014073	0.001499	0.00239	0.003575	0.002511	0.002491	0.003256
p Value (z)	0.775	0.219	0.576	0.000	0.485	0.741	0.874	0.779	0.772	0.979
p Value (t(29))	0.777	0.229	0.580	0.000	0.490	0.743	0.875	0.781	0.775	0.979

Mouse	21	22	23	24	25	26	27	28	29	30
Intensity	2662.382	7616.871	2595.819	2708.454	1472.886	3412.169	3493.19	6025.117	15089.88	7479.101
Row scaled	0.001333	0.003102	0.001777	0.00535	0.002491	0.001728	0.001376	0.002071	0.003066	0.003876
p Value (z)	0.443	0.971	0.559	0.373	0.773	0.546	0.454	0.644	0.959	0.777
p Value (t(29))	0.449	0.971	0.564	0.380	0.775	0.551	0.460	0.647	0.959	0.779

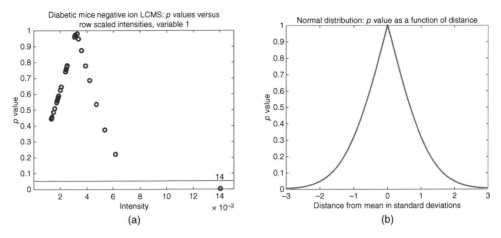

FIGURE 5.5 (a) Left case study 10 (30 diabetic mice analysed using negative LCMS) p values of row scaled intensity of variable 1 and $p = 0.05$ critical limit, (b) Right p values of a theoretical normal distribution.

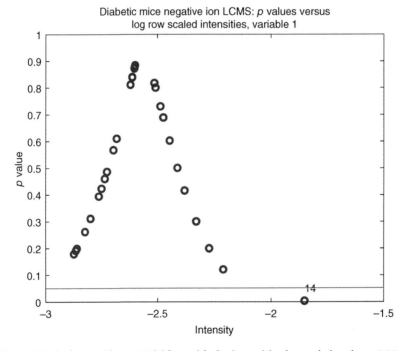

FIGURE 5.6 Equivalent to Figure 5.5(a) but with the intensities log scaled and $p = 0.05$ critical limit.

In most examples in the later chapters, we will not log transform the mouse data, because although this does seem a good transformation for variable 1, it is not necessarily appropriate for other variables in the data. In practice, a comprehensive analysis of the information would try out different approaches for transforming the data to find an optimum. However, most methods will identify significant metabolites or outlying samples, it is just that the p values may differ according to how the data are transformed. It is important to use p values as a guide rather than having an exact physical meaning and often several approaches to calculations can be compared to provide a consensus.

5.2.3 t-Distribution and Degrees of Freedom

Although the discussion above is based on the samples being normally distributed, in practice this is an approximation it is more common model to use the t-distribution. This book is not about statistical theory, but a major reason why the normal distribution might not be most appropriate for experimental data is that the experimentally estimated standard deviation is only an approximation of the underlying (population) standard deviation and is usually lower.

In strict statistical terms:

- The population standard deviation, usually denoted by σ, is a parameter, and a feature of the population (i.e. of all possible samples with a given characteristic) defined by $\sigma = \sqrt{\sum_{i=1}^{I}\left(x_i - \bar{x}\right)^2 / I}$.

- The sample standard deviation, usually denoted by s, is a statistic and is the estimate of the population standard deviation. For example, if we sample 30 diabetic mice, we are taking only a small portion of all possible diabetic mice, in order to estimate the variability or standard deviation of our measured variable in the population, and define this by $s = \sqrt{\sum_{i=1}^{I}\left(x_i - \bar{x}\right)^2 / (I-1)}$.

However, the population standard deviation can also be used for scaling data, and when we want to compare the z- and t-distribution directly, it is easier to use the same definition of the standard deviation. When there are several samples (10 or more) this would make no meaningful observable difference, and for purpose of comparison we will use the population standard deviation throughout this text although strictly speaking we should use the population standard deviation when describing the z-distribution and sample standard deviation when describing the t-distribution.

Nevertheless, when the number of samples is relatively modest (less than about 20) it is usual to correct for the small sample size by modelling the data using the t-distribution rather than the z-distribution. This should be a slightly better fit to the experimental data. Still, it assumes the underlying population is normally distributed. In practice, although both the z- and t-distributions differ a bit, this depends on many different assumptions about the underlying population of measurements, such as normality, which are unlikely to be correct, but it is more usual to use the t-distribution.

As the z- and t-distributions differ, we will look at this difference.

- In this book, we will use the concept of degrees of freedom, often denoted ν, in several places. If there are I samples, the data are characterised by $\nu = I - 1$

degrees of freedom. Formally, the number of degrees of freedom indicates the number of independent values that can vary in an analysis without breaking any constraints. If the sample size is 10, 1 degree of freedom is lost by standardisation; if we are provided with 9 out of 10 measurements from a standardised distribution the 10th must be fixed so that the overall mean of 0 and standard deviation of 1 is achieved.

- The shape of the t-distribution differs according to the sample size. There is a differently shaped t-distribution according to the number of degrees of freedom, which we denote $t(\nu)$.

- As I increases, the t-distribution approaches the z-distribution in shape. A t-distribution with, for example, 100 degrees of freedom or $t(100)$, is in practice indistinguishable from a z-distribution.

- For the simulated case study of Table 5.1, there are 10 samples from class A (blue class), and so 9 degrees of freedom.

- In Figure 5.7, we superimpose the t-distributions for both 4 and 9 degrees of freedom and the z-distribution together with the corresponding two-tailed cdfs. As the number of degrees of freedom decreases, the shape of the t-distribution diverges from the z-distribution and becomes flatter with wider wings.

- This is better illustrated in Figure 5.8 with a $p = 0.05$ critical limit indicated. Although the variation in p value against distance from the mean does not look very different, in fact using a normal distribution, $p = 0.05$ is reached at about 2 standard deviations and for a t-distribution at $p = 0.05$ at about 3 standard deviations. There is not much difference around the mean, but further out, for example, at critical values of $p = 0.05$, 0.01, etc. there can be a substantial difference.

In our simulated case study, we have 9 degrees of freedom, so can calculate the p values assuming a t-distribution with 9 degrees of freedom. These are presented in Table 5.2. The higher the p value (the closer the measurement is to the mean of the data), the closer the p values calculated using both types of distribution.

- Hence, for sample H, the p value calculated using the two distributions is almost identical.

- For sample J, which has the lowest p value, the difference is largest and increases from 0.034 (z-distribution) to 0.063 (t-distribution with 9 degrees of freedom). If we set a critical value of $p = 0.05$, we would now estimate that sample J is above the critical p value and so not an outlier or extreme sample if using the t-distribution.

Of course, although by statistical theory there is a difference between z- and t-distributions, especially when sample sizes are low, in practice this depends on the underlying population being normally distributed, which is unlikely to be obeyed perfectly in experimental samples. It would be possible to perform tests of normality, but this would usually require a very large number of samples, which would be impractical and so there is usually no correct answer as to what statistic should be employed. In general, the t-distribution is used in most practical cases but both would

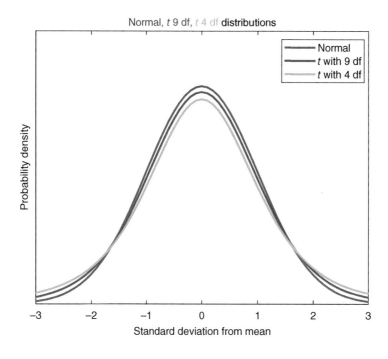

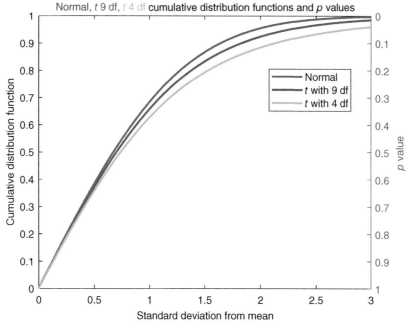

FIGURE 5.7 Normal (z) distribution, compared to the t-distribution with 9 and 4 degrees of freedom together with the corresponding cdf.

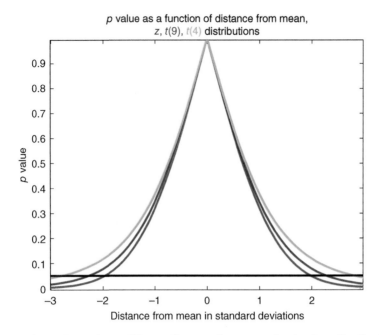

FIGURE 5.8 Change in *p* value at different distances from mean for the three distributions of Figure 5.7, with $p = 0.05$ critical value indicated.

identify outlying samples and if the sample size increases to say 20 or 30, there would be no observable or practical difference between *p* values estimated using the *z*- or *t*-distribution.

We will illustrate by a real example, the 30 diabetic mice from case study 10, using negative ion LCMS and row scaling each sample to 1.

- We can calculate the *p* values using the *t*-distribution with 29 degrees of free-dom for variable 1 of the diabetic mice of case study 10, as presented in Table 5.3 and compared to the *z*-distribution.
- When *p* values are close to 1, there is no difference to 3 significant figures between the estimated *p* values using both models. Even when *p* is between 0.7 and 0.8 the difference is relatively small. It is largest, in our samples when *p* values are around 0.2, for example, mouse 12.
- For mouse 14, the *p* value is the same to 3 significant figures, and using either measure we would conclude this is an outlying sample.

The main use of *p* values in this situation is to detect unusual measurements or samples. The statistics cannot say why but can certainly lead a user to home in on samples or measurements that are unusual, and if lots of variables have been meas-ured and many samples, for example, 100 samples and 1000 variables, it can be tricky to detect outliers within such a large dataset of 100,000 measurements, but statistical methods can be used to red light any that are extreme and can be faster than manually examining large datasets.

- As an example, we can estimate those measurements that are below critical values of p of 0.05, 0.01 and 0.001 for the 30 mice diagnosed as diabetic and the 32 metabolites identified from negative ion LCMS of case study 10.
- These are represented as heat plots in Figure 5.9.

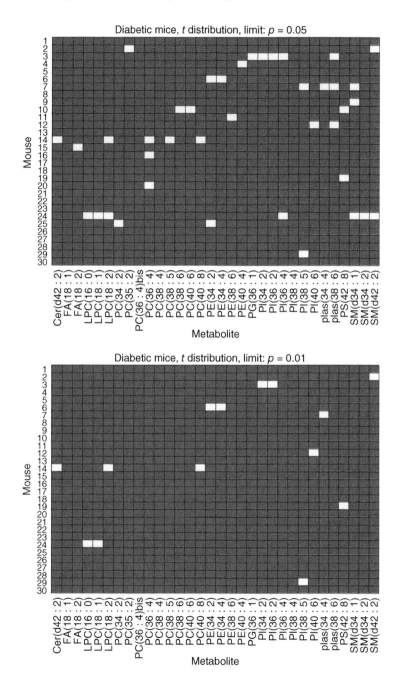

FIGURE 5.9 *P* values below different critical limits for 32 metabolites detected in the 30 diabetic mice by negative ion LCMS (case study 10).

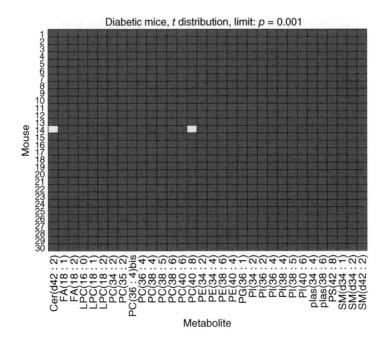

FIGURE 5.9   (Continued)

- There are $30 \times 32 = 960$ elements in the matrix. So, if the data can be modelled by a t-distribution, we would expect approximately to the nearest integer 48 measurements to have $p < 0.05$, $10 \ p < 0.01$ and $1 < 0.001$.

- Looking at the figures, we can count 40 have $p < 0.05$, 14 have < 0.01 and 2 have $p < 0.001$. This suggests our distribution is a reasonable model. Of course, we would need a very large sample size to be sure, and this type of experiment is not likely to be possible as it is not feasible to keep instrumental conditions stable over very long periods when performing chromatography of a very large number of samples, so we can only expect an approximate agreement.

- Mouse 14 and to an extent 24 seem the least conforming to the model, based on the number of metabolites below the critical thresholds, but no sample sticks out as a major outlier, suggesting that the population sampled and the analytical procedure are robust and the diabetic mice can be safely considered as a homogeneous group, although a careful investigator may want to check these two samples to be sure.

Sometimes we can spot more obvious outliers.

- Figure 5.10 is obtained from the NMR profiling of the maize (case study 5) and just those 54 samples harvested at 8.5 °C and measurements for which $p \leq 0.05$ are indicated.

- 73 of the 1836 (corresponding to 54 samples $\times$ 34 variables) have $p < 0.05$, or 0.040 of the samples, which suggests the overall model is quite a reasonable fit.

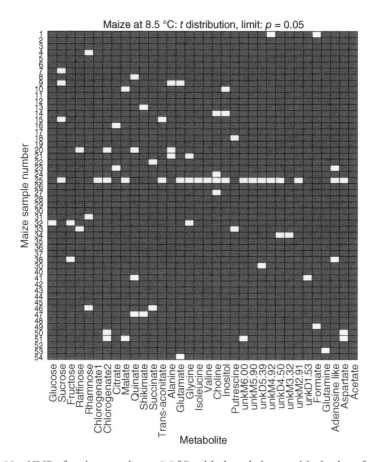

FIGURE 5.10   NMR of maize samples, at 8.5 °C, with those below a critical value of $p = 0.05$.

- However, sample 25 has 19 out of the 34 metabolites with a $p < 0.05$ and is so probably an outlier. This does not of course necessarily imply there is a problem with its metabolic signature but could be with the analytical procedure, we would not know without further work which is beyond the scope of this text.

The conclusions from the t-value evaluation could be compared to those obtained from the PC plots. The scores plot of PC2 versus PC1 for the standardised maize NMR data is presented in Figure 5.11(a) with sample 25 (8.5 °C) marked. Looking at this graph, we would not identify this sample as an outlier. In Figure 5.11(b) PC3 is represented by one of the axes. Sample 25 might now be potentially viewed as a possible outlier but not very strongly and still could be missed. In Figure 5.11(c) we perform PCA just on the 54 samples at 8.5 °C growth temperature. In this case, the metabolite contents based on NMR intensities are standardised just over these samples, and not the entire dataset. For brevity we will not describe this procedure in numeric detail, but sample 25 now clearly appears to be an outlier in the PC scores plot of PC2 versus 1, having a prominently negative score along PC1.

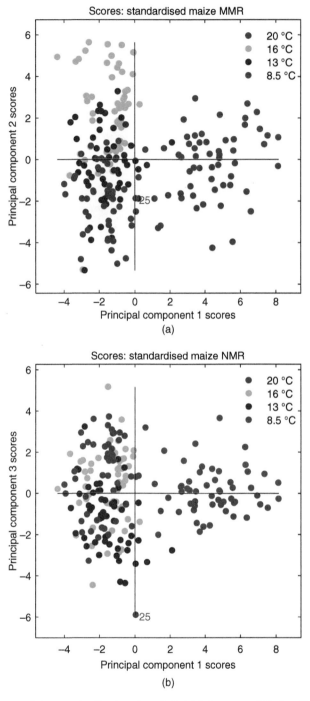

FIGURE 5.11   Scores of the standardised maize NMR (a) PC2 versus PC1 for all the data (b) PC1 versus PC3 for all the data (c) PC2 versus PC1 just for the 8.5 °C data, with sample 25 indicated.

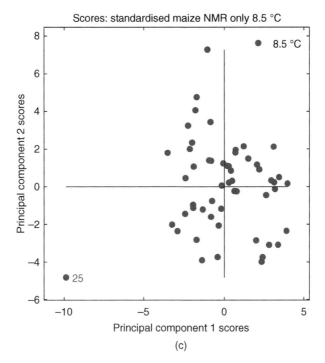

FIGURE 5.11   (Continued)

Of course, although we may have spotted this if we had originally performed PCA just on the data at 8.5 °C, often there are so many possible PC plots, that investigators only visualise a few, and in most cases, this would be from the overall dataset. So, p values, whilst they are only an indication of the significance of a variable, can provide useful complementary evidence and would highlight that sample 25 is a bit unusual. On the basis that several metabolites have $p \leq 0.05$ for this sample (19 out of 34), it sug gests there are problems which could be investigated further. These problems could be analytical rather than biological and so should not without further investigation neces sarily imply unusual metabolism.

5.2.4 χ^2-Distribution

When there is only 1 variable, χ^2 is simply the square of z. It is characterised by 1 degree of freedom, with $\nu = 1$ in this case.

- The area to the right of the χ^2 is illustrated in Figure 5.12 at a variance of 2.25 or standard deviation of 1.5.
- If we denote cdf as the cumulative distribution function for measurements obtained less than a specified value, we see that $2 \times (1 - normcdf(x)) = 1 - \chi^2$ $cdf(x^2, 1)$. Remember to use 1 degree of freedom if comparing to the normal distribution.
- This can be checked for any number, for example, if $x = 1.5$, $normcdf(1.5) = 0.933$.
  - This is the left-hand area corresponding to all values less than 1.5 standard deviations from the mean.

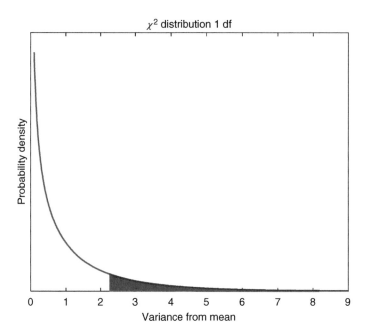

FIGURE 5.12 χ^2-distribution for 1 degree of freedom with the right-hand area at a variance of 2.25 (or squared standard deviation) from the mean illustrated.

- This means that $1 - 0.933 = 0.067$ of the data lies to the right of 1.5 (>1.5 standard deviations) or $2 \times 0.067 = 0.134$ on either side (two-tailed).
- Hence, the probability (or p value) of being more than 1.5 standard deviations away from the mean on either side (two-tailed) is 0.134 based on an underlying normal distribution.
- To calculate χ^2, square 1.5 to give 2.25.
- We now find $\chi^2 cdf(2.25,1) = 0.866$ and $1 - 0.866 = 0.134$ or the same.
- Hence, when there is 1 variable both the z- (normal) and χ^2-distribution provide the same answer.

Where χ^2 is useful is when there is more than one variable or dimension. In such a case there is no positive or negative direction from the centre of a series of measurements, and it is necessary to calculate the square of the distance from the centre. χ^2 with more than 1 degree of freedom can be used to model or estimate p values for multidimensional datasets, whereas the z-distribution is only applicable in 1 dimension.

The shape of the χ^2-distribution is dependent on the number of degrees of freedom and the pdf is illustrated in Figure 5.13 for different values of ν. The mode of $\chi^2(\nu)$ is at $\nu - 2$ if $\nu > 2$, otherwise 0. So, a dataset characterised by 4 variables would be approximated by $\chi^2(4)$ with a mode at 2. Of course, this depends on the underlying population being normally (or more accurately multinormally) distributed.

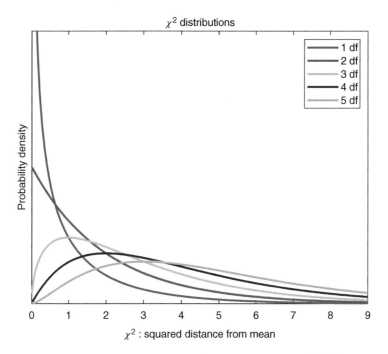

FIGURE 5.13 χ^2 probability density distributions for different degrees of freedom.

The distance in 1 dimension is in units of standard deviations, so the squared distance (which is conventionally used) is in units of variance. Hence, for sample A and variable 1 of Table 5.1, the distance from the centre of class A is -0.486 standard deviations or a squared value of 0.236 units.

Note that once we exceed 1 degree of freedom (or 1 dimension) the units of distance are not usually measured in standard deviations or variances, but a new measure called the Mahalanobis distance which we will discuss in Section 5.3. For this section, we simply denote this as a squared distance, although it can be denoted just as χ^2.

In chemometrics, we usually use multivariate datasets so if using χ^2 often need to use a high number of degrees of freedom.

Note interestingly that the probability of finding samples near the centre of a distribution is low if there are more than 2 degrees of freedom, which may at first be counterintuitive. It can be explained, in three dimensions, by considering the surface area of a sphere. As the distance from the centre increases, the surface area increases, and although the density of points may decrease, there is a greater chance of finding points due to a larger surface area. As the distance increases even more, the density reduces more rapidly than the surface area increases. So, there is an optimum distance from the centre where it is most likely to find a sample.

- In Figure 5.14 we convert the pdf of Figure 5.13 into a cdf of χ^2.
- We can define $p = 1 - cdf$, as on the right-hand axis. It can be defined as the probability of finding a sample at a given distance from the centre if the underlying distribution is multinormal.

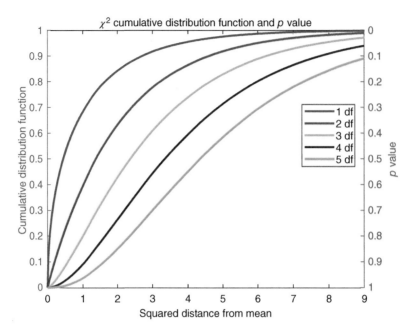

FIGURE 5.14 χ^2 cumulative distribution function and p values for varying degrees of freedom, plotted against squared distance from mean.

- As an illustration, we can plot $\chi^2 cdf(2)$ against distance from the centre of distribution in two dimensions, representing the concentrations of 2 variables or metabolites as in Figure 5.15(a), the shape is analogous to a Gaussian distribution in three dimensions. A corresponding cloud of 100 points representing simulated experimental values is also indicated, obtained using a multinormal distribution. Note that we are representing the distance rather than the squared distance on the axes in this case.
- The cdf can be converted to a p value as $p = 1 - cdf$. For clarity, we take the logarithm of p and visualise this in Figure 5.15(b).
- A plane through $\log10(p) = -1$ would correspond to a critical value of $p = 0.1$ and any point below the area the plane and surface intersected would be likely to be obtained one time in 10 or less if a member of the parent population. This is illustrated in Figure 5.15(c).
- Finally, we could project the p values onto the original two dimensions as contours as in Figure 5.15(d). Being 100 data points, normally distributed, we expect 1 outside $p = 0.01$ limit (which there is), 5 outside $p = 0.05$ (there are 4), and 10 outside $p = 0.1$ (there are 8, and this corresponds to the cut-off in Figure 5.15(c)). Obviously, there will be some small deviations from a perfect fit to a χ^2 distribution, but the approximation is quite a reasonable one.

We will discuss in more detail multivariate use of χ^2 in Section 5.3.
Meanwhile, we return to our univariate examples.

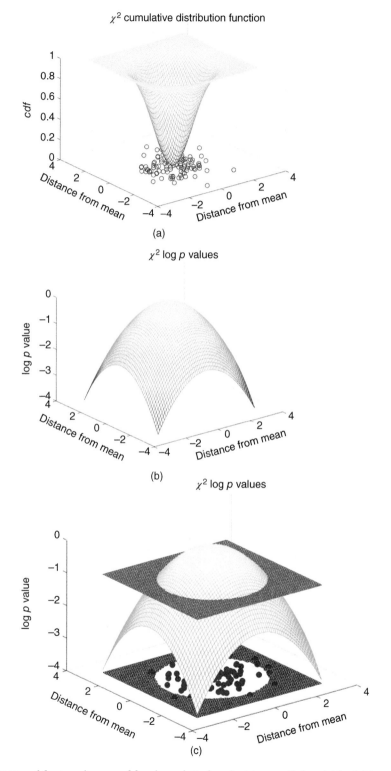

FIGURE 5.15 χ^2 for two degrees of freedom plotted against two spatial variables (a) top cdf together with a cluster of data points, (b) middle top log p (to the base 10) and (c) middle bottom critical value of $p = 0.1$ corresponding to a value of $\log(p) = -1$ and a plane at -1 (d) bottom projection of 5.14(a) with p value contours indicated.

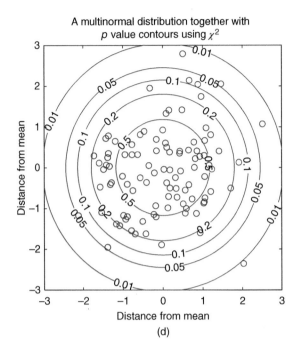

FIGURE 5.15   (Continued)

- p values as calculated using a χ^2-distribution with 1 degree of freedom, are identical to p values calculated using a z-(or normal) distribution.
- As an example, consider the first sample from the simulated case study in Table 5.1.
- This is 0.486 standard deviations below the mean of class A using the sample standard deviation.
- When using the z-distribution we find $p = 2(1 - normcdf(0.486)) = 0.627$ as listed in Table 5.2.
- The same result is obtained by $p = 1 - \chi^2\ cdf(0.486^2,1) = 0.627$ where the '1' denotes the number of degrees of freedom or variables.

As p values using the χ^2-distribution with 1 degree of freedom are identical to those obtained using the z-distribution, we will not discuss this further in this section but will introduce this distribution again in the context of multivariate datasets below.

5.2.5   F-Distribution and Hotelling's T^2

The F-distribution relates to the t-distribution in an analogous way the χ^2-distribution to the z-distribution.

The F-distribution is characterised by two degrees of freedom, $F(\nu_1,\nu_2)$. In our case

- The first degree of freedom ν_1 corresponds to the number of variables, and for a univariate dataset will equal 1,

- The second degree of freedom ν_2 corresponds to the number of samples minus ν_1. In the case of class A from the dataset of Table 5.1 it equals 9.

When there is only 1 variable, $F(1,\nu_2)$ is simply the square of $t(\nu_2)$. Similar to the χ^2-distribution, the p value is the proportion of the area under the pdf to the right of the variance from the mean, analogous to Figure 5.12.

In order to understand better how the F distribution relates to the t- and χ^2-distributions when there is just 1 variable (i.e. $\nu_1 = 1$) we illustrate this graphically in Figure 5.16.

- Figure 5.16(a)–(c) represent the F-, χ^2- and t-distributions as traditionally presented.
- For the F-distributions, $\nu_1 = 1$ and ν_2 varies in our example. An F-distribution of $\nu_1 = 1$ and $\nu_2 = 9$ would correspond to a distribution arising from a series of univariate measurements with 10 samples. For the χ^2 distribution $\nu = 1$. For the t-distribution ν varies in the same way as F.
- Although the curves for the F-distribution appear to differ only slightly when ν_2 varies, in fact, if the figure were expanded (to keep this text brief we leave this exercise to the reader) there is some difference between the curves, which becomes more evident in the cdf below.
- In Figure 5.16(d) the horizontal scale for the t-distribution has been squared so as to be comparable to the other distributions.
- Figure 5.16(e) and (f) represent the cdf and corresponding p values for the F- and χ^2-distributions. As the number of degrees of freedom (ν_2) for F increases, the curve converges onto χ^2. However, although the different cdf curves for F seem quite similar if a given critical value of p were defined, the critical distance from the mean will differ quite considerably when the sample size is relatively small. We leave the reader to do these calculations if interested.
- Figure 5.16(g) and (h) are the corresponding cdfs and p values for the t-distribution. Although the cdf for the squared distance to the mean appears similar to the corresponding values of F, the lowest probability is 0.5, because the t-distribution is symmetrical around the mean.
- Figure 5.16(i) and (j) represent the corresponding two-tailed cdfs, and we can see that the t cdfs and F cdfs are identical for $t(\nu)$ and $F(1,\nu_2)$ where $\nu = \nu_2$, providing the distance from the mean is squared, and a two-tailed distribution is used.

In practice, this means we obtain identical p values if we correctly use F or t.

- Consider the case study of Table 5.1 and just the first 10 samples representing Class A.
- Sample A, variable 1, has a value of 22.84, which if standardised over Class A using the population standard deviation becomes -0.486.
- The one-tailed t cdf with 9 degrees of freedom is 0.319, so the two-tailed p value is 0.638.

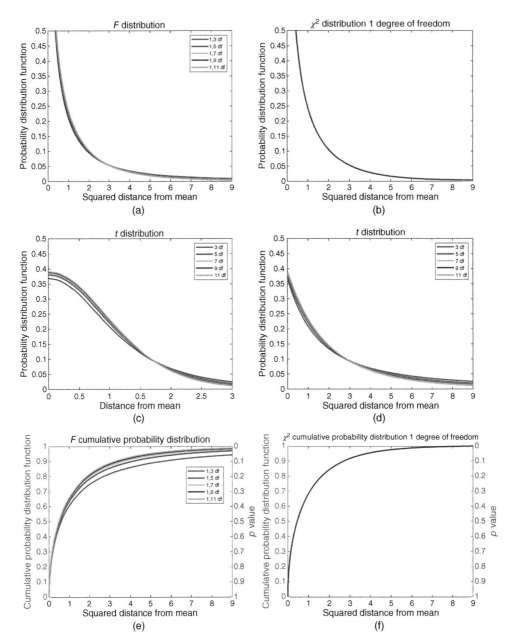

FIGURE 5.16   Illustration of how p values for the F-distribution with $\nu_1 = 1$ relate to the t and χ^2-distributions, (a) to (j) as discussed in the text.

- To calculate the equivalent value of p using F, compute $1 - Fcdf(0.486^2, 1, 9) = 1 - 0.361 = 0.638$, or the same (there is a small rounding approximation).
- The value of p using either the z- or χ^2-distribution is 0.627 as already discussed.

Hence, if correctly implemented we could use either t or F when there is one variable and come to identical conclusions about p. Using z or χ^2 also provides identical p values to each other but differ from those obtained using t and F.

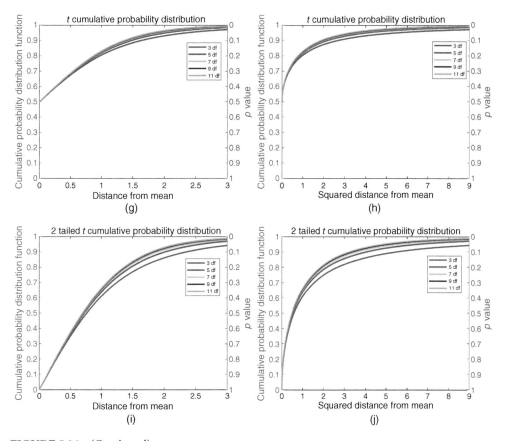

FIGURE 5.16   (Continued)

When there is more than one variable the situation becomes somewhat more complicated.

The F-distributions for $\nu_2 = 20$ and various values of ν_1 are illustrated in Figure 5.17 together with the corresponding cdf.

- Note that the horizontal axis is labelled 'F value' rather than squared distance from mean. The F-distribution is used in many situations such as ANOVA, which we will discuss in Chapter 10, but a scaling factor is necessary if used to calculate p values for a defined distance from the mean when there is more than 1 variable ($\nu_1 > 1$).

- If $\nu_1 > 2$ there is zero intensity when $F = 0$. When extended to multivariate distributions, this implies that we do not expect to find samples in the centre of a distribution if characterised by more than two variables.

- Once ν_1 increases, the mode of the distribution converges on $F = 0.909$ for this case when ν_1 is large, meaning that it is most likely samples are found at around this value rather than (as when there is just one variable) in the centre. The mode of an $F(\nu_1,\nu_2)$ distribution is given by $(\nu_1 - 2)/\nu_1 \times (\nu_2/(\nu_2 + 2))$ or in our case where $\nu_2 = 20$, it equals $(\nu_1 - 2)/\nu_1 \times 0.909$. However, when we are dealing with distances in multivariate space, as is common in chemometrics, we need to scale the F values to give a new statistic, Hotellings T^2 as described below.

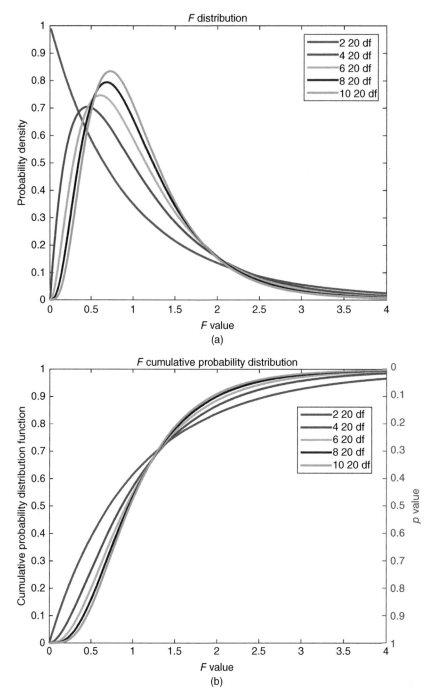

FIGURE 5.17   (a) top F probability density distribution, (b) bottom F cumulative probability distribution and p values, for various degrees of freedom.

- For T^2 if there are J variables and I samples (usually represented by a $I \times J$ matrix) mutlinormally distributed
- then the statistical definition is $T^2 = F \times (I-J)/(J \times (I-1))$.
- This follows the F-distribution with $\nu_1 = J$ and $\nu_2 = I-J$.
- So, if there are 10 samples and 5 variables, as in the data for Class A of Table 5.1 it follows an $F(5,5)$ as $J = 5$ and $I = 10$ providing we assume the variables are multinormally distributed.
- Note that if there is just one variable $T^2 = F \times (I-1)/(1 \times (I-1)) = F$, so the two distributions are identical for univariate datasets.
- As we will see below, this definition only gives a reasonable answer if $I-J$ is large compared to J.

We can now illustrate the T^2 distribution. For brevity, we will only present a few key graphs.

- Figure 5.18 presents the T^2-distribution together with the pdf when $I = 100$, that is, a sample size of 100, and varying values of J.
- So, if $I = 100$ and $J = 5$, there are 5 and 95 ($= I-J$) degrees of freedom, respectively.
- If we compare the cdf and p values with the equivalent for χ^2 in Figure 5.14 we see they are very similar as expected, hence if there are several samples both models will give almost identical p values. Experimentally it should not be possible to differentiate.
- The pdf appears superficially somewhat different to Figure 5.13 but this is only in the vertical scale, for brevity we leave this and let the interested reader work this out, the shape is the most important feature of a pdf.
- In Figure 5.19 we present the corresponding T^2-distributions but with $I = 20$. They should be compared to Figure 5.17, but note the horizontal scale has been stretched. The graphs are not exactly comparable because I is not equal to the number of degrees of freedom, but the adjustment is only very minor. For conciseness, we select only a few figures for illustrative purposes.
- What has happened is that the curves have been stretched compared to the F-distribution curves. So, for example, if there are 10 variables, and 20 samples a squared distance from the mean of 16 gives a p value of 0.6 this implies if multinormally distributed, 60% of samples lie farther out.
- The transformation from F to T^2 is essential to make a meaningful interpretation of the distance of a sample from the centre of a group in multivariate space.

We will illustrate the use of T^2 and F when there is more than one variable, in the next section.

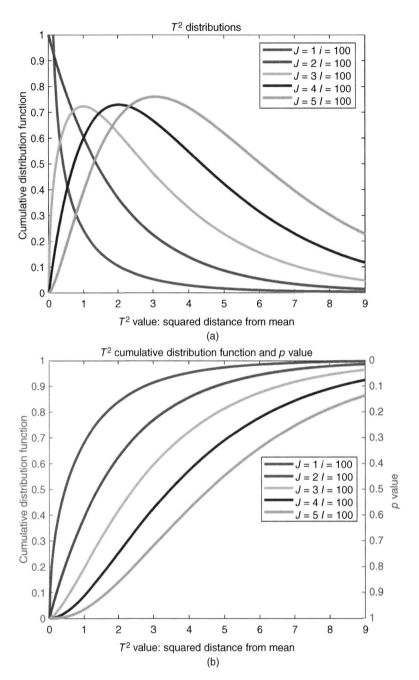

FIGURE 5.18 T^2 (a) top probability distribution, (b) bottom cumulative distribution function, when $I = 100$ equivalent to a 100 sample dataset.

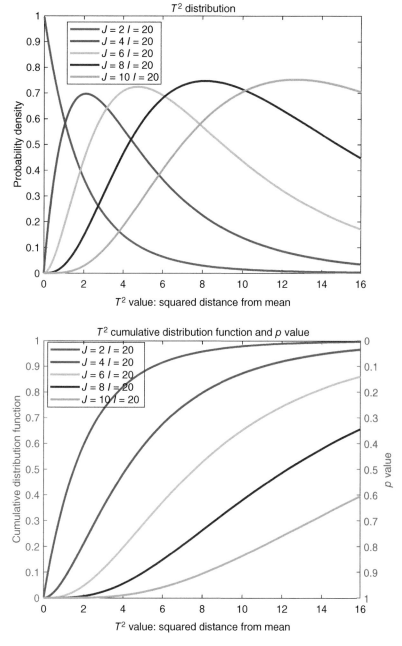

FIGURE 5.19   As the previous figure but with $I = 20$.

5.3   MULTIVARIATE CALCULATION OF *P* VALUES AND THE MAHALANOBIS DISTANCE

Most metabolomics datasets are multivariate, that is several measurements are obtained for each sample, such as chromatographic peak heights, NMR intensities at different frequencies or spectroscopic intensities. Above we focussed primarily on univariate measurements. It is still useful for example to calculate the *t* statistic for the intensities of each peak separately, but more commonly we use multivariate measures.

When there is more than one variable, we can no longer calculate *t* or *z* except when studying one variable at a time, and we have to use χ^2, T^2 or *F* as appropriate.

We will initially illustrate the use of *p* values in two dimensions using just the first two variables of class A for the data in Table 5.1.

- We can calculate the Euclidean distance of each sample from the centre of the data as follows:
    - Calculate the centroid $\bar{x} = \begin{bmatrix} 28.021 & 28.310 \end{bmatrix}$
    - Then for each sample calculate the vector distance $d_i = x_i - \bar{x}$, so for sample A it becomes
      $[-5.183 \ 5.007]$ and the geometric (or Euclidean) distance, $d_i = \sqrt{d_{i1}^2 + d_{i2}^2} = \sqrt{d_i d_i'}$ so for sample A this is 7.206.
- In order to calculate the *p* values using either χ^2 or T^2, we have to standardise this distance. The population standard deviation of the 10 samples from the centre is 5.320 (for statistically trained readers this may irritate and formally one should use the sample standard deviation when calculating T^2 but the population standard deviation for χ^2 however for simplicity, we will just use one definition, as this makes numerical comparisons easier).
- Hence, the distance corrected for the standard deviation is 7.206/5.320 = 1.355 for sample A.
- We can then visualise these distances as in Figure 5.20(a). We can see that the contours representing distance from the centre do not model the distribution of the samples very well. This is because x_2 has a much lower variance to x_1 so the vertical axis is flatter. The distances are listed in Table 5.4.
- We can calculate the *p* values from the squared distances, in this case using Hotelling's T^2 with $\nu_1 = 2$ and $\nu_2 = (10-2) = 8$ as in Table 5.4 and Figure 5.20(b).
- We can see that the *p* values are not very realistic, for example, we would normally expect around 1 out of 10 to have a value less than 0.1, and around half have values more than 0.5.

The main problem is that our distance measure is not very useful. We can of course use any distance measure, the assumption being that samples are multinormally distributed in the direction of the distance contours.

A solution is to change the distance measure. There are two problems with using the Euclidean distance measure above.

TABLE 5.4   Euclidean and Mahalanobis distance of samples from the centre of class A of the data from Table 5.1 together with *p* values calculated using Hotelling T^2 using only the first 2 variables.

	A	B	C	D	E	F	G	H	I	J
Stand euc distance	1.355	2.504	1.731	1.803	0.669	1.306	2.764	1.022	2.325	4.308
p value using T^2 (Euclidean)	0.476	0.121	0.317	0.291	0.824	0.499	0.086	0.645	0.152	0.011
Mahalanobis distance	1.099	2.157	1.169	1.008	0.347	0.731	1.524	1.028	1.647	2.239
p value using T^2 (Mahalanobis)	0.604	0.189	0.568	0.652	0.948	0.794	0.399	0.641	0.349	0.170

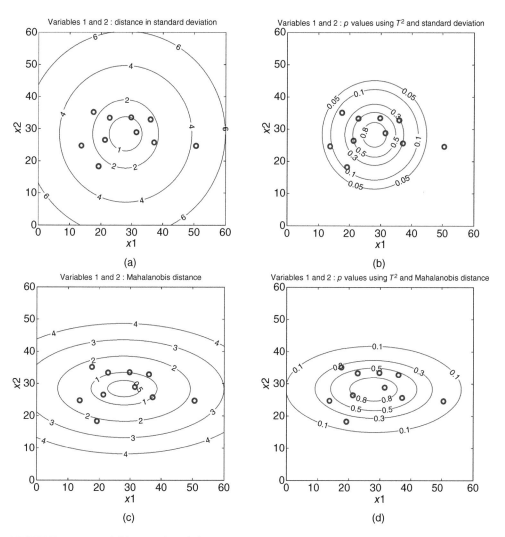

FIGURE 5.20   Variables 1 and 2 of class A in Table 5.1 as described in the text (a) Euclidean distance (b) *p* values using Euclidean distance (c) Mahalanobis distance (d) *p* values using Mahalanobis distance.

- Variables can vary over very different ranges. For example, we might measure the length of people's noses and of their legs. Variation of just 1 cm in leg length would be of equal significance to variation of 1 cm in nose length unless the two types of measurements are scaled so that their variation (often expressed in the form of variance) is considered equally important.
- Some variables can be correlated. For example, the length of arms and legs might be correlated, but if we are interested in differences in shape, we are often more interested in uncorrelated variables.

The Mahalanobis distance is a common alternative to the Euclidean distance and is defined by $d_i = \sqrt{(x_1 - \bar{x}) S^{-1} (x_1 - \bar{x})'}$ where S is the variance–covariance matrix.

- In our case for the first two variables $S = \begin{bmatrix} 113.48 & -0.45 \\ -0.45 & 25.62 \end{bmatrix}$ (there is very little correlation between the first two variables)

- so $S^{-1} = \begin{bmatrix} 0.0088 & 0.0002 \\ 0.0002 & 0.0390 \end{bmatrix}$

- and for the first sample, $d_1 = \sqrt{[-5.183 \quad 5.007] \begin{bmatrix} 0.0088 & 0.0002 \\ 0.0002 & 0.0390 \end{bmatrix} \begin{bmatrix} -5.183 \\ 5.007 \end{bmatrix}} = 1.099.$

- The p values can be then calculated using T^2 as usual (or of course χ^2 but for brevity, we illustrate just one approach) using the squared distances.
- The Mahalanobis distance and p values of all 10 samples from class A are tabulated in Table 5.4.

There are several things to note.

- First note that the sum of squares of all the Mahalanobis distances $\sum_{i=1}^{i=10} d_{iA}^2 = 20 = 10 \times 2 = I \times J$. This is a universal property so long as the population variance–covariance matrix is used in the calculations (otherwise it will equal $(I-1) \times J = 18$ in our case as the reader can verify).
- Sample J is now, although still far from the centre, not an extreme outlier, with a $p = 0.170$ compared to 0.011 where the Euclidean distance is used.
- It can be seen clearly in the plots of Figure 5.20(d) compared to Figure 5.20(b), that the Mahalanobis distance models the direction of variability much better than the Euclidean distance.
- Note that for a full component model, the Mahalanobis distance using all the original variables will be the same as the Mahalanobis distance using all non-zero PCs.

Where the χ^2- and T^2-distributions are useful is when there are several variables. Most chemometrics datasets are multivariate. The data of class A in Table 5.1 can be organised as a 10×5 matrix.

We can then calculate p values that any individual sample is a member of this class as follows using either χ^2 or T^2-distributions and using the Mahalanobis distance measure.

For brevity, we tabulate the values in Table 5.5 and leave the reader to reproduce the full calculations if desired.

- Note the sum of squares of Mahalanobis distances for the 5 variable model equals 50 ($=10 \times 5$) as expected (using the population variance–covariance matrix).
- We see sample F has the lowest p value and sample H the highest, using both criteria.
- The p values for both criteria differ but show the same trend; however, the p values for T^2 are always larger. This is illustrated in Figure 5.21.
- A problem with T^2 when the ratio of samples to variables is small, it is that it is based on $F(J, I - J)$. If $I - J$ is close in size to I, the estimate of $I - J$ is very critical. In our case when there are five variables $I - J = 5$ which is the same as J. The problem here relates to how accurately the number of degrees of freedom can be estimated. For a theoretical distribution, it is assumed all variables are multinormal. If there are some correlations, the dimensions (or relative eigenvalues) of later PCs are small. So is $\nu_2 = 5$ or not, if there are 3 large dimensions in space and the final 2 are very small is it appropriate to use a five-dimensional or three-dimensional normal distribution? This can make a considerable difference to the F value, and may result in overestimates of p. This problem is not important if I is much greater than J.
- If we use less variables, by PCA, using centred data we can reduce the matrix to a 10×2 scores matrix. The p values are also tabulated in Table 5.5. Note that now the two p values are much closer, although T^2 will always result in a higher p value. The relationship is presented in Figure 5.22.

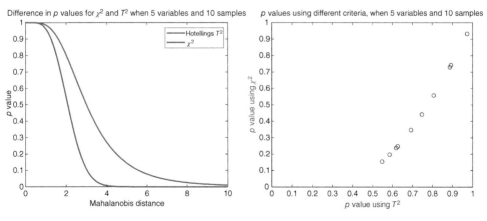

FIGURE 5.21   Variation of p value against Mahalanobis distance for class A of Table 5.1 and all 5 variables or PCs using different criteria.

TABLE 5.5 Mahalanobis distance and p values using both χ^2 and Hotelling's T^2 for the 10 samples of class A, using both 5 and 2 variables (or PCs), for the case study of Table 5.1.

	A	B	C	D	E	F	G	H	I	J
Mahalanobis distance (5 variables)	1.651	2.366	2.190	2.580	1.675	2.835	1.987	1.152	2.602	2.709
p Value (χ^2) (5 variables/PCs)	0.742	0.347	0.441	0.248	0.730	0.154	0.557	0.932	0.238	0.197
p Value (T^2) (5 variables /PCs)	0.892	0.692	0.747	0.626	0.887	0.548	0.807	0.972	0.619	0.586
Mahalanobis distance (2 variables)	0.862	1.314	2.046	0.907	1.203	0.553	1.720	0.636	1.640	2.171
p Value (χ^2) (2 variables/PCs)	0.690	0.422	0.123	0.663	0.485	0.858	0.228	0.817	0.260	0.095
p Value (T^2) (2 variables /PCs)	0.728	0.496	0.217	0.705	0.551	0.875	0.321	0.839	0.351	0.186

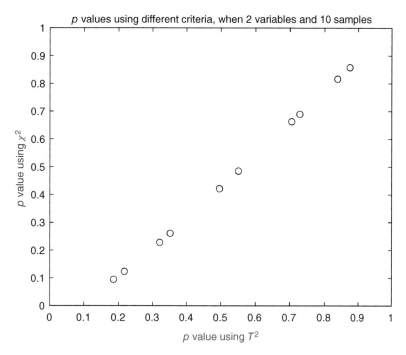

FIGURE 5.22 Variation of p value against Mahalanobis distance for class A of Table 5.1 and a 2 PC model using different criteria.

- However, the order of significance of the samples will always be the same for both criteria, providing the same number of PCs are used in the model. Note how some samples though show different ranks using both a 5 PC and 2 PC model, suggesting that some later PCs are characteristic of specific samples.

Although Hotellings T^2 is implemented in many software packages, it can provide misleading results if the number of variables is relatively high compared to the number of samples. It is safe, for example, when using a 2 PC model for a sample size of 100, but if the number of variables increases substantially, $I-J$ becomes close in size to J and can result in considerable distortions in the p values, so in general χ^2 is recommended. Both criteria of course can be used for ranking samples or variables in order of significance however but so could the Mahalanobis distance without calculating p values.

We will discuss the use of p values for classification in more detail in Chapter 7, but we can calculate the p values for the 10 samples of class B using the Mahalanobis distance to the centre of class A so the Mahalanobis distance and T^2 or χ^2 measures can be used to assess how well samples that were not part of the original training set fit the model. This is discussed extensively in Chapter 7.

We now move onto a real case study, case study 5 (NMR of maize) harvested at 8.5 °C. Refresh by examining Figure 5.10. We will initially calculate the p values using all 34 original variables.

- First calculate the 34×1 centroid of the data $\bar{x}$.
- and the 34×34 variance–covariance matrix S using the population standard deviation/variance.
- The squared Mahalanobis distance of sample i to the centroid is given by

$$d_i^2 = \left(x_i - \ddot{x} \right) S^{-1} \left(x_i - \ddot{x} \right)'$$

- The squared Mahalanobis distance should follow approximately χ^2 with 34 degrees of freedom.
- The p values together with the Mahalanobis distance are presented in Table 5.6.
- These can also be visualised graphically as in Figure 5.23. Samples 25 and 24 have low p values. A threshold of $p = 0.1$ is illustrated. However, we would expect some samples to have p values below 0.1 (in fact 5 if the measurements were normally distributed so this number is lower than expected) and by this criterion, although both samples might be suspicious they are nothing out of the ordinary. It however might in practice be worth checking them if other analytical data were available. If samples have low p values using several independent analytical techniques, this is good evidence they are outliers.
- Because all 34 variables (or non-zero PCs which will result in the same p values) were used in the model, the Hotellings T^2 criterion will not be successful, for reasons discussed above, resulting in very high p values all apparently over 0.9. This problem has been discussed above and illustrated on the small 10-variable simulated dataset. We recommend using χ^2 unless the number of variables is very much smaller than the number of samples.

It is of course not always necessary to include a model with all 34 variables or non-zero PCs, and a common alternative is to reduce the model by PCA. There are endless combinations and so to keep this chapter brief, we illustrate just one possibility.

TABLE 5.6   Mahalanobis distance using all 34 variables and corresponding p values using χ^2 for the 54 samples of NMR extracts of maize harvested at 8.5 °C.

Sample	d	p	Sample	d	p	Sample	d	p
1	0.959	0.285	11	0.962	0.837	21	3.212	0.567
2	1.178	0.867	12	0.617	0.257	22	0.668	0.332
3	1.903	0.171	13	2.589	0.538	23	0.504	0.097
4	0.773	0.342	14	3.268	0.221	24	2.016	0.510
5	0.353	0.810	15	3.639	0.493	25	4.002	0.073
6	0.895	0.634	16	1.064	0.623	26	0.332	0.711
7	4.513	0.167	17	0.975	0.876	27	1.345	0.753
8	0.882	0.931	18	1.758	0.571	28	0.699	0.502
9	3.776	0.404	19	0.701	0.982	29	0.393	0.726
10	0.792	0.662	20	1.401	0.301	30	1.627	0.264

Sample	d	p	Sample	d	p	Sample	d	p
31	1.455	0.396	41	0.461	0.653	51	2.287	0.493
32	2.278	0.430	42	1.503	0.476	52	0.803	0.476
33	0.777	0.306	43	0.710	0.296	53	0.412	0.298
34	0.735	0.242	44	0.735	0.627	54	1.644	0.516
35	1.150	0.819	45	0.697	0.192			
36	1.513	0.658	46	0.821	0.236			
37	0.471	0.660	47	1.541	0.121			
38	1.451	0.277	48	1.753	0.324			
39	2.393	0.466	49	1.809	0.280			
40	0.376	0.747	50	1.505	0.426			

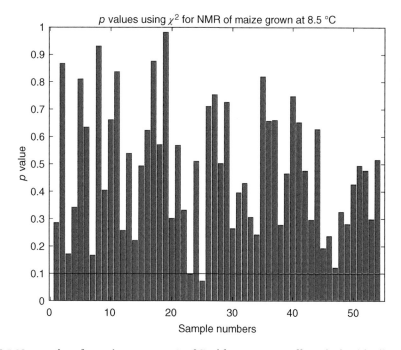

FIGURE 5.23 p values for maize grown at 8.5 °C with $p = 0.1$ cut-off marked, with all variables in the model and no standardisation.

- The maize NMR data are standardised over the 54 samples harvested at 8.5 °C.
- PCA is performed and the first 5 PCs are retained.
- p values are calculated using the χ^2 criterion and illustrated in Figure 5.24. The appearance is a little different from Figure 5.23; however, sample 25 is even more clearly an outlier.

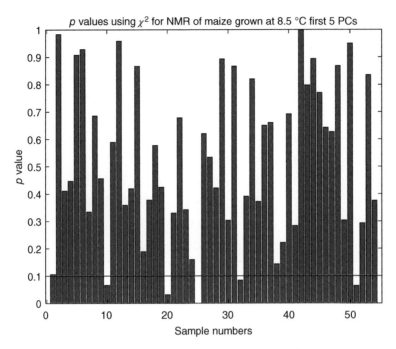

FIGURE 5.24 p values for maize grown at 8.5 °C with $p = 0.1$ cut-off marked, with first 5 PCs in model, data standardised prior to PCA.

- Of course, interpreting the differences between a 5 PC model and a full 34 PC model could fill an entire chapter, or more, but we leave to the reader, the main message is to try different approaches to get an overall consensus view, and that each case study will be different.

In this section, we have extended the discussion of statistical distributions and criteria to multivariate datasets, as is common in chemometrics. We will build on these approaches in later chapters. Univariate measures are still regularly employed as complementary to the multivariate approaches.

5.4   DISCRIMINATORY VARIABLES

In metabolomics, one of the most important jobs is to determine which variables are significant discriminators between two or more groups of samples. These are often called markers. One group, for example, may consist of diseased donors, and the other healthy donors. Chapters 9 and 10 will extend these concepts, but in this section, we illustrate the use of simple univariate statistics based on the distributions introduced in Section 5.2.

Many common approaches look at univariate differences between the mean value of each variable in different classes.

- In Table 5.1, we have two classes, Class A (blue group) consisting of 10 samples and Class B (red group) likewise.
- Of the 5 variables, which are the most discriminating or in statistical terms, most significant, variables?
- For each variable, we can calculate how different the means are relative to the standard deviation, and often call this a t statistic (although it can be used in other tests).
- There are several different definitions of t, but a common definition is $t = \left(\bar{x}_A - \bar{x}_B\right) / \sqrt{\left(s_{pool}\left(1/I_A + 1/I_B\right)\right)}$ where
  - $\bar{x}_A$ is the mean value of the relevant variable over class A
  - I_A the sample size of class A
  - and $s_{pool} = \sqrt{\left(\left(I_A - 1\right)s_A^2 + \left(I_B - 1\right)s_B^2\right) / \left(I_A + I_B - 2\right)}$ is often called the pooled standard deviation where s_A is the sample standard deviation of class A.

This definition models the standard deviation as being approximately the same for each class, which is not always the case, and there are more complicated equations if the difference in standard deviation between two groups is taken into account. However, for simplicity, we only use one (the most widely employed) definition in this chapter. Note also we will use the sample standard deviation rather than population standard deviation in this section. This fine distinction may irritate the more statistically trained readers, but in practice makes very little difference. In the previous section, we were more concerned with multidimensional scaling for which the population definition is more suitable, whereas this section is primarily about estimation.

- To calculate t for variable 1, we have
  - $\bar{x}_{1A} = 28.02$ and $\bar{x}_{1B} = 14.17$
  - $I_A = 10$ and $I_B = 10$
  - $s_A = 11.23$ and $s_B = 4.98$, so $s_{pool} = \sqrt{\left(9 \times 11.23^2 + 9 \times 4.98^2\right)/18} = 8.69$
- Hence, $t_1 = \left(28.02 - 14.17\right) / \left(8.69 \times \sqrt{2/10}\right) = 3.567$
- We can now calculate p values using either the two-tailed normal (z) distribution ($p = 0.00036$) or two-tailed t-distribution with 18 degrees of freedom ($p = 0.00221$). Although called a t value, it can equally be used to calculate a p value using the z-distribution.
- Squaring $F_1 = t_1^2 = 12.72$ and we find the same p values via χ^2 with 1 degree of freedom ($p = 0.00036$) or F with $\nu_1 = 1$ and $\nu_2 = 18$ ($p = 0.00221$).
- Most people use the t-distribution, but all four distributions (as above) can be used to calculate a p value, if desired.

We calculate the values of t and F together with the corresponding p values in Table 5.7 for the 5 variables and two classes of the simulated dataset of Table 5.1.

TABLE 5.7   Calculation of the t and F statistic for the 5 variables and two classes in Table 5.1, together with the relevant p values for the significance they are different between classes A and B.

	1	2	3	4	5
t	3.57	1.03	2.47	−9.46	−0.10
F	12.72	1.06	6.10	89.58	0.01
p (using t or F)	0.00221	0.31584	0.02372	2.07E-08	0.92432
p (using z or χ^2)	0.00036	0.30217	0.01350	2.94E-21	0.92326

We can see that variables 2 and 5 are not good discriminators (or markers), whereas the other three variables are likely to be good markers or discriminatory variables.

As an example, consider case study 1, of the control and pre-arthritic donors, involving a 49×81 matrix, of which the first 19 samples are from the control class and the next 30 from pre-arthritic donors.

- Initially standardise the data.
- Perform PCA. The scores plot of PC2 versus PC1 was presented in Figure 4.14(a), which does not look very encouraging, similarly discouraging plots can be obtained for all the earlier PCs. As we will see in Chapter 7 this could be improved using PLS though.
- We can alternatively calculate the t statistic for each variable, the first 19 samples belonging to class A (controls) and the remaining 30 samples to class B (pre-arthritic). Note that there is no need to standardise the variables for this purpose although the same result would be obtained if we did. Class A is denoted as positive and class B as negative.
- For readers that like to perform these calculations, variable 1 is Methionine, the t value is 0.065 and its corresponding p value is 0.948, so not likely to be a discriminatory variable.
- The p values for all 81 variables using the t-distribution are presented in Figure 5.25, with $p = 0.05$ criterion indicated. The t value is determined using class A as the positive class. This makes no difference to the p value, but does affect the sign of the t statistic.
- Five variables have a p value of less than 0.05, according to the t-test and are tabulated in Table 5.8.
- Note that four have a negative t value, these will have a lower level in class A (controls) than class B (pre-arthritic) and so are markers elevated for donors that are likely to develop arthritis.
- Of course, it is not necessary to use $p = 0.05$ as the cut-off criterion and all markers with low p values are potentially interesting. We will see that other indicators such as PLS loadings and VIP values (Chapter 9) provide alternate views. Also, the calculations above assume that the distribution of the metabolites is normal: in practice this is unlikely; however, the shape of the distribution

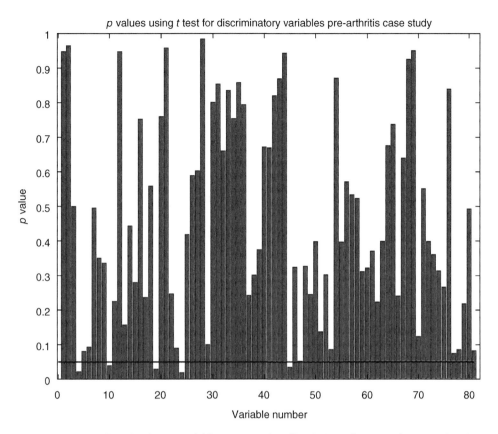

FIGURE 5.25 p values for the 81 variables measured to discriminate between the control and pre-arthritis groups, using the t statistic with $p = 0.05$ cut-off indicated.

TABLE 5.8 5 variables from the pre-arthritis case study 1 with p values < 0.05 using the t statistic.

Variable	4	10	19	24	45
Metabolite	Kynurenine	Hypoxanthine	Lyso-PC(16:0)	Lyso-PC (14:0)	Oleic acid
t	−2.382	−2.115	−2.256	−2.428	2.177
p	0.021	0.040	0.029	0.019	0.035

can sometimes be improved via scaling such as log scaling or using alternate tests. However, p values using the t statistic are good for exploratory analysis for which compounds are likely to be markers and variables with extreme p values are definitely of interest and suitable for ranking of metabolites.

The p values calculated using methods in this section should not be interpreted literally, the numerical values should be taken with a pinch of salt, but they can be employed to sort which metabolites are potential markers, as a first step for further investigation. Most metabolomic studies should be viewed as exploratory, to narrow down a large number of potential markers to those that deserve further study, often using independent methods.

5.5 CONCLUSIONS

In this chapter, we have introduced the idea of statistical distributions, p values, and their application both to identify if samples belong to a given group and whether variables are likely to be markers. As metabolomics has its application in biomedical science, many papers and organisations and funding bodies and degree conferring bodies demand the use of p values, and as such the chemometrician needs to talk the language of their collaborators.

We will introduce p values in several different contexts later in this text so this chapter is primarily an introduction to statistical concepts and computations. However, as p values are increasingly used in metabolomics, it is important that all workers in the field have a basic understanding of the background to this measure.

CHAPTER 6

Choosing Samples

6.1 MOTIVATION

In metabolomics, it is often necessary to plan how many and what types of samples are appropriate for subsequent analysis.

There are two main needs:

The first involves studying the effects on metabolism in a controlled (usually a laboratory) environment. An example is environmental toxicology, where the effects of various conditions or potential toxins on metabolites of interest as in case study 8 (effect of environmental conditions on Daphnia). In these situations statistical Design of Experiments (DoE) is usually employed, resulting in a list of conditions that the experimenter then performs.

The second involves selecting samples from a large population or database. It is important to avoid bias, for example, if one is studying the possible effects of a disease it is important to balance other factors such as age or gender to ensure that the effects studied are genuinely due to the factor of direct interest and are not confounded with uninteresting factors; also sometimes a decision may be to restrict the target population, for example, just to males within a defined age group and ethnicity. Often there are many samples available, for example, from a hospital, and the investigator picks a representative set in a systematic way. Sometimes due to analytical restrictions (e.g. the stability of LCMS or GCMS instruments and columns) only a limited sample size of 50–100 samples can reasonably be analysed, so the job is to pick a subset of often hundreds of samples for further analysis.

We will look at both approaches below.

Data Analysis and Chemometrics for Metabolomics, First Edition. Richard G. Brereton.
© 2024 John Wiley & Sons Ltd. Published 2024 by John Wiley & Sons Ltd.
Companion website: www.wiley.com/go/Brereton/ChemometricsforMetabolomics

TABLE 6.1   Simulated case study – concentration of a metabolite when a plant is grown under different light conditions and temperature (arbitrary units).

		Daylight hours				
		8	10	12	14	16
Temperature	14	18.2	18.6	19	19.4	19.8
	16	18	19.6	21.2	22.8	24.4
	18	17.8	20.6	23.4	26.2	29
	20	17.6	21.6	25.6	29.6	33.6
	22	17.4	22.6	27.8	33	38.2

6.2   DESIGN OF EXPERIMENTS

When experiments can be performed under controlled conditions usually in a laboratory environment, classical DoE (Design of Experiments) is often employed. The prime aim is to study the effect of several factors on the change in concentrations of metabolites.

There are several motivations for using DoE, but a major one is because the different factors often interact, that is, they are not independent. Table 6.1 represents a small simulated case study, representing the concentration of a compound in plants grown at different temperatures and with different daylight hours. Can we relate daylight hours and temperature to the concentration of the metabolite?

A traditional experimental approach is called one factor at a time.

- Keep one factor constant, for example, 12 daylight hours.
- Vary the other factor, temperature, and we see that concentration increases from 19 to 27.8, and will conclude that the concentration of the metabolite increases gradually as temperature increases.
- Then keep the temperature constant, for example, 18 °C, and we see a similar gradual increase in the concentration of the metabolite.

This apparently simple answer is flawed.

- As an example, if we grow the plants at 8 daylight hours, we see that the concentration of the metabolite actually decreases as temperature increases.
- At 16 daylight hours, the increase in concentration with temperature is much more marked than at 12 hours.

Hence, we could come to different conclusions according to how we perform the experiment, and how we interpret the results, if we are not careful. The relationship between concentration, temperature and daylight hours is illustrated in Figure 6.1 together with the 25 experimental points. It can be seen that the relationship is curved, so that apparently different trends will be obtained according to which cross sections are sampled.

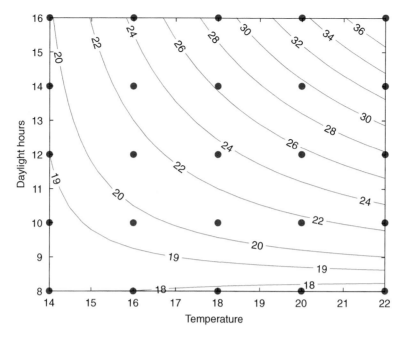

FIGURE 6.1   Experimental points and relationship of concentration as a function of temperature and daylight hours.

This is where systematic experimental design comes into play.

- The reason for this strange behaviour is that there are interactions between the variables.
- We can form a model between x (concentration), c_1 (temperature) and c_2 (daylight hours) as follows:

  In this case, $x = 51.6 - 2.5c_1 - 4.0c_2 + 0.3c_1c_2$, Note: in most statistics texts y is used for the response and x the factor, but as explained in Chapter 10 we use different notation here to have compatibility with multivariate chemometrics terminology.
- In real cases the model is usually an approximation, so there is an additional error or residual term and may contain other terms, such as squared terms. It is rare to model squared terms in the analysis of metabolomic data (although possible) but interaction terms are often of interest.
- However, the coefficients provide a numerical indication of the effects of concentration and temperature, which as described in Chapter 10, and can be interpreted as a p value to provide an estimate of the significance of each term.

We will see that if we aim to form a numerical model such as the relationship above, we can cut down the number of experiments also, and do not need 25 measurements saving time and resources.

Statistical DoE consists of a number of steps. In Chapter 10, we will discuss how to analyse this type of data. The main aim of this chapter is to describe the main steps involved in designing experiments.

6.2.1   Factors, Response and Coding

The first step is to define the factors we want to study. In the simulated case study, these were temperature and daylight hours. The usual aim of experimental designs is to see how the changes in these factors influence a response, often in the form of a chromatogram or spectrum which often then is interpreted in terms of changes in concentrations of metabolites as one or more external factors are varied. Chemometrics cannot provide information about causality. For example, we cannot tell whether the increase in the concentration of a metabolite is caused by a disease or whether the disease is caused by build-up of the metabolite. We can only tell whether the factors and response are related or correlated. In some cases, such as temperature, it will be obvious that it is the temperature (factor) that causes metabolic changes (response) but this is not universally self-evident.

In case study 8 (environmental toxicology of *Daphnia*), the factors are salinity, temperature and hypoxia, and the response is a set of GCMS chromatograms.

In some literature the factors are also called variables, but in chemometrics, we often use the term variables to refer to the information obtained from analytical measurements such as chromatograms. The relationship between statistical and chemometric terminology is further discussed in Chapter 10. Figure 6.2 illustrates the distinction between factors and variables.

Rather than using original data, the values of the factors are usually transformed or coded, for two principal reasons:

- Statistical designs are common to many different types of experiments, using coded variables, so it is only necessary to use standard designs.
- The mathematical models using coded data can be interpreted better, and each factor is equally important in the model.

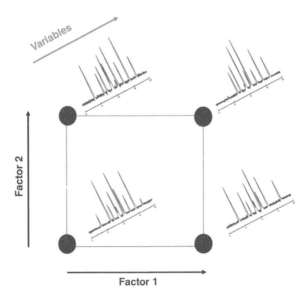

FIGURE 6.2   Factors and variables for a two-level, two-factor design.

A common approach is to study factors at two levels.

- If the factors are quantitative, these correspond to high (coded +1 or +) and low (coded −1 or −) values, for example, a high and low temperature or high and low pH.
- In some cases the factors are categorical, for example, different genotypes, and have no quantitative relationship. It is still possible to code these, for example, +1 corresponds to genotype A and −1 to genotype B, so the interest may be whether the genotype has an effect on the response or whether there is an interaction between genotype and another factor, for example, treatment (is a treatment more effective on a specific genetic group?).
- It is also possible to mix quantitative and categorical factors.

In the case of the *Daphnia* (case study 8) the factors are quantitative.

- Salinity at a low level (−1) was $0\,g\,L^{-1}$ and high level (+1) $5\,g\,L^{-1}$.
- Temperature at a low level (−1) was 20 °C and at a high level (+1) was 25 °C.
- Hypoxia at a low level (−1) was $9\,mg\,L^{-1}$ and high level (+1) was $2\,mg\,L^{-1}$ (note +1 = hypoxic and involves a low oxygen concentration in this coding).

There is of course no restriction to just two levels for each factor, for example, one factor may relate to a time course, and it could be desirable to study the process at five points in time rather than only two, for example, a treatment or progress of disease. However, it is important to realise that this involves a considerable increase in experimentation compared to a two-level design and most preliminary (sometimes called scoping or screening) designs are at two levels. Sometimes a second more detailed stage at several levels can be performed only on those factors that are considered to be important.

Once factors are coded we can usually use standard designs. In chemometrics, rather than having a single response (often denoted y in classical statistical texts but x in most chemometrics texts) we often have many responses or variables, such as GCMS peak heights or spectral intensities.

6.2.2 Replicates

Replication can be an important feature of many designs. Usually, there is uncertainty (or errors) in any measurement, that is when repeating an experiment, there will be slightly different results each time. This may be due to sampling errors or due to instrumental irreproducibility.

There are several reasons for replication. The classical statistical motivation is to compare the predicted size of an effect to the experimental error. Often these are calculated as a ratio, and the magnitude of this ratio can be assessed using statistical tests, often the F test, by methods such as ANOVA as discussed in Chapter 10. This results in a p value and statisticians then often talk about the significance of a particular factor or term in an equation. We will illustrate the calculations in Chapter 10. The primary aim of this section is to guide the investigator as to how to decide which experiments to perform.

A second and practical reason is that many analytical measurements can often be subject to quite unpredictable problems with reproducibility. Occasionally, a chromatogram is of unexpectedly low quality. So, by taking several replicates, we can be confident (unless the experiments were really difficult) that there will be some useful data at each set of experimental conditions, even if one analysis or sample needs to be rejected for analytical reasons. For case study 8, experimentally, 5 replicates were recorded at each of 8 unique combinations of conditions, resulting in 40 samples in total, and then at each of the 8 experimental conditions, 3 representative chromatograms were selected, resulting in 24 samples for further analysis. This allowed rejection of any unusual chromatograms, and means that all experimental conditions will correspond to 3 high-quality chromatograms.

Case studies 5 (NMR of maize) and 6 (effect of nitrate on wheat) involve performing 3 replicates for each set of experimental conditions, in addition for case study 6 for each of 2 conditions and 3 parts of the plant, there were six replicate plants and further each plant was analysed 3 times. Due to greater reproducibility of spectroscopy compared to chromatography and very controlled growth conditions, additional measurements were not obtained beyond the usual 3 replicates for case study 5.

When performing experiments under controlled laboratory conditions, it is often useful to build in replication. In some traditional experimental designs, sometimes only one or two sets of conditions are replicated, often in the centre of the design, and it is assumed errors are uniform throughout different experimental conditions. However, in many situations in metabolomics, where it is possible, and resources are available, it is more usual to perform replicates at every unique set of experimental conditions. Much of traditional statistics was developed when experimentation was expensive, and so replicating every condition would have involved extensive resources, whereas using modern instrumentation this can be done faster, especially if the replication is analytical rather than involving resampling. Hence, modern experimentation has more flexibility in the number and nature of replication compared to experiments reported in the classical literature.

Note that replicates can sometimes be of different types, according to what sources of error are expected, for example, sampling and analytical replicates. In very controlled experiments such as plant growth experiments, where the conditions and genetics can be carefully controlled, and just one or a small number of factors vary, we might consider different plants as sampling replicates. In areas such as human or animal metabolomics where such control is not often possible, different donors would not usually be viewed as replicates, and replication is primarily in the analytical procedure, such as extraction or instrumental analysis although if there are different stages in the analysis each could be replicated. Any design can include replicates.

6.2.3   Statistical Designs

There are a very large number of formal experimental designs. However, many are not common in metabolomics, so we restrict this chapter to describing just a few designs that are commonly reported. Classical statistics was primarily concerned with a univariate response and relationships between the response and value of each factor, forming mathematical models often with squared and interaction terms such as $x = b_0 + b_1 c_1 + b_2 c_2 + b_{11} c_1^2 + b_2 c_2^2 + b_{12} c_1 c_2 + e$ where x is a univariate response, and the c's are the values of the factors, often as coded levels (see Section 6.2.1 for a discussion about coding).

- The term b_0 is an intercept term.
- Terms such as b_1 are linear terms.
- Terms such as b_{12} are two-factor interaction terms; when there are more than two factors, of course, multifactor interactions can be envisaged such as three-factor interactions; in practice, it is hard to estimate or attach much meaning to very high factor interactions and interactions above two are unlikely to be reported in metabolomic studies.
- Terms such as b_{11} are quadratic terms.
- The term e represents a residual or error.

In metabolomics, some of this detail is often unachievable, especially as the response is normally multivariate. Usually, non-linear terms such as quadratic terms are neglected in most chemometric models, so simpler designs can be employed. To study quadratic interactions, there must be at least three levels for each factor, so two-level designs cannot be used for quadratic models. The aim is not always to produce a quantitative physical estimation, but to see which metabolites are markers, even if modelling their functional relationship to the factors being studied is not of direct interest. Interactions though may provide valuable information in certain situations, although it is rarely necessary to study more than two-factor interactions. However, biological processes do interact and the effect of a factor may differ according to the level of other factors, for example, the effect of a treatment might be different on children as compared to adults, or on people of different ethnicities.

Approaches such as ANOVA and ASCA (Chapter 10) can be used quantitatively to analyse the resultant data in some cases and can provide p values that the concentration of a given metabolite is affected by a specific factor, or, graphical methods such as PCA can be used to illustrate the main trends. Replication is usually included on top of the basic design. This chapter will be exclusively involved in describing the designs whereas analysis of the resultant data will be described in later chapters.

6.2.3.1 Fully Crossed Designs

The most basic involves completing a grid representing all factors at all levels. In areas such as plant metabolomics, this is often possible, as experiments are relatively cheap and easy to perform, so a comprehensive grid of conditions can be studied. These are called either fully crossed designs or full factorial designs.

Under such circumstances, after identifying each factor, we can study each at several levels. The number of levels can differ for each of the factors.

If there are F factors and each are studied at K_f levels, the experiment consists of $M = \prod_{f=1}^{F} K_f$ unique experimental conditions.

For case 6 (effect of nitrate on wheat):

- There are two levels of nitrate,
- and three parts of the plant
- making $F = 2$, $K_1 = 2$ and $K_2 = 3$, so $M = 2 \times 3 = 6$.

If we code the levels for the nitrate -1 (without) and $+1$ (with), and for the leaf -1 (base), 0 (mid) and $+1$ (tip) then the design is presented in Table 6.2, in the form of a list of unique experimental conditions. In addition, each unique set of conditions is further replicated in the case reported involving 6 (plants) $\times$ 3 (extracts) or 18 replicates, with $18 \times 6 = 108$ analyses overall. Note that the three levels for the leaf could be considered categorical, that is, they do not form a quantitative sequence, rather like three unrelated genotypes or different properties such as type of disease. In Chapter 10, we will introduce the ANOVA test, which does not require the levels to have a quantitative relationship, unlike regression modelling. Multilevel designs can be analysed either by the ANOVA test, assuming they are unrelated or by regression, assuming the levels are related. When there are two levels, these methods give the same answers, but for more than two levels, they differ.

6.2.3.2   Two-level Full Factorial Designs

A very common type of design is a two-level factorial design, which is the simplest type of fully crossed design.

In order to determine the effects of different factors, the factors are studied just at two levels. If there are F factors and two levels, this requires $\prod_{f=1}^{F} 2 = 2^F$ experiments.

TABLE 6.2   Experimental design for case study 6.

Experiment number	Coded value: nitrate	Coded value: leaf
1	+1	+1
2	−1	0
3	+1	−1
4	−1	+1
5	+1	0
6	−1	−1

Coded values: nitrate: +1 = with, −1 = without, Leaf: −1 = tip, 0 = mid, +1 = base.

- For case study 8, there are three factors, and so $2^3 = 8$ different experimental conditions.
- The design is given in Table 6.3. Note that as the design is in coded values, the same design could be used for any two level three factor experiment, for example, the factors could be two genotypes, two genders and control/treated groups.
- Samples (*Daphnia*) at each unique set of experimental conditions are replicated 3 times also, making a total of $2 \times 8 = 24$ factors.

Designs of this type can be set up for any number of factors, although after four factors (16 unique conditions), the number of experimental conditions may become quite large. It is often important to build in replicates, and as such, having too many unique sets of conditions or unique factors can make the experiments cumbersome if these are in addition to replicates. For practical purposes, for example, instrumental stability or resources, very large experimental designs are often not feasible as it is often useful to ensure replication in addition to studying different factors.

A very important feature of factorial designs is that conditions are chosen to avoid confounding. This problem occurs if the effects of two or more factors cannot be distinguished. Table 6.4 illustrates a design where two factors (1 and 3) are confounded.

The correlation between each column of Table 6.3 is 0, whereas the correlation between columns 1 and 3 of Table 6.4 is −0.5. This means that for the design of Table 6.4 it will not be possible to distinguish the effects of these two factors, despite the fact the number of experiments, and levels of each factor, are the same for both designs. Hence, it is important to choose designs for which none of the factors are confounded, ensuring the factors are uncorrelated. These designs are also uncorrelated for all possible interaction terms, such as $x_1 x_2$ with each other and the single factors.

TABLE 6.3   Experimental design for case study 8.

Experiment	Factor 1	Factor 2	Factor 3
1	+1	+1	+1
2	+1	+1	−1
3	+1	−1	+1
4	+1	−1	−1
5	−1	+1	+1
6	−1	+1	−1
7	−1	−1	+1
8	−1	−1	−1

Coded values: Factor 1 – salinity: −1 = 0 g L⁻¹ and +1 = 5 g L⁻¹; Factor 2 – temperature: −1 = 20 °C and +1 = 25 °C; Factor 3 – hypoxia: −1 = 9 mg L⁻¹ and +1 = 2 mg L⁻¹; note that +1 for hypoxia means low oxygen levels.

TABLE 6.4   An inappropriate design where factors 1 and 3 are confounded.

Experiment	Factor 1	Factor 2	Factor 3
1	1	1	1
2	−1	1	1
3	−1	1	−1
4	1	1	−1
5	−1	−1	1
6	1	−1	−1
7	1	−1	−1
8	−1	−1	1

6.2.3.3   Fractional Factorial Designs

Sometimes the number of experiments required to perform full factorial designs is larger than the capacity of a laboratory or a funded study. Whereas usually feasible to perform full factorial designs when performing plant growth experiments, some clinical studies, for example, a for animal models, can be expensive.

- A fractional factorial two-level design involves performing 2^{F-P} experiments.
- So, if $P = 1$ (a half factorial), and $F = 3$, we need to perform $2^{3-1} = 4$ experiments.
- Can we find a subset of the four original experiments be selected that allows us to study all three factors?
- Rules have been developed, to produce these fractional factorial designs obtained by taking the correct subset of the original experiments. Table 6.5 illustrates a possible fractional factorial design that enables all factors to be studied. There are a number of important features.
- Every column in the experimental matrix is different.
- In each column, there are an equal number of '−1' and '+1' levels.
- For each experiment at level '+1' for factor 1, there are equal number of experiments for factors 2 and 3, which are at levels '+1' and '−1', and so for all other combinations of factors.
- The correlation coefficients between the columns are 0.

TABLE 6.5   Fractional 3 factor 2-level factorial design.

Experiment	Factor 1	Factor 2	Factor 3
1	1	1	1
2	−1	1	−1
3	1	−1	−1
4	−1	−1	1

- The full 3-factor factorial 2-level design is illustrated in Figure 6.3, with each axis representing one factor, together with a corresponding half factorial design; readers should note that there are two possible half factorial designs in this case.
- In classical statistics, a full 3 factorial 2-level design can be used to form a model between the response x and the three factors of the form $x = b_0 + b_1c_1 + b_2c_2 + b_3c_3 + b_{12}c_1c_2 + b_{13}c_1c_3 + b_{23}c_3c_3 + b_{123}c_1c_2c_3 + e$ where there are 8 coefficients which can be calculated using 8 unique experimental conditions.
- When reducing to 4 experiments only 4 terms can be calculated by changing the model to $x = b_0 + b_1c_1 + b_2c_2 + b_3c_3 + e$. The terms for the reduced design are confounded with terms for the full factorial model, as follows:
  - b_0 with b_{123}; b_1 with b_{23}; b_2 with b_{13}; b_3 with b_{12}.
  - This can be checked by multiplying the relevant columns in Table 6.5, for example, multiplying the columns for Factor 1 and Factor 2 representing the interaction between these factors gives the column for Factor 3, so the interaction between the first two factors is confounded with the third factor.
- This means that it is impossible to distinguish interactions from linear effects, and we lose a little additional information.

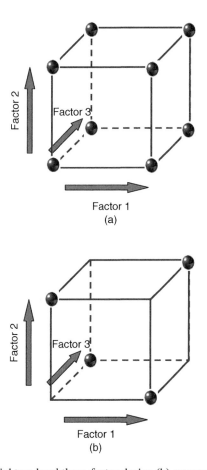

FIGURE 6.3 (a) Full factorial two-level three-factor design (b) corresponding half factorial design.

In the case of a half factorial 3 factor 2-level design, all information about interactions is lost (or more accurately confounded). However, a half factorial 4 factor 2-level design, involving $2^{4-1} = 8$ experiments, will retain some information about interactions as 5 experiments are needed for modelling the 4 factors and intercept, so there are 3 degrees of freedom remaining to obtain information on some of the interactions, dependent on design.

When performing a detailed metabolic study, it is sometimes important to study interactions, for example, it may be desirable to determine whether a specific treatment affects different phenotypes in different ways, in which case treatment and phenotype interact, and enough experiments are required to check if this is so. However, a preliminary study might just involve trying to find out whether different possible factors have any significant effect on the metabolic profile and eliminate those that seem irrelevant, for example, genetic makeup may not have a significant effect on a treatment and as such could be eliminated from further investigation. Under such circumstances, partial factorial designs are adequate as a first step. Since further experiments may be expensive, sometimes a smaller, scoping, study is sufficient for a preliminary project and can answer questions using limited resources.

When there are a large number of possible factors, there are yet more efficient designs such as Plackett–Burman designs, but it would be rare that a meaningful metabolomic study is performed with more than 4 or 5 factors, so full factorial or partial factorial two-level designs are usually sufficient.

6.3　SAMPLING DESIGNS

A different problem arises when selecting samples from a population, for example, individuals in a hospital with a certain disease. Which individuals should be selected for analysis?

Statistical sampling methods have an important role in selecting a representative set of samples from a population. Consider predicting the result of an election, out of an electorate of a million. Going to an upmarket shopping centre on a Wednesday afternoon and asking a hundred (or even a thousand) people their opinions are likely to result in a biased estimate, as the sort of people identified will not be representative of the overall electorate. Therefore, it is important to find a method to identify a balanced (representative) subsection of a population in order to reduce bias.

In clinical studies, it is often difficult to find a perfectly balanced cohort, but some guidance is provided. The first step is to define the candidate population.

- Choose which factors to exclude. For example, for case study 1 (pre-arthritis) excluded individuals with BMI (body mass index) > 30, smokers and those who did not fast for a defined period prior to sampling; all donors were female (so males were excluded).
- In some cases, choose which factors are required for inclusion. For example, for case study 3, NMR of triglycerides in school children, the criteria are as follows: All children are in fifth grade (10 year olds) from Sogn and Fjordane County

in western Norway. Inclusion criteria were that schools should have at least seven children in fifth grade; and that children were healthy (with no serious or chronic illnesses) and able to participate in daily physical activity and physical education (PE). Participants had to be able to complete standard academic performance tests.

The results of this will narrow down the number of samples and reduce variability due to extraneous factors. Sometimes the number is still very large. For example, when performing analyses by LCMS a sample size of 100 is about the maximum given problems of instrumental stability, so if the cohort identified is 1000 individuals the number of samples needs to be reduced tenfold, whereas if using a spectroscopic method like NMR larger sample sizes are often viable.

If the sample sizes need to be reduced further, there are two principal approaches:

- Simple random sampling, for example, if there are 500 possible individuals and we need a sample size of 100, randomly select one-fifth of the samples. If various exclusion and inclusion criteria have already been applied previously, then the remaining cohort may be sufficiently homogeneous for further meaningful study. Often an independent neutral selector can be asked to select the samples or a simple computerised random number generator could be employed,

- Systematic methods: There are a large number of such methods. In many metabolomic studies, though, it is often necessary to compare two or more groups. As an example, in case study 1 (pre-arthritis), individuals are tested several years before some will develop arthritis. However, only a small proportion will develop the disease, so the pre-arthritis group may be a relatively small fraction of the population. The rest are available as controls, but we want to select only a small number as controls. Although some exclusions have already been applied, there still will be a wide range of characteristics within these controls, and a common approach is to narrow these down by balancing the characteristics of the two groups (pre-arthritic –a small group, and controls – potentially large) by selecting a subset of the controls to have similar characteristics to the constrained pre-arthritic group. The controls were chosen to have a similar mean age (51.5 years) to the pre-arthritic group (54.1 years) and after that were randomly selected.

There are many other systematic approaches to sampling clinical data. Normally, in large clinical studies, ideally the chemometrics expert in a team works with clinicians who decide on sampling design, when there are large and complex studies, but only has limited influence on the overall sample design as such large experiments often involve agreement by a large consortium and may involve sampling over many years. Sometimes, clinical studies will result in a large cohort of individuals, and it is up to the chemometrician often in collaboration with an analytical chemist, to decide on a subset of samples from the database for further investigation. In smaller, more controlled, laboratory-based studies, it is possible for the chemometrician to be in control of the selection, for example, in model mouse studies, where phenotype, age, sex, diet, etc., can be built in at an early stage.

One important consideration in most statistical sampling is the sample size, that is, how many donors you need to obtain a result with a given level of confidence. However, in many chemometrics studies, this is often limited to resources and instrumental capabilities. Also, it also is often not possible to make statistically robust predictions to a given degree of confidence, especially with human donors. In model animal studies, it is possible to use more controlled conditions and so there are less problems with reproducibility. Most of metabolomics involve making preliminary predictions, for example, that a metabolite is a candidate marker of a disease at an earlier stage before physical symptoms are present, and that provides the basis for future, often more costly, research that would lead to a more definitive answer.

Another consideration in many metabolomics studies is to define a control group that has characteristics that are similar to the group that is being studied, for example, diseased donors, or donors subjected to a specific treatment.

Finally, some types of experiments can be run in replicates, for example, plant experiments such as case study 5 (NMR of maize), in which there are 3 replicate plants for each of 18 different maize types. However, although this could be regarded as a type of sampling design, it can also be viewed as a fully crossed design (Section 6.2.3.1).

Although most chemometricians will not be directly involved in sample design for large studies, we list several of the most common approaches below. As most readers of this text will not be actively involved in designing large clinical experiments, we will not describe these in detail but it is useful to gain an understanding from collaborators and if there is a chance, in preliminary project meetings and proposals to ensure these issues have been carefully planned in advance. For extensive and very expensive clinical studies, some taking several years, usually there has been a clinical design, sometimes many years before the analytical chemistry and chemometrics are performed. On the other hand, even if this has happened, the chemometrician may be able to select a subset of samples and so participate in the later stages of selection. Some formal named methods are listed below.

6.3.1   Simple Random Sampling

As described above, we may take 1000 school children, and if we want a sample size of 100, randomly select 1 in 10.

6.3.2   Systematic Sampling

Similar to above, we choose the samples systematically. For example, if our 1000 children were numbered from 1 to 1000, we choose children numbered 1, 11, 21, etc. In contrast, for simple random sampling, any child can be selected, as example, they might be numbered 39, 517, 322, 892 and so on until 100 are identified.

6.3.3   Stratified Sampling

This involves, as a first step, dividing the population into groups, as an example we might have one group of males and another females. We might further have groups of people aged between 21 and 40, 41–60 and 61–80. From each of these six subgroups (sex and age), we then pick individuals by one of the methods described in 6.3.1 or 6.3.2.

6.3.4   Cluster Sampling

This involves dividing the entire population into subgroups of approximately similar characteristics, for example, school children as described in case study 3 (NMR of triglycerides), and a few of the schools are selected. In its simplest form, all donors are sampled from each school (or hospital or district as appropriate).

6.3.5   Multi-stage Sampling

This is an extension of cluster sampling. For example, we might decide to sample 5 out of 20 hospitals in a city. Within these hospitals, we cannot sample every patient with a particular characteristic, so choose one of the methods above to select a subsection of patients.

CHAPTER 7

Determining the Provenance of a Sample

7.1 PATTERN RECOGNITION

Pattern Recognition (PR) now largely embraces classification. Many decades ago, PR had a much wider remit, for example, handwriting analysis or facial recognition, but especially in the area of chemometrics it relates primarily to methods for classification.

PR can answer many questions, for example, as follows:

- Is a sample a member of a predefined group? For example, is an extract of blood serum diagnostic of a disease, defined by a group of donors known to have a particular disease or predefined characteristic?
- Is a sample an outlier, not a member of any predefined groups? For example, can we determine when taking extracts of plants whether a sample is not a member of any predefined cultivar?
- Is there enough evidence in the analytical data to be able to distinguish two or more postulated groups? For example, if we record the LCMS (liquid chromatography tandem mass spectrometry) of extracts of plasma of two groups of donors with different genetic characteristics, is there adequate biological and analytical information to distinguish these groups?
- Which variables (or markers) are most characteristic of a group or best able to discriminate between different groups? Which metabolites are the most likely biomarkers?

Data Analysis and Chemometrics for Metabolomics, First Edition. Richard G. Brereton.
© 2024 John Wiley & Sons Ltd. Published 2024 by John Wiley & Sons Ltd.
Companion website: www.wiley.com/go/Brereton/ChemometricsforMetabolomics

The last question will be the focus of Chapters 9 and 10. We will primarily be concerned with answering the first three questions in this chapter using classification methods.

Classification is usually considered a method of supervised PR, where some preliminary structure, normally by dividing the samples into groups or classes, is imposed on the data. Unsupervised PR consists of methods such as cluster analysis and dendrograms, which are rarely used in metabolomic studies and which we will not describe in this text.

7.2 PRELIMINARY PROCESSING PRIOR TO CLASSIFICATION

Before applying classification algorithms, it is always necessary to look at the data and prepare it. This text will focus on multivariate classification problems where several measurements are made on each sample. Simple univariate approaches might involve classifying donors into two or more groups according to the concentration of a single compound, for example, in the urine or blood, with often one group defined when its concentration is high and the other low. Metabolomics methods primarily involve looking at a profile of many different metabolites and so we need to prepare multivariate datasets before classification algorithms.

Some of the common steps are as follows:

- First, pre-process the data, if justified common approaches as discussed in Chapter 4 include
  - row scaling often to a constant total but sometimes ratioed to a standard,
  - logarithmic scaling,
  - column centring almost always,
  - and column standardisation.
- Sometimes discard uninformative variables and samples, and fill in missing data, according to specific problems, as described in Chapter 4.
- Perform PCA (Principal Components Analysis) and examine the scores and loadings graphically.
- Remove any unusual samples, variables, or even unexpected groupings if necessary.
- Pre-process the new data again if necessary and perform PCA again.
- Look again at the scores and loadings to see if the result appears sensible. If not return to repeat pre-processing or removal of redundant variables or samples, until you are satisfied the data are sensible.

As these procedures have been described in detail in Chapter 4, we will not illustrate them again, except to note that pre-processing of columns such as mean centring has to be performed a second time if one removes any suspicious or outlying samples. PCA plots also often have a different appearance once outlying samples have been removed.

Often this is done on a training set. We will discuss validation in Section 7.7.2, where the data are divided into a training set (to form a model) and test set (to see how well the model performs), but it can also be done by auto-prediction on an entire experimental dataset. In some cases, datasets are repeatedly divided into training and test sets via an iterative method (sometimes 100 times or more) in which case graphical inspection each time is not possible, so the preliminary inspection and preparation are often best performed using an auto-predictive model, for example, whether to log scale or standardise is determined on the entire dataset, and then this procedure is repeated each time on the subset of samples forming the training set. There is no universal rule.

7.3 SIMULATED CASE STUDIES

In addition to illustrating the methods using experimental metabolomics datasets introduced in Chapter 3, we will also employ two simulated case studies.

- The main case study, consisting of two classes is presented in Table 7.1 and will be used in several sections of this chapter.
- It consists of a 41×12 matrix; 20 samples have been assigned to class A and 21 to class B.

For this dataset, the only pre-processing we will do is to mean centre the data, prior to PCA. Obviously, other types of pre-processing may be necessary for different situations as discussed in Chapter 4.

- Figure 7.1(a) presents the scores of the first 2 components of the centred data. There seems quite good separation between the two classes in the direction of PC2 with some overlap. There is an obvious outlier (sample B21 in this case). PC1 is mainly dominated by the outlier. This suggests sample B21 has a large influence on the PCs.
- Figure 7.1(b) and (c) are the corresponding t values and p values (using the t statistic as described in Section 5.4) for the 12 variables. We see that variables 11 and 12 have very low t and so high p values.
- Figure 7.1(d) represents the loadings for PC2, as this is the main direction of discrimination. In this case, the magnitude of the loadings represents how important a variable is, and the sign, whether it is most associated with class A or class B.
- If the scores and loadings are compared, we see negative scores for PC2 are most characteristic of class A and vice versa for class B, so variables with the most pronounced negative loadings are characteristic of class A and vice versa for class B. This would suggest that variables 6 and 8 may be the best markers for class B, and variable 10 for class A. Note that the sign of each component may be different if using different software and cannot be controlled.
- P values confirm which variables are likely to be most important as discriminators, low p values being the most useful. The magnitude and signs of the t values also provide fairly similar information to the loadings.
- We do not expect an exact correspondence between the conclusions drawn from p values and loadings first because the latter look at a multivariate viewpoint whereas the p values are based on a univariate t test, and second because PC2 is not a perfect discriminator.

TABLE 7.1 Simulated case study consisting of a 41 × 12 matrix of measurements from two classes A (samples 1 to 20) and B (samples 21–41).

Samples	1	2	3	4	5	6	7	8	9	10	11	12
A1	32.41	42.14	77.24	33.87	31.25	133.89	25.18	100.10	72.03	36.34	46.32	54.21
A2	51.76	65.42	96.58	42.83	54.83	175.63	49.07	117.20	99.09	64.25	64.64	85.20
A3	31.25	42.26	66.84	35.80	36.75	125.38	21.07	88.44	73.81	41.17	46.19	61.49
A4	45.75	55.55	77.56	31.08	49.83	142.86	37.33	99.53	87.78	61.44	56.14	71.75
A5	55.00	68.97	97.94	45.08	64.51	178.18	54.65	121.69	108.16	76.24	71.30	93.75
A6	74.05	70.31	89.17	41.59	80.53	178.67	60.06	117.25	105.66	98.33	79.26	101.03
A7	35.28	49.52	73.89	35.71	38.99	136.81	36.11	95.44	82.35	44.18	48.58	62.88
A8	85.80	81.33	105.16	46.80	92.76	198.24	72.66	128.92	123.11	112.84	96.57	117.07
A9	106.85	82.42	98.05	43.87	104.73	187.54	91.95	114.29	115.70	148.13	100.87	121.98
A10	53.98	54.21	78.91	32.85	59.50	154.97	47.54	103.87	89.54	74.13	59.51	77.24
A11	25.73	45.51	70.84	33.91	31.20	138.67	26.40	97.76	74.83	32.02	41.10	56.30
A12	44.49	55.93	86.84	41.45	51.12	161.20	42.12	115.01	88.73	57.98	61.63	77.49
A13	71.96	65.85	87.33	34.66	77.57	165.52	62.90	104.31	101.69	92.94	76.93	93.36
A14	53.19	60.03	84.23	38.63	63.12	170.02	47.97	112.85	92.21	74.68	63.79	85.29
A15	65.31	70.69	93.38	44.73	72.10	184.76	60.98	126.16	110.42	84.78	74.31	101.28
A16	82.65	79.61	96.10	43.88	79.23	184.22	66.13	118.75	115.68	116.18	84.81	109.80
A17	89.11	73.69	84.42	46.83	92.38	173.92	70.20	113.20	112.41	121.47	92.32	104.75
A18	74.15	77.21	100.47	52.55	82.94	198.70	61.82	121.48	117.44	103.88	86.39	105.75
A19	101.64	80.39	103.89	49.48	97.76	191.83	83.89	123.27	117.18	138.28	103.17	116.29
A20	63.75	77.11	102.19	53.97	72.48	190.94	44.68	134.92	109.29	74.44	75.46	97.82

(continued)

213

TABLE 7.1 (*Continued*)

Samples	Variables											
B1	47.92	67.42	100.34	51.47	51.12	190.58	34.97	136.34	102.68	47.85	62.90	90.85
B2	35.76	67.58	107.94	50.94	49.39	200.88	36.01	143.00	114.36	43.06	62.32	85.23
B3	28.17	74.32	124.36	58.29	44.73	225.71	31.93	164.22	123.01	34.35	58.82	91.05
B4	48.92	70.44	92.80	48.72	53.56	184.38	44.68	128.90	101.17	62.38	62.81	87.62
B5	36.62	57.63	83.38	39.80	49.42	169.21	36.22	116.27	90.93	50.18	54.86	77.08
B6	67.18	92.63	135.88	67.56	72.63	255.67	58.32	175.91	137.52	80.27	91.76	119.52
B7	29.90	61.61	95.02	51.40	49.52	184.39	37.26	130.74	98.30	35.91	60.09	78.97
B8	51.00	78.84	106.99	55.86	51.43	214.21	39.57	150.62	116.93	57.88	66.68	90.44
B9	47.60	83.81	125.94	61.30	58.96	233.40	49.55	172.60	133.06	60.85	76.09	108.57
B10	22.90	47.46	70.23	32.67	24.05	132.89	17.92	97.59	70.74	17.90	37.49	52.88
B11	53.08	68.42	108.39	57.05	57.20	192.16	42.44	133.56	106.89	62.00	69.94	92.54
B12	52.01	64.30	91.57	44.05	59.20	173.69	44.06	117.53	103.78	74.51	70.26	90.62
B13	32.53	56.90	93.74	48.07	41.36	169.12	26.64	123.19	91.34	38.67	54.98	71.32
B14	47.23	74.39	109.88	55.66	55.80	207.65	44.47	146.59	116.68	54.52	65.26	84.37
B15	61.43	76.45	117.77	50.32	67.74	204.98	54.55	139.04	116.64	72.97	80.73	102.54
B16	40.15	62.38	92.63	45.72	52.31	180.87	34.70	127.14	97.85	49.25	53.69	76.34
B17	26.22	46.66	73.68	34.11	38.40	144.50	23.18	99.12	74.08	27.34	39.74	63.48
B18	56.93	87.66	129.62	61.68	71.01	243.76	50.15	176.18	138.36	67.03	80.95	107.46
B19	44.12	81.53	122.27	59.62	51.47	227.22	39.21	157.33	116.69	52.33	70.02	98.50
B20	63.71	75.66	115.55	51.25	69.81	216.49	57.86	152.89	119.15	85.28	86.36	103.63
B21	180.82	145.27	167.06	83.30	187.22	325.72	148.28	197.96	202.31	251.43	183.79	209.90

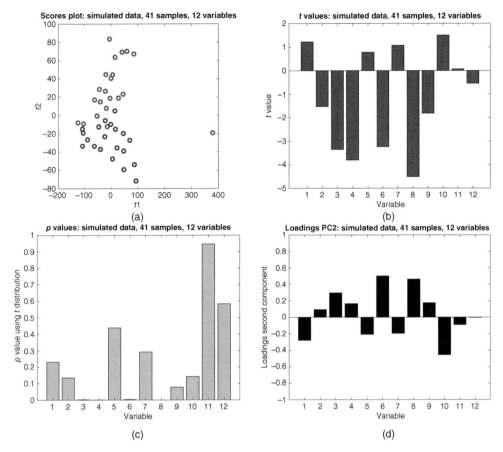

FIGURE 7.1 (a) Scores of first 2 PCs (b) t values (c) p values and (d) loadings of PC2 for data of Table 7.1; Class A blue symbols, Class B red symbols.

However, for the purpose of preliminary analysis, we can see that sample B21 is an outlier and probably is best not to include in a class model. Removing this sample to create a 40×12 matrix makes a significant difference.

- Figure 7.2 presents the equivalent graphs to Figure 7.1 after removal of B21.
- The scores cover the PC space more evenly. Note that the axis for PC2 is inverted, in that class A is represented mainly by positive and class B negative scores. This is an unavoidable consequence of the algorithm and has no significance, as the sign of PCs cannot be controlled.
- The t values and loadings for PC2 (which is still the most discriminatory) are therefore also inverted in sign.
- Note that the pattern of the p values appears quite different between Figure 7.1(c) and Figure 7.2(c). Variables 12 still has a very high p value, and therefore is unlikely to be useful as a discriminator. However, variables 5 has now a much lower p values after B21 has been removed, suggesting that it might have some discriminatory power. Variables 3, 4, 6 and 8 have low p values and high magnitude t values and loadings using both datasets. Variable 11 appears

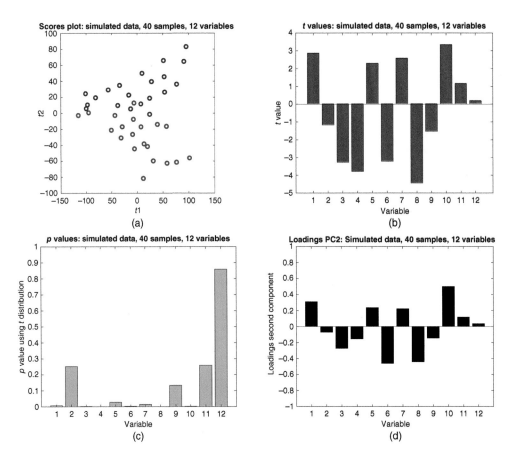

FIGURE 7.2   (a) Scores of first 2 PCs (b) t values (c) p values and (d) loadings of PC2 for data of Table 7.1 after sample B21 (outlier) has been removed; Class A blue symbols, Class B red symbols.

more significant than when the outlier is included, but using evidence from both figures together we will highlight variables 11 and 12 as probably the least informative.

Removing outliers can therefore change conclusions, especially about which variables (or markers) are likely to be most significant and is a useful first step.

It is also possible to remove redundant variables. Sometimes these are very obvious, for example, spectroscopic noise in NMR (nuclear magnetic resonance), and can seriously distort analysis, especially if columns are standardised. We have addressed this already in Section 4.7.3, so this is primarily a reminder to look at this prior to forming classification models. Whether this step is necessary is very much dependent on the characteristics of the analytical technique employed.

At preliminary inspection, it looks as if variables 11 and 12 play little role in discrimination, and we will look at what happens if we remove them.

- Figure 7.3 illustrates the effect when variables 11 and 12 are removed.
- By definition, the *t* and *p* values of each variable will remain unchanged as they are univariate measures.

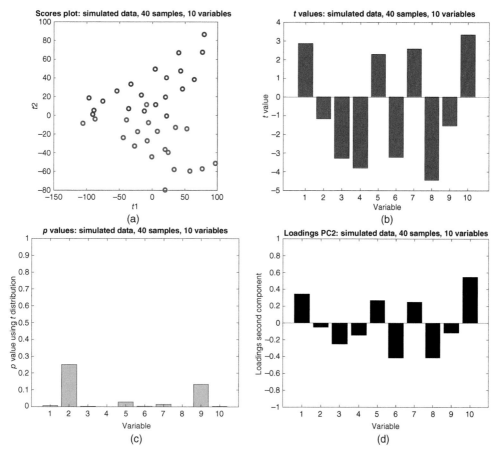

FIGURE 7.3   (a) Scores of first 2 PCs (b) t values (c) p values and (d) loadings of PC2 for data of Table 7.1 after sample B21 (outlier) has been removed and variables 11 and 12 have been removed; Class A blue symbols, Class B red symbols.

- Interestingly, there is hardly any change in the scores, as the two variables removed have very little influence on the pattern. This is not always so, for example, if there are a lot of uninformative spectral features such as in many NMR spectra, and all the variables are standardised, removing these can clarify trends.
- There will be some change in the pattern of the loadings, but this is quite small.

As an illustration, in this chapter, we will reduce the 41×12 data matrix of Table 7.1 to a 40×10 data matrix, excluding sample B21 and variables 11 and 12, when subsequently illustrating classification methods. We could also have also removed variable 2, but for illustrative purposes, we will follow calculations through with the first 10 variables.

A smaller 30×2 case study consisting of 30 samples from three classes will be employed to illustrate the methods of Section 7.6 and will be introduced in that section.

7.4   TWO-CLASS CLASSIFIERS

This chapter will primarily be concerned with classification methods. Two-class classifiers are the simplest to understand.

Two class classifiers aim to produce a boundary between two groups of samples. This is often called a hard boundary and is illustrated symbolically in Figure 7.4. Objects to the left of the boundary represent Class A illustrated by circle symbols and to the right Class B illustrated by diamond symbols. For a new sample, we can predict which class it belongs to according to which side of the boundary it lies.

In Figure 7.4, we assume that the boundary between different classes is linear and that they can be exactly separated. However, this does not need to be the case. Two other situations are illustrated in Figure 7.5. Figure 7.5(a) illustrates a non-linear boundary. There is no requirement that classes are best separated by a linear model, although most chemometric methods involve finding linear boundaries. Figure 7.5(b) illustrates two classes that cannot be linearly separated and often happens.

Finally, there are numerous different mathematical models or algorithms. These all result in different types of boundaries between the classes. Figure 7.6 illustrates three different possible boundaries. Both boundaries A and B perfectly separate two classes using linear models. Boundary C does not succeed. Each boundary can be formed using a different computational model or statistical criterion. There are hundreds of such models or approaches available and an experienced chemometrics expert would spend time choosing which he or she feels is most suitable for a particular problem. In this text, we will describe just a few of the most common approaches for two-class classifiers.

Of course, most datasets involve recording several different variables such as chromatographic peaks or spectral intensities. Hence, a dataset where each sample is measured at 100 different spectral wavelengths could theoretically be represented using a 100-dimensional graph. We cannot, of course, visualise this on paper or on a computer screen, but the boundary, in this case, could be represented by a 99-dimensional hyperplane. The computer can of course calculate which side each sample is to this hyperplane to give a prediction as to each sample's provenance but to better understand the

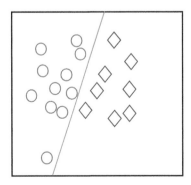

FIGURE 7.4   Principle of hard boundary in a two-class classifier.

methods, we use simpler graphical approaches in this text to provide a preliminary picture using 2D projections: the principles can be extended to imaginary multidimensional space.

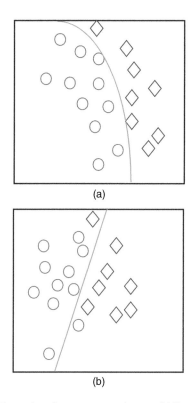

FIGURE 7.5   (a) Non-linear boundary between two classes, (b) linearly inseparable classes.

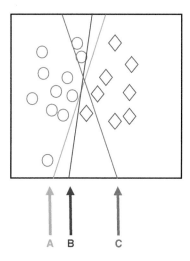

FIGURE 7.6   Three different types of linear boundaries.

7.4.1   Linear Discriminant Analysis

LDA (Linear Discriminant Analysis) is one of the simplest and oldest approaches for multivariate classification. It is best understood as a two-class classifier, although can be extended to more than two classes.

We will start with the 40×10 simulated data matrix of Table 7.1 excluding sample B21 and variables 11 and 12.

- In Section 5.3, we discussed how to calculate the Mahalanobis distance from the centroid of a group but to a single predefined class.
- We will initially use all the 10 variables in our model.
- For LDA we calculate the Mahalanobis distance to the centre of each class but using a variance–covariance matrix based on all the samples in the data matrix, 40 in this example, S_{all}. For readers that wish to follow the calculations in this chapter, S_{all} is a 10×10 matrix, and using the population standard deviation, for example, $s_{all3,4} = 144.74$ is the population covariance between variables 3 and 4.
- We compute the Mahalanobis distance from the centre of each class

$$d_{Ai} = \sqrt{\left(x_i - \bar{x}_A\right) S_{all}^{-1} \left(x_i - \bar{x}_A\right)'}$$

  - where S_{all}^{-1} is the 10×10 inverse of the variance–covariance matrix for all 40 samples, so that $s_{all7,3}^{-1} = -0.023$ for readers following this text numerically.
  - and $\bar{x}_A$ is the mean of the 20 samples that belong to class A in the form of a 1×10 row vector (using notation in this book).
- The distance to the centre of class B against the distance to the centre of class A for all 40 samples is presented in Figure 7.7(a). The equidistant line is drawn, and samples are assigned according to which side of the line they fall.
- Notice two samples are misclassified, which are A20 and B12, and are marked.
- As a check, the distance of sample A13 to the centroids of each class is $d_{A,A13} = 3.02$ and $d_{B,A13} = 3.65$, assigning it to class A.
- Note that identical results would be obtained if the scores of all 10 non-zero PCs were used instead, and identical results are also obtained whether the data are centred of not.
- Note that there are no samples very close to the origin. This intuitively strange result can be explained by the shape of the χ^2 distribution with 10 degrees of freedom (see Section 5.2.4); when there are more than 2 degrees of freedom, the probability density at the origin is 0.

Mahalanobis distances can only be calculated in this way if the variance–covariance matrix has an inverse. For this to happen:

- The number of variables must be less than the number of samples in the overall dataset; in our case $10 < 40$ so we can safely compute the Mahalanobis distance.

- None of the variables can be correlated.
- If the number of variables is close to the number of samples, or some variables whilst not completely correlated have correlation coefficients very close to ±1, for example, >0.95, then although it is possible to obtain Mahalanobis distances, they may be a bit difficult to interpret and it is best to reduce the number of variables.

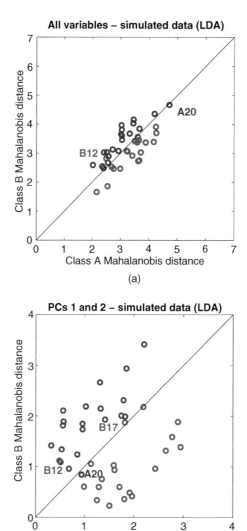

FIGURE 7.7   Graph of Mahalanobis distance for class B versus class A for 40×10 simulated case study of Table 7.1 and LDA using (a) all variables (b) PCs 1 and 2 (c) PCs 1 to 5 with misclassified samples labelled.

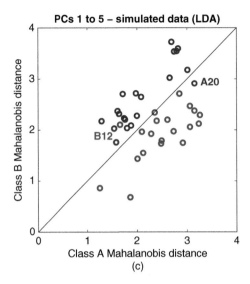

FIGURE 7.7   (*Continued*)

Often to overcome this limitation, instead of calculating the Mahalanobis distance using the full dataset (or scores of all non-zero PCs) it is first reduced by PCA and the first few PCs are retained.

- Figure 7.7(b) and (c) present the equivalent plots for a two PC and a five PC model. The data are centred prior to PCA (although this does not affect the results).
- As a check, the score $t_{A7,2} = 15.06$ and the distance using a 5 PC model of sample A9 to class A = $d_{A,A9} = 2.68$ and to class B is $d_{B,A9} = 3.72$, assigning this correctly to class A.

We can identify the misassigned samples for each model, as presented in Table 7.2.

- First, if we compare to Figure 7.7 we can check the number of misassigned samples in each class is the same in the table and the figure.
- We also see from the figure as can be verified by numerical calculations, that misassigned samples lie very close to the border between the groups, and so it is possible that the linear boundary between the classes could be refined to get a more accurate model.

TABLE 7.2   Misclassified samples using different models and LDA for 40×10 simulated case study.

Model	Misclassified samples
PCs 1 and 2	A20 B12 B17
PCs 1–5	A20 B12
All data (or PCs)	A20 B12

- At first glance, it seems that the more the PCs or variables in the model, the better the classification, as the model with 2 PCs misclassifies 3 samples, compared to the other models.
- However, this is not necessarily the most appropriate answer. In Section 7.7.2, we will discuss validation. Using more PCs can overfit a model, because the original (training set) may not be perfect, for example, we may be studying a disease, and some diseased donors may have been misdiagnosed or at different stages of progression, or contain a few samples with very different genetics or other features, but increasing the number of PCs may try to fit these 'false' samples.

We can understand the two PC model graphically as in Figure 7.8. Figure 7.8(a), the LDA contours are marked for each class. Figure 7.8(b), the equidistant line $d_A = d_B$ is drawn, samples being assigned to their corresponding class according to which side of the line they lie. The three samples assigned to their incorrect classes are marked with closed circles and labelled, they should be compared to the 2 PC Mahalanobis class distance plot in Figure 7.7.

Obviously, once there are more than 2 variables, the divisor becomes a hyperplane and after 3 variables cannot be visualised, but the principle illustrated in Figure 7.8 using 2 variables can easily be extended.

We now illustrate these principles using case study 10, using the negative ion LCMS data of mice diagnosed diabetic and non-diabetic, involving a 71×32 matrix, consisting of class A (30 diabetic mice) and class B (41 non-diabetic) and 32 variables. The data are row-scaled and then standardised prior to further analysis as discussed in Chapters 3 and 4.

- Figure 7.9 presents the plots of the Mahalanobis distance to the centroid of class B against class A, using models involving the scores of 2, 5 and 31 PCs.
- Although there are 32 variables, because the data have been row-scaled, there are only 31 degrees of freedom: this is caused by a property called closure, and is the maximum number of non-zero PCs we can use in the model; note if row-scaled we could not use the full raw data as there will be correlation between the variables since the value of the 32nd variable will be obtained from the remaining 31 variables.
- If you are following this numerically, remembering to use the population standard deviation, we find using a 2 component model $d_{A2} = 1.71$ and $d_{B2} = 1.34$, that is, mouse number 2 is assigned to the non-diabetic group even though it was originally thought to be diabetic; in contrast $d_{A2} = 4.98$ whereas $d_{B2} = 5.00$ using a 31 component (full) model, and so is narrowly assigned to the diabetic class.
- Table 7.3 lists the misclassified samples as the number of PCs in the model is increased. The number of misclassified samples decreases with increasing number of PCs. Note that the samples misclassified may vary, for example, mice number 2 and 7 change around as the number of PCs are increased from 10 to 20, but they will be very close to the border.

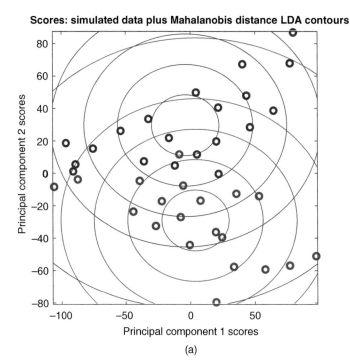

(a)

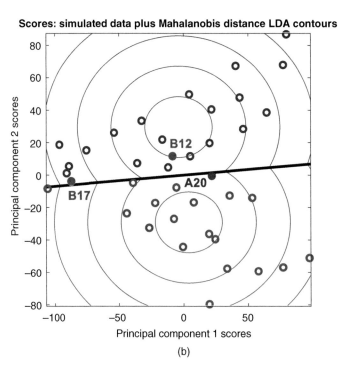

(b)

FIGURE 7.8   Simulated data, 2 PC model, showing LDA Mahalanobis distance contours at distances of 0.5, 1, 1.5, 2 and 3, with the three misassigned samples marked in the bottom figure.

- Figure 7.10 illustrates the LDA Mahalanobis distance contours and misclassified samples for the 2 PC LDA model.

The first impression is that the more the PCs the better the classification. However, this may involve over-fitting, and as we will discuss later, whether the answer is appropriate is best checked using validation as discussed in Section 7.7.2. Clearly, using a

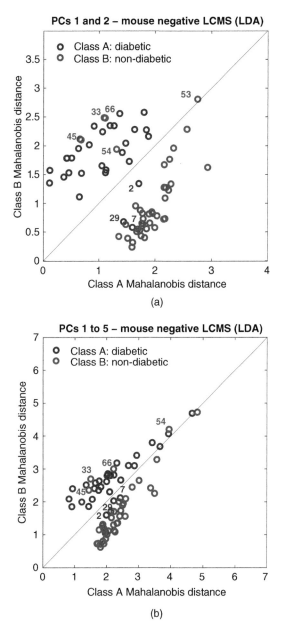

FIGURE 7.9 Mahalanobis distance using LDA for dataset of case study 10 LCMS: Class A diabetic, Class B non-diabetic; misassigned samples are labelled.

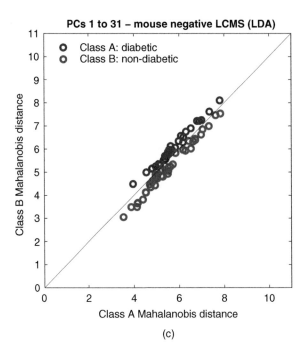

(c)

FIGURE 7.9   (Continued)

TABLE 7.3   Misclassified samples using different models and LDA for 71×32 model of diabetic and non-diabetic mice.

Model	Misclassified samples
PCs 1 and 2	Diabetic: 2 7 29; Non-diabetic: 33 45 53 54 66
PCs 1–5	Diabetic: 2 7 29; Non-diabetic: 33 45 54 66
PCs 1–10	Diabetic: 7; Non-diabetic: 33 45 54 66
PCs 1–20	Diabetic: 2; Non-diabetic: 33 45
All non-zero PCs (31)	None

full 31-PC model results in a meaningless classification model in this case; however, a 2-PC model may be a little too few. In practical terms, it is probably best to use predictions employing just the first few PCs. Some samples may appear to be misclassified using a 2-PC model and although there is no hard rule, one can look at how they were originally diagnosed, and whether they are best removed from further analysis. If the aim is to see which metabolites are the best markers for diabetes, it might be a good policy to remove samples whose provenance is slightly doubtful, and use chemometrics as a second step to narrow down the mice of interest. In any experimental situation there will be variability and the first, often clinical or observational, classification can be refined in a second chemometric step.

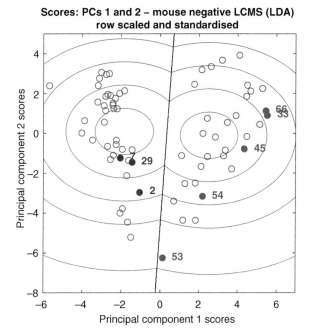

FIGURE 7.10   Illustration of LDA boundaries and the misclassified samples for a two PC model and the negative ion LCMS data of case study 10, the diabetic mice.

7.4.2   Partial Least Squares Discriminant Analysis

One of the most widespread methods employed in metabolomics is PLSDA (Partial Least Squares Discriminant Analysis).

The PLS method has a long vintage in chemometrics circles. Like LDA, PLSDA in its basic form is a linear method, and we will show that under certain circumstances it gives identical results to LDA. Although it is possible to include squared terms, this is rarely done, and in this book, we will restrict to the linear models.

The major advantage in PLSDA comes from its flexibility, especially in coping with situations, unforeseen in traditional statistics, where the number of variables far outnumber samples. Traditional PR studies are primarily concerned with whether two groups can be separated rather than why. In a traditional problem, for example, we may be interested to see if we can separate two groups of faces that are visually distinct using a computational algorithm for image processing: we are not primarily interested in whether people can be distinguished from the size of a pimple on the left of their face, but more whether we can produce an automated model to determine the differences between two or more groups. In metabolomics we are often interested in further diagnostics, for example, we measure a large number of metabolites, which correspond to markers that distinguish between two or more groups, and hence if we know the identities of these markers, which metabolites cause or are a consequence of a disease or other difference between groups of samples. In addition to classification abilities similar to LDA (the differences primarily relating to data scaling) there is a wealth of further diagnostics and a large number of statistics that have been historically developed in the chemometrics literature, especially for determining which variables are most signification for discrimination as discussed in Chapter 9. Many historic classifiers, for example, in traditional biology, were developed in an era when the number of samples usually exceeded the number of variables substantially, which is not always the case in metabolomics, hence the need for different ways of handling the data.

In fact if one performs PLSDA on data retaining all non-zero PLS components, it gives identical results to LDA for the classifier if group sizes are equal, as we will see below but additional information could be obtained about the variables. An analogy in Chapter 10 is a comparison between the ANOVA test and ANOVA followed by regression modelling: the former will provide a p value as to whether there is a significant difference between different groups but the latter will provide additional diagnostic information, although the p values will be the same.

7.4.2.1   PLSDA for Equal Class Sizes

We will initially describe PLSDA for balanced data, that is when each class is of the same size and consists of two classes. Although there are several algorithms, the main ones due to Wold and Martens, the overall principles are quite straightforward. In this text, we will restrict to the Wold NIPALS algorithm. In this section, we restrict our discussion to the PLS1 algorithm.

- PLS tries to relate two types of variables, called the X block (in our case the experimental measurements) and the c block (in our case the classifier).
- The c block consists of a single vector, which in our case consists of two classes which we label $+1$ and -1 according to whether a sample is a member of class A or class B, respectively. Some authors label classes using 0 and 1 but our description by default centres the c block when the data are balanced. For brevity, we will not compare other approaches to labelling the classifier.
- Hence, for the 40×10 case study derived from Table 7.1 (excluding B21 and variables 11 and 12), the X block consists of the 40×10 measurements, and c (sometimes alternatively denoted y) is a vector of dimensions 40×1 which consists of the first 20 numbers equal to $+1$, and the second 20 equal to -1. The principles are illustrated in Figure 7.11.

PLS using the NIPALS algorithm forms two sets of equations, as follows:

$$X = TP + E$$
$$c = Tq + f$$

- Centring is an important first step. We will initially assume that X is centred down the columns, whereas if there are equal numbers of samples in two classes, c is also centred if we label class A by $+1$ and class B by -1. We will discuss unequal class sizes (unbalanced data) in Section 7.4.2.2.

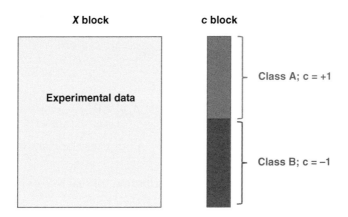

FIGURE 7.11 Principles of PLSDA calibration; balanced two-class dataset.

- The dimensions of T, P and q are $I \times A$, $A \times J$ and $A \times 1$ where A is the number of PLS components used in the model. Using the definitions in this chapter, the p vector for each component is a row vector, although some authors prefer it to be defined as a column vector (which we will not use so as to avoid confusion, in this text).
- The matrix T is called the scores matrix.
- There are two types of loadings, the x loadings (P) and the c loadings (q) unlike PCA where there is only one type of loadings. Using the algorithm described in this book there is one matrix T common to both blocks called the scores.
- The product of T and P approximates to the experimental data and the product of T and q to the c vector, the common link being T.
- Note that T and P for PLS are different in nature to T and P obtained in PCA as discussed in Chapter 4; historically, the terms loadings and scores have been used both for PCA and PLS although this can cause confusion. It is important to recognise that there are several algorithms for PLS available in the literature, and although the predictions of c should be the same in each case, providing the data have been suitably pre-processed in the same way, the scores and loadings depend on algorithm; we only illustrate one (the NIPALS algorithm) in this book for brevity.
- Although the scores are orthogonal (as in PCA), P (or x loadings) are not (which is an important difference to PCA), and, furthermore, the x loadings are not normalised, so the sum of squares of each p vector does not equal 1, using the algorithm described in this section. Different algorithms may result in other properties. If you are using a favourite package, it is a good idea to check. This differs from PCA where the only difference between algorithms is in scale.
- In addition to scores and loadings, a further matrix W of dimensions $J \times A$, also called a PLS weight matrix, is obtained as a consequence of the algorithm. We define w_a as a column vector for each component a in this text.

- The weights vectors are orthonormal unlike the x loadings, so that

  - $\sum_{j=1}^{J} w_{ja}^2 = \boldsymbol{w}_a' \boldsymbol{w}_a = 1$ and $\sum_{j=1}^{J} w_{ja} w_{jb} = \boldsymbol{w}_a' \boldsymbol{w}_b = 0$

  - whereas $\sum_{j=1}^{J} p_{aj}^2 = \boldsymbol{p}_a \boldsymbol{p}_a' \neq 1$ and $\sum_{j=1}^{J} p_{aj} p_{bj} = \boldsymbol{p}_a \boldsymbol{p}_b' \neq 0$

- An important step involves predicting $\boldsymbol{c}$, in our case the classifier. This can be done in one of two ways:

  - The first is by summing the contribution of each component so that

    $\hat{c}_{iA} = \sum_{a=1}^{A} t_{ia} q_a$ assuming the mean value of c is 0 (which it is for equal class sizes)

  - The second is via a regression vector $\boldsymbol{b}$ where $\boldsymbol{b} = \boldsymbol{W}(\boldsymbol{P}\,\boldsymbol{W})^{-1}\boldsymbol{q'}$ in our notation so that $\hat{c}_i = \boldsymbol{x}\boldsymbol{b}$; unlike for loadings and weights, the regression vector incorporates all information from past components, so for an A component model although there will be an $A \times J$ loadings matrix, there will be a single $J \times 1$ regression vector $\boldsymbol{b}$.

    In both cases, it is assumed $\boldsymbol{X}$ is centred for simplicity.

- The latter equation $\hat{c}_i = \boldsymbol{x}\boldsymbol{b}$ is very useful because it can also be employed to predict the value of c for unknown samples as well as samples used to form the model (training set samples).

- Additionally, the size of each PLS component can be calculated, in analogy to the eigenvalue of a PC, by multiplying the sum of squares of both $\boldsymbol{t}_a$ and $\boldsymbol{p}_a$ together, so we define the magnitude of a PLS component as:

$$\kappa_a = \left(\sum_{i=1}^{I} t_{ia}^2 \right)\left(\sum_{j=1}^{J} p_{aj}^2 \right).$$

This is sometimes incorrectly called an eigenvalue but does not have the same properties. Note that in contrast to PCA, the size of successive values of κ_a does not necessarily decrease as each component is calculated. This is because PLS not only models the $\boldsymbol{X}$ data, but is a compromise between $\boldsymbol{X}$ and $\boldsymbol{c}$ block regression. There is of course no universal agreement as to how to determine the size of each successive PLS component, but the equation above is a common and well-accepted one.

- The final step is to use the value of c to assign a sample to one of the classes. The simplest rule is to assign all samples with a positive value to class A and negative to class B. We will see later that this is not always the ideal solution, but for this example, we will use $c = 0$ as the decision criterion.

There are various ways of determining A, the optimal number of PLS components. In classical chemometrics, this occupied a great deal of energy, but much of classical chemometrics was concerned with analytical and physical chemistry where accurate quantitative predictions were important; in metabolomics, the predictions are usually more qualitative and indicative in nature so although it is desirable to choose the number of PLS components using a systematic approach, more often it can be done empirically. We will not discuss the numerous approaches for choosing the number of PLS components in detail in this text but readers are referred to Section 8.4 for an overview.

The method will be initially illustrated by the 40×10 reduced case study of Table 7.1.

The scores, loadings, weights and predictions of c are presented for a one-component model in Figure 7.12 and two-component model in Figure 7.13.

- The decision criterion is 0, so all samples with a predicted value of +1 are assigned to class A, otherwise class B. The number of misclassified samples are those with the incorrect signs for the predictions by the relevant model.
- For a one-component model, the predictions are simply proportional to the scores.
- The weights are similar but not identical to the loadings.

For readers who are following the calculations numerically, remember that different algorithms give different values of loadings and scores, but identical predictions.

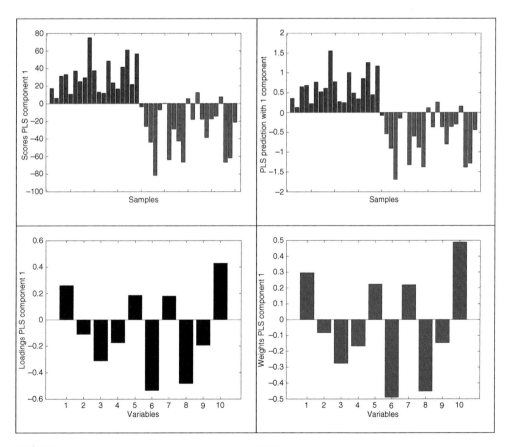

FIGURE 7.12   Simulated data: one-component PLSDA.

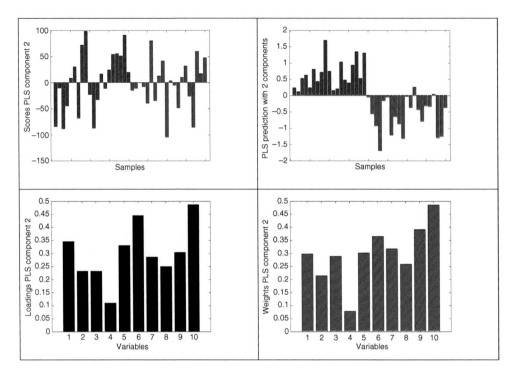

FIGURE 7.13 Simulated data: two-component PLSDA.

If you are following the calculations in this chapter, check the following

- $t_{12,2} = -32.81$, $w_{6,2} = 0.364$, $p_{2,6} = 0.446$.
- The predicted value of sample A6, $\hat{c}_{A6} = 0.764$ after one component, and 0.806 after two components.
- The regression coefficients for the first 3 components are presented numerically in Table 7.4. The predicted values $\hat{c}_a$ for the training set for an a component model can be calculated by $\boldsymbol{X}\,\boldsymbol{b}_a$ where $\boldsymbol{b}_a$ is the ath column of the matrix of regression coefficients.

The misclassified samples are listed in Table 7.5 for various models. The 2 PLS component model identifies the same samples as LDA using 2 PCA components, and the model with all non-zero components identifies A20 and B12 as does the LDA model.

We can understand this better by looking at a two PLS component model for our 40×10 simulated dataset.

- Figure 7.14 presents the scores of the first two PLS components with the contours for c indicated. Note that the contours are linear in contrast to the Mahalanobis distance of the PC scores (Figure 7.8).
- Whereas in this case the PLS scores plot is quite similar to the PC scores plot but rotated (the relative positions are slightly different using the NIPALS algorithm), the main separation between classes is in PLS component 1 for this data.

TABLE 7.4 Matrix of regression coefficients for the first 3 PLSDA components of the simulated case study.

Variable	b_1	b_2	b_3
1	0.0061	0.0066	−0.0069
2	−0.0017	−0.0014	−0.0062
3	−0.0057	−0.0054	0.0148
4	−0.0034	−0.0033	−0.0136
5	0.0046	0.0051	−0.0023
6	−0.0101	−0.0097	−0.0343
7	0.0045	0.005	0.0176
8	−0.0093	−0.0091	−0.0051
9	−0.003	−0.0025	0.0286
10	0.0101	0.0109	0.0141

- The decision threshold or separator is when $c = 0$ which means that if the predicted value of c is positive, we assign the sample to class A, otherwise class B; there can be other decision thresholds, for example, if there are unequal numbers of samples in each class or unequal variance, but the correction in Section 7.4.2.2 is usually sufficient for unequal class sizes and a decision threshold of $c = 0$.

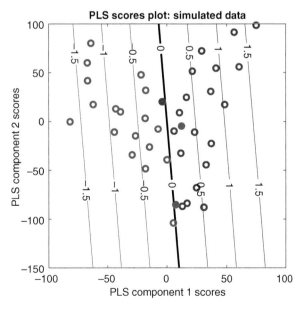

FIGURE 7.14 Two-component PLSDA model with predicted values of c indicated by contour levels and misclassified samples indicated with closed circles.

- Three samples are misclassified as tabulated in Table 7.5. There are three misclassified samples, 1 from class A, and 2 from Class B.

When there are more than 2 PLS components, the decision threshold becomes a hyperplane. For example, if there are 3 PLS components in the model, the decision threshold is a plane separating the two groups and so on.

As a method for determining the provenance of samples, PLSDA has no advantage over LDA. LDA can be performed using a varying number of Principal Components as PLSDA can be with a varying number of PLS components. LDA has the advantage that groups can be of different sizes, and the model is unchanged, whereas the PLS model depends on group size as discussed in the next section.

However, there are two advantages that make PLS useful in metabolomics.

- The loadings and weights provide insight into which variables are markers for each group: Usually, the first PLS component is most discriminatory, so the loadings and weights of PLS component 1 provide predictions as to which are most likely to be markers.
- PLS can be done when the number of variables far exceeds the number of samples, as often happens, for example, in chromatography or spectroscopy. For LDA we have to reduce the variables either using PCA or by eliminating some variables. Of course, an LDA model can be computed just using the first few PCs but then loses interpretation.

7.4.2.2 PLSDA for Unequal Class Sizes

It is common that the number of samples in each group differs. For LDA, this does not matter, it is only necessary to determine the centre of each class and the overall standard deviation. However, this is slightly more complicated for PLSDA. For example, if there are 20 samples in class A and 10 in class B, the c vector will not be centred. If we do centre this vector, what is the decision criterion? Centring would change c for class A from +1 to +0.67 and for class B from −1 to −1.33. Do we still use a decision threshold of 0 (the mean) or −0.33 (the average of the two class centroids)? When using PLSDA for unequal class sizes, always check how the data are transformed and what decision threshold has been used.

TABLE 7.5 Misclassified samples using a number of PLS components for 40×10 simulated case study.

Model	Misclassified samples
1 component	A20 B5 B10 B12 B17
2 components	A20 B12 B17
3 components (and more)	A20 B12

There are many ways of overcoming this, but a simple approach is weighted centring of the X matrix.

- If there are I_A samples in class A and I_B in class A, for each variable j calculate $\overline{\overline{x}}_j = \left(\overline{x}_{Aj} + \overline{x}_{Bj} \right) / 2$.
- Then, for each column j transform $^{wcen}x_{ij} = x_{ij} - \overline{\overline{x}}_j$
- Label c as normal even with unequal class sizes, and no need to centre.
- Perform PLSDA on this new weighted centred matrix.

As an example, we take the first 25 samples of the simulated case study of Table 7.1 using the first 10 variables.

- This can be represented by a 25×10 matrix of which
- Samples 1–20 are A1–A20 and members of class A.
- and sample 21–25 are B1–B5, members of class B.
- The mean $\overline{x}_3$ of variable 3 over all samples is 96.73, so the overall mean centred value for sample 3 $^c x_{8,3} = 105.16 - 96.73 = 8.43$.
- The mean of $\overline{x}_{A,3}$ of variable 3 over the 20 samples of class A is 88.55, and the mean of $\overline{x}_{B,3}$ of variable 3 over the 5 samples of class B is 101.84 so the weighted mean $\left(\overline{x}_{A,3} + \overline{x}_{B,3} \right) / 2 = 95.19$.
- Hence, the weighted mean centred value for sample 3 is $^{wmc}x_{8,3} = 105.16 - 95.19 = 9.97$. The reader can check these numerically if desired.

The effect is illustrated in Figure 7.15. When just centring the columns, the boundary shifts away from the smaller class, meaning that six class A samples are misclassified. When using weighted mean centring, there are only 2 misclassified samples, one from each class, as the boundary is equidistant between the two classes.

There are other ways of solving this, for example, using different decision criteria for c to assign samples into their respective classes, or using intercept values for both X and c; however, for brevity, we illustrate one simple approach. The importance is always to check how data have been transformed prior to PLSDA to ensure sensible results.

We will now illustrate the approach using case study 1 (LCMS of control and pre-arthritic donors).

- The data consists of a 49×81 matrix, consisting of 19 controls (class A; samples 1–19) and 30 pre-arthritic donors (class B; samples 20–49).
- The control group is labelled $+1$ and the pre-arthritic group is labelled -1.
- These data are initially standardised down each column for this case study.
- We then illustrate two sets of results (1) when the data are mean centred down each column – as a consequence of standardisation and (2) when the data are weighted mean centred down each column after standardisation, so they are first scaled and then recentred.

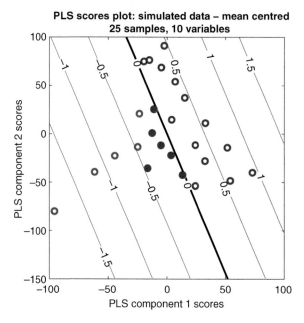

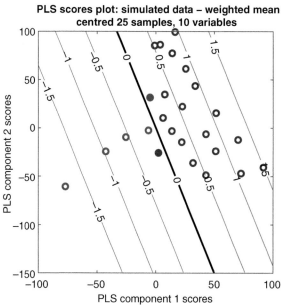

FIGURE 7.15   Effect of weighted mean centring on PLS predictions and boundaries. misclassified samples marked with closed symbols, and values of c indicated in the contour levels.

The plot of the scores of the first two PCs and first two PLS components are illustrated in Figure 7.16 when the data are standardised and so column centred. It can be seen that the two groups appear better separated using PLS. It is important to remember that this is an auto-predictive (or training set) plot, and so the algorithm attempts to separate the groups as well as it can using the combined X and c block data. If we are certain that the metabolomic data really can separate the groups, the PLS model is of course clearer, but if not it could involve over-fitting.

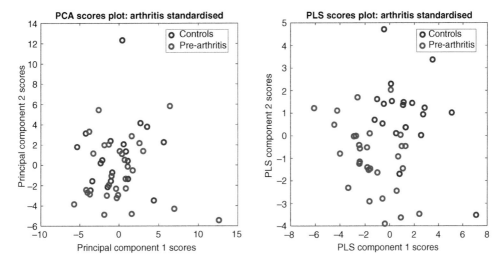

FIGURE 7.16   Case study 1 (pre-arthritis) after standardising and the scores of the first 2 PCs and PLS components compared, both mean centred.

The predicted values of c together with the decision boundary and misclassified samples are illustrated in Figure 7.17 for a 2 PLS component model using both mean-centred and weighted mean-centred model. Although relative positions of the samples in the PLS scores plots are identical in nature, the position of the boundary differs slightly.

If you are checking numerically,

- $t_{12,2} = 1.46$ for the weighted mean centred and $t_{12,2} = 1.70$ for the overall mean centred dataset; if the two scores plots of Figure 7.17 are compared note that whereas the relative position of all the data points is identical, the c values are shifted as a consequence of the shift in column centre.

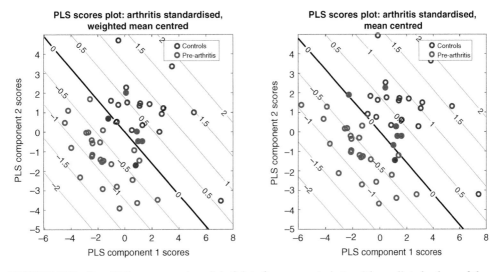

FIGURE 7.17   Two PLS component model of data from case study 1, with predicted values of the classifier c, using both centred and weighted centred method and misclassified samples marked with closed circles.

- The prediction of sample 12 (from the control group) $\hat{c}_{12} = 0.267$ for a one-component model, 0.687 for a two-component model, 0.691 for a three-component model and 0.957 for a four-component model when using weighted mean centring.
- The corresponding predictions are $\hat{c}_{12} = 0.330$, 0.809, 0.848 and 1.126 for the overall mean-centred model. Notice that although the value of $\hat{c}_{12}$ for the overall mean-centred model is always greater than for the weighted mean-centred model (in this case), the difference is not constant.
- The number of misclassified samples for different numbers of PLS components and column centring is presented in Table 7.6.
- In most models except for the one-component model, the overall centred method is biased towards the larger class (B) and so misclassifies a higher proportion of class A.
- For a two-component model, 6/49 (12.2%) are misclassified overall, of which 2/19 (10.5%) are from class A and 4/30 (13.3%) from class B, using the weighted mean centring. The proportion of samples misclassified from both classes is very similar.
- The two-component models can be compared visually to Figure 7.17 where the corresponding misclassified samples are marked.

We can also look at which metabolites are characteristic of each group, via the loadings of PLS component 1 and will use the weighted centred model only for brevity.

- They are presented visually in Figure 7.18.
- Loadings greater than 0.2 are marked in colour, and are suggestive of variables characteristic of each group; 0.2 is an arbitrary threshold but does distinguish the most extreme variables and so most likely markers.

TABLE 7.6   Misclassified samples for the arthritis dataset (case study 1) using different number of PLS components.

Number of PLS components	Class A misclassified (controls)	Class B misclassified (pre-arthritis)
(a) Weighted mean centring		
1	5 samples: 3 5 9 10 16	9 samples: 23 27 28 32 36 40 41 46 48
2	2 samples: 1 9	4 samples: 27 28 46 48
3	1 sample: 1	1 sample: 28
4	n/a	1 sample: 28
(b) Overall mean centring		
1	5 samples: 3 5 9 10 16	9 samples: 23 27 28 32 36 40 41 46 48
2	1 sample: 1	6 samples: 27 28 41 44 46 48
3	n/a	3 samples: 28 41 46
4	n/a	2 samples: 28 46

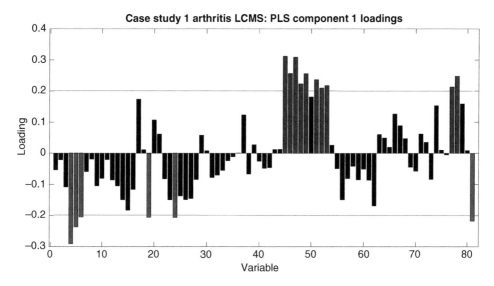

FIGURE 7.18 Loadings of the first PLS component for case study 1 LCMS of pre-arthritic and control donors, using weighted mean centred data, with loadings greater than ±0.2 marked, and provisionally assigned to their group (control = positive / blue, pre-arthritic = negative / red).

- Often we called the loadings characteristic of class B (pre-arthritic donors) as potential markers for the disease, whereas loadings characteristic of class A (controls) suggest metabolites that are suppressed in individuals that carry the disease.
- The identities of these metabolites are presented in Table 7.7.
- This approach is effective in this case because the first PLS component is a good separator, and is not a universally recommended method.

TABLE 7.7 Metabolites characteristic of each group for case study 1 (LCMS of pre-arthritic and control donors) based on loadings > 0.2 using weighted mean centred PLSDA.

Class A (controls)		Class B (pre-arthritic)	
45	Oleic acid	4	Kynurenine
46	Linolenic acid	5	Tryptophan
47	Cis-gondoic acid	6	Phenylalanine
48	Docosahexaenoic acid	19	Lyso-PC (16 : 0)
49	Docosapentaenoic acid	24	Lyso-PC (14 : 0)
51	Palmitic acid	81	3-Indolelactic acid
52	Cis-9-palmitoleic acid		
53	Stearic acid		
77	B-Hydroxypalmitic acid		
78	Beta-hydroxylauric acid		

Of course, we could also use weights instead of loadings and 0.2 is an arbitrary criterion: it mainly identifies the most characteristic compounds identified in each group. There are many other approaches for determining which variables are most significant which is the basis of Chapter 9, several based on PLS especially where the first component is not sufficient to separate the groups. Metabolites that are characteristic of the controls are likely to be suppressed with the pre-arthritic donors. Of course, the interpretation is for the clinicians and is not discussed in this text.

7.4.2.3   OPLS

A recent modification of the PLS algorithm called Orthogonal Partial Least Squares (OPLS) has been described and is available in several software packages. Many readers of this text may have come across the method as it is commonly encountered.

OPLS is most usually employed as a way of visualising differences between groups and makes the distinction clearer than either PCA or PLS alone.

The principle is that there are two types of variation:

- Correlated variation between the data X and the classifier c, this improves the separation.
- Non-correlated (orthogonal) variation between data X and c. This could be considered a form of noise, that if removed would make the remaining data better correlated against c. This is sometimes alternatively called within group variation or occasionally could be viewed as error.

The algorithm finds and removes this undesirable orthogonal variation.

- After pre-processing, the first step is to find orthogonal components characterised by scores t_{ortho} and loadings p_{orth} using the OPLS algorithm.
- After choosing the number of orthogonal components, sometimes just 1 or 2, the orthogonal adjustment can be calculated as $E_{ortho} = T_{ortho} P_{ortho}$.
- Now adjust $X_{adjust} = X - E_{ortho}$ and perform normal PLS on this new filtered or adjusted dataset to obtain T_{adjust} and P_{adjust}.
- The number of PLS components used for calculating the orthogonal components and in the second (main) PLSDA calculation does not need to be the same.

If used for visualisation, often the scores are plotted for the orthogonal component against the first adjusted component, t_{1ortho} versus $t_{1adjust}$.

- The scores $t_{1adjust}$ should have a good correlation against c (the classifier) and so show good discrimination.
- The scores t_{1ortho} have a correlation of 0 against c and so remove the relationship to the classifier.

Of course, the vertical axis is uninformative and this is really a way of presenting a 2D plot, it would be possible to present the adjusted scores as a 1D graph instead. OPLS is primarily an approach for visualisation.

The procedure is best illustrated by a case study.

- We will use case study 10, using the positive ion LCMS of the control and diabetic mice, organised as a 71×146 matrix, with 30 mice in class A (diabetic) and 41 class B (controls).

- The first step is to row-scale the data to a constant total. If you wish to check this, $x_{17,30} = 2.76 \times 10^4$. The sum $\sum_{j=1}^{146} x_{17,j} = 9.73 \times 10^7$. Hence, $^{rs}x_{17,30} = 2.84 \times 10^{-4}$.

You should also check that the sum of each row is 1 after this transformation.

- The data are next standardised, in this case, we use the population standard deviation for scaling. For column (or variable) 30, $\bar{x}_{30} = 7.53 \times 10^{-4}$ and $s_{30} = 4.93 \times 10^{-4}$. Hence, the standardised value is -0.953. These data are used for PLS.

- In this text, we recommend weighted mean centring for PLSDA. For variable 30, $\bar{x}_{A30} = 3.54 \times 10^{-4}$ and $\bar{x}_{B30} = 1.05 \times 10^{-3}$ so the weighted mean $\bar{x}_{30} = 7.00 \times 10^{-4}$. Hence, the standardised and mean weighted centred value $x_{17,30} = -0.844$. We will perform PLSDA on this dataset. For readers without this capacity, there will be a small shift in the horizontal axis if using overall rather than weighted centring. For brevity, we do not illustrate results both using overall and weighted mean centring, and stick to the latter but the principles are the same.

- In Figure 7.19, we illustrate the scores of the first two components using PCA and PLSDA (as described in Section 7.4.2.2). The Mahalanobis distance of 2 from the centre of each group is illustrated as an indication of the size and orientation of each group: equally a different Mahalanobis distance or a p value could have been chosen. The overlap is shaded using this criterion.

- It can be seen that the overlap is slightly less for PLS than PCA as expected. More importantly, there is a better separation along the first component when using PLS.

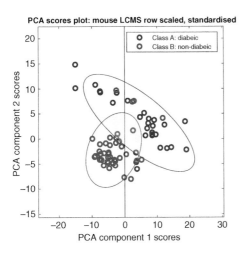 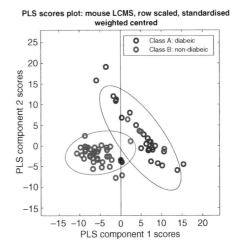

FIGURE 7.19 PCA and PLSDA scores plot of the mouse positive ion LCMS data of case study 10; Mahalanobis distance of 2 for each group is drawn, as indicative, with overlap shaded.

- We now calculate the orthogonal components and reduce the original data by subtracting varying numbers of orthogonal components. The orthogonal components are representative of variation that is unrelated to c and so not directly interesting to the problem in hand.
- If you are checking your calculations, we present the scores of the first component of sample 17 in various ways as follows:
  - For PCA $t_{17,1} = 8.519$.
  - For PLS $t_{17,1} = 6.997$.
  - The first orthogonal component $t_{ortho17,1} = 3.416$.
  - The adjusted components after 1 orthogonal component has been removed is $t_{adjust17,1} = 5.134$; after 2 have been removed $t_{adjust17,2} = 5.891$; after 3 have been removed $t_{adjust17,3} = 4.676$.
- The graphs of the scores of the first orthogonal component against the first adjusted component after a differing number of orthogonal components are removed, are presented in Figure 7.20.

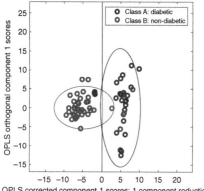

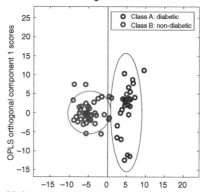

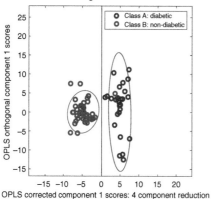

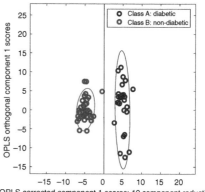

FIGURE 7.20   OPLS performed on the data of Figure 7.19 as described in the text.

- When one component has been removed the amount of overlap between the two classes reduces. This continues as more components are removed. Perfect separation is achieved when 10 components have been removed. This procedure forces a solution that appears eventually to be perfect, even though there are some misdiagnosed samples in the original dataset, and should be viewed primarily as a way of presenting data to look better.

When the number of variables is less than the number of samples, of course, it is not always possible to reach a situation with perfect separation.

The first component obtained on the reduced data provides a better graphical separation, and can also be used for loadings as well. Figure 7.21 illustrates the scores and

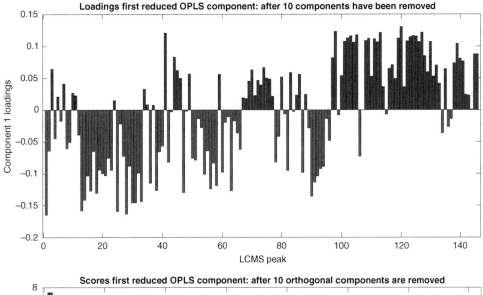

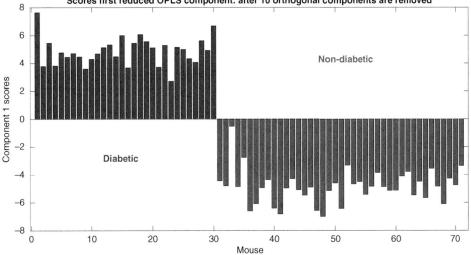

FIGURE 7.21   Loadings and scores of the full LCMS of the first component of the mice of case study 10 after OPLS and 10 orthogonal components have been removed.

loadings for OPLS component 1 after 10 orthogonal components have been removed to give a perfect separation. Positive loadings are coloured in blue and relate to class A (diabetic), and negative loadings are in red (non-diabetic).

This gives a clue as to which variables or metabolites are most characteristic of each group; however, it is important not to over-interpret such graphs. They force a solution, and this may not necessarily provide a true picture. If there are more variables than samples it is always possible to find a way to force a perfect separation, whether this makes sense or not. In Section 7.7.2, we will discuss validation. However, one advantage is that it can suggest markers for each group, and provide first clues as to which metabolites are most significant.

It is important to understand that OPLS should not be primarily employed as an approach for improving the prediction of class membership or the provenance of a sample, but mainly as a way of presenting information to emphasise the distinction between predefined groups. The loadings though do give clues as to which metabolites may be best for distinguishing the groups, although should be complementary to other approaches as described in Chapters 9 and 10.

Finally, although many readers may have been familiar with OPLS because it is part of common and widespread software, a method called Target Projection (TP) has also been proposed with significant similarities to OPLS. There is not space in this text to compare both approaches in depth in this chapter; however, TP is used for calculation of selectivity ratios as described in Section 9.4, but some readers will come across the latter method, which was reported independently in the literature several years before OPLS.

7.5   ONE-CLASS CLASSIFIERS

A one-class classifier involves creating a boundary around each predefined group of samples. Figure 7.22 illustrates two one-class classifiers. Instead of creating a boundary between two or more groups, a one-class classifier encloses each group separately.

One-class classifiers can be more flexible than two-class classifiers, as they allow other additional types of decision. In Figure 7.23, we illustrate some different situations.

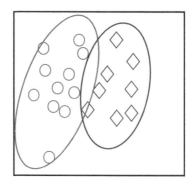

FIGURE 7.22   Two one-class classifiers.

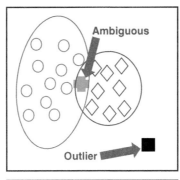

FIGURE 7.23   Different situations for one-class classifiers.

- Outliers can be identified as members of no predefined groups.
- Ambiguous samples can be identified as having characteristics of more than one class (either due to analytical ambiguity or genuine physical characteristics).
- There can be more than two classes.

Two-class models try to assign samples into one of two predefined classes, whereas one-class classifiers are not so constrained. In this section, we will illustrate one-class classifiers when there are only two groups, but the data can be extended to any number of groups each with their own decision criteria.

7.5.1   Quadratic Discriminant Analysis

QDA (Quadratic Discriminant Analysis) is closely related to LDA, the main difference is in how the variance–covariance matrix of each class is calculated.

We will start with the 40×10 simulated data of Table 7.1 excluding sample B21 and variables 11 and 12.

- In Section 5.3, we discussed how to calculate the Mahalanobis distance from the centroid of a group, but to a single class.
- For QDA we calculate the Mahalanobis distance to the centre of each class but using a variance–covariance matrix based only on all the samples of each class independently, rather than LDA which uses an overall variance–covariance matrix. In our example, there are 20 samples for both classes A and B, which

are used to obtain two variance–covariance matrices S_A and S_B. For readers that wish to follow the calculations in this chapter, S_A is a 10×10 matrix, and using the population standard deviation, the element of the third row and fourth column $s_{A3,4} = 61.9$; and for class B, using S_B the corresponding value is $s_{B3,4} = 147.4$.

- We compute the Mahalanobis distance from the centre of each class

$$d_{Ai} = \sqrt{\left(x_i - \bar{x}_A\right) S_A^{-1} \left(x_i - \bar{x}_A\right)'}$$

  - where S_A^{-1} is the 10×10 inverse of the variance–covariance matrix for the 20 samples of class A in this case, so that $s_{A7,3}^{-1} = -0.031$ for readers following this text numerically.
  - and $\bar{x}_A$ is the mean of the 20 samples that belong to class A in the form of a 1×10 row vector (using notation in this book).

with similar calculations for class B.

- The distance to the centre of class B against class A for all 40 samples is presented using all 10 variables (or non-zero PC scores) in Figure 7.24.
- This looks quite different from the corresponding plot for LDA in Figure 7.7(a). In fact the spread of Mahalanobis distances for samples within each class is very similar to LDA but the spread between classes is quite different.
- As a check, the distance of sample A13 to the centroids of each class is $d_{A,A13} = 3.31$ and $d_{B,A13} = 6.05$, assigning it to class A. Note that compared to LDA, the distance to the centroid of class A is quite similar but to class B is very different.

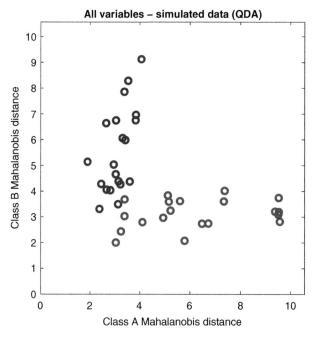

FIGURE 7.24 Graph of Mahalanobis distance for class B versus class A for 40×10 simulated case study of Table 7.1 and QDA using all variables.

- Note as an additional check, if you use the population standard deviation for the standardisation,

$$\sum_{1}^{20} d_{iA}^2 = I_A \times J = 20 \times 10 = 200,$$ and the same for class B (if using the sample standard deviation it would equal 190). This property is not so for the equivalent LDA models because the variance–covariance matrix for LDA is calculated using all 40 samples, whereas the mean is based on 20 samples for each class.

- Note that the Mahalanobis distance using all 10 variables is identical to that obtained using all the scores of all 10 PCs and centred data. For this comparison, PCA must be done on the entire dataset rather than each class individually (which we will discuss as an alternative approach of SIMCA (Soft Independent Modelling of Class Analogy) in Section 7.5.2).

Unlike a two-class model, we have to define the boundaries for each class separately.

- It is usual to define boundaries using either a limiting Mahalanobis distance or a p value.
- To illustrate we will compute a critical p value using χ^2 (as discussed in Chapter 5). It is possible to use Hotelling's T^2 instead but as previously discussed (Section 5.2.5) when the ratio of samples to variables is low the p values can be misleadingly high. The sample rather than population standard deviation and covariance could also be used in the definition of the variance–covariance matrix; however, differences are very small and unlikely to be experimentally meaningful.
- If we set $p = 0.05$, then the critical value of χ^2 for 10 degrees of freedom is 18.31, so the limiting Mahalanobis distance is $\sqrt{18.31} = 4.28$.
- In Figure 7.25(a) we mark this limit. We can see:
  - All samples are correctly assigned to their classes.
  - but 11 samples are ambiguous by this criterion, that is they fall within both classes.
- The predicted group memberships are presented in Table 7.8, with ambiguous samples highlighted in bold. To check numerically, $d_{A,A12} = 2.38$ and $d_{B,A12} = 3.30$, so although closer to the centroid of class A, both are less than the critical value 4.28 and this sample is deemed to be ambiguous.
- QDA can also be implemented as a two-class classifier, as illustrated in Figure 7.25(b). In this case, we assign each sample unambiguously to one of the two predefined classes but identify one possible misclassified sample, which is B12. This can be compared to Figure 7.7(a). This sample is identified as ambiguous in our one-class QDA model. It is also misclassified in the LDA model using all variables and full component PLSDA models.

In this text, we recommend using QDA as a one-class modelling technique and LDA for two-class models, although there is no fundamental issue against employing QDA as a two-class technique instead.

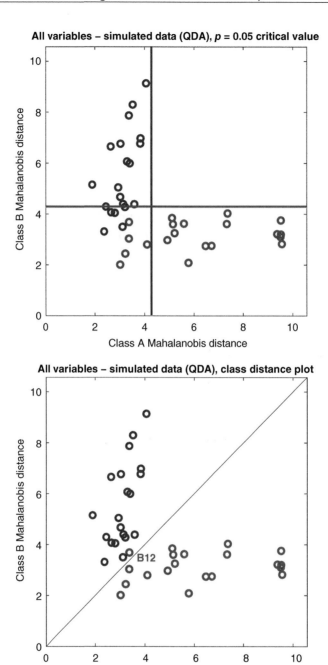

FIGURE 7.25   Graph of Mahalanobis distance for class B versus class A for 40×10 simulated case study of Table 7.1 and QDA using all variables with (a) top critical distance of $p = 0.05$ marked and (b) bottom the equidistant line marked representing QDA used as a two-class classifier (not described in detail in this text).

The method can be better understood by reducing this to a two-variable problem by calculating the first two PCs of the centred data and performing QDA on just these two PCs.

- After centring the scores can be calculated to give a 40×2 matrix T if retaining just the first two PCs. If you are checking your calculations, you should find $t_{B6,2} = -51.41$.

- The variance–covariance matrix for class A is $S_A = \begin{bmatrix} 2996.5 & 890.5 \\ 890.5 & 565.1 \end{bmatrix}$ and for class B is $S_B = \begin{bmatrix} 2434.7 & -658.7 \\ -658.7 & 528.4 \end{bmatrix}$ and the means are $\bar{t}_A = -\bar{t}_B = \begin{bmatrix} 3.950 & -29.331 \end{bmatrix}$ (note that because the data are centred and the class sizes are equal the means are the negative of each other).

- Calculate the distances to the centroids of each class and check the distance of sample A13 to the centroids of each class is $d_{A,A13} = 1.032$ and $d_{B,A13} = 4.22$. Note that these are quite different from the distances calculated by LDA.

- Using $p = 0.05$ and the χ^2 distribution with 2 degrees of freedom, the critical distance is $d = \sqrt{5.99} = 2.45$; hence, any samples with a distance greater than this are not consider members of the parent class.

- The plot of Mahalanobis distance from the centre of class B versus class A is presented in Figure 7.26(a).

- We see that only sample B3 is outside the $p = 0.05$ critical limit for its own class. This is a little lower proportion expected than (we would anticipate 2 out of 40 samples if the data followed a perfect normal distribution) but not unreasonable.

- More significantly though, 9 samples for class A and 6 from class B are now ambiguous. This does not mean there is a problem with our classification model but that the classes overlap using 2 PCs. The unusually assigned samples are listed in Table 7.8. This does not imply an error in our model, just that we now have more choices.

- We can gain additional insight from Figure 2.26(b) which shows the $p = 0.05$ contour (at a Mahalanobis distance of 2.45). We can see that the single sample at $p < 0.05$ distance is only just on the border and that there is a great deal of overlap between the classes, resulting in 15 samples that appear ambiguous.

- The proportion of ambiguous samples could suggest that 2 PCs are not enough to distinguish the two classes and that more PCs should be included in the model.

Comparing Figure 7.25(a) and Figure 7.26(a) it can be noted that there are no samples close to the centre (Mahalanobis distance of 0) for the 10 PC model, whereas for the 2 PC model samples are distributed close to the centre (for the modelled class). This can be understood by the probability density function in Figure 7.27. When there are 10 degrees of freedom, the mode of $\chi^2(\nu)$ is at $\nu - 2$ (see Section 5.2.4), of 8, so the square root is 2.83, meaning that most samples are found at around this distance from the centre. When there are only 2 degrees of freedom, the mode is at the centre. The pdfs correspond well to the distribution of Mahalanobis distances both when we use 2 and 10 PCs.

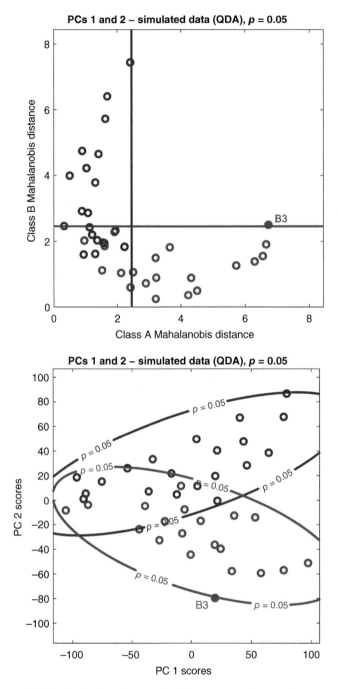

FIGURE 7.26   Graph of Mahalanobis distance for class B versus class A for 40×10 simulated case study of Table 7.1 and QDA using all variables after PCA with (a) top critical distance of $p = 0.05$ marked and (b) bottom represented as a contour.

TABLE 7.8 QDA predictions using $p = 0.05$ criterion for simulated case study and various models.

	All variables	PCs 1 and 2
Sample	Prediction $p = 0.05$	Prediction $p = 0.05$
A1	A	AB
A2	AB	AB
A3	AB	AB
A4	A	AB
A5	A	AB
A6	A	A
A7	A	AB
A8	A	A
A9	A	A
A10	AB	A
A11	AB	AB
A12	AB	AB
A13	A	A
A14	AB	A
A15	A	A
A16	A	A
A17	A	A
A18	A	A
A19	A	A
A20	A	AB
B1	B	B
B2	B	B
B3	B	None
B4	B	AB
B5	AB	AB
B6	B	B
B7	B	B
B8	B	B
B9	B	B
B10	B	AB
B11	B	B
B12	AB	AB

(*continued*)

TABLE 7.8 *(Continued)*

	All variables	PCs 1 and 2
Sample	Prediction p = 0.05	Prediction p = 0.05
B13	AB	B
B14	B	B
B15	B	B
B16	AB	AB
B17	AB	AB
B18	B	B
B19	B	B
B20	B	B

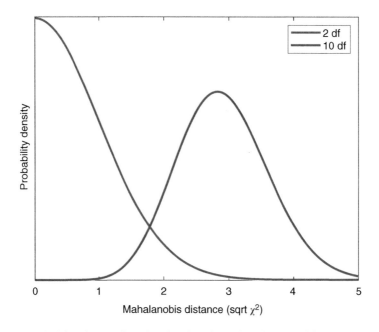

FIGURE 7.27 Probability density function for χ^2 and 2 and 10 degrees of freedom versus the square root of χ^2.

It is important to remember that p values do represent just an approximation and most real data will not be modelled perfectly using a normal distribution. There are many statistical tests for normality, if the data approximately obey a normal distribution, then the results provide reasonable guidance for further investigation. A very simple approach is to calculate the p values for each of the samples in each class separately. Their p values should be linearly related to rank, apart from any obvious outliers. For example, if there are 20 samples in a class, we expect 10 to have a p value < 0.5 and 5 to have a p value < 0.25 and so on. For brevity, we will not illustrate this calculation in full, but recommend this approach as a simple graphical guide, to check that the model meaningfully approximates the desired probability distribution. For a 2-component model and the simulated dataset, we illustrate the corresponding rank plot in Figure 7.28 as an example. Sometimes increasing the number of PCs in the model results in better linear relationship between rank and p value. As discussed in Chapter 5, p values should not be taken literally and used more as guidance and a common scale to compare different methods and indices.

- If we feel there are too many ambiguous samples, we could either change the number of PCs in the model or change the critical p value.
- If we use a critical p value of 0.2, the value of the Mahalanobis distance using χ^2 cut-off criterion is 1.79 for 2 degrees of freedom.
- The corresponding graphs are in Figure 7.29; the reader can compare the results to those obtained using $p = 0.05$, and find that there are less ambiguous samples, more misclassified samples and one outlier. Although it is common to use a low p value, which is more appropriate depends on the aim of the analysis, and is not a hard rule.

To illustrate this further we will discuss a real two-class model, using as an example the 71×32 dataset of negative LCMS of the diabetic and non-diabetic mice of case study 10. As described elsewhere, the data are first row-scaled and then standardised. For brevity, we will illustrate mainly the resultant graphics rather than provide full numeric calculations.

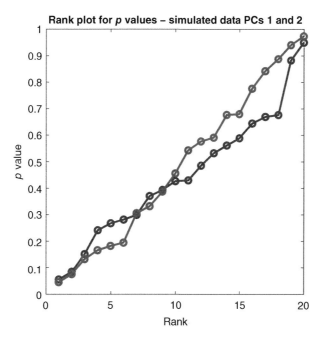

FIGURE 7.28   Example of rank plot as discussed in the text.

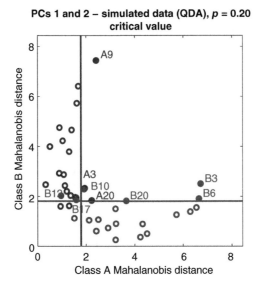

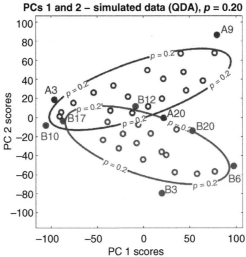

FIGURE 7.29 As Figure 7.26 but using a p value of 0.2 as discussed in the text, with samples outside the boundaries marked.

- The 2 PC model is presented in Figure 7.30, both via a Mahalanobis distance and a contour plot.
- The $p = 0.05$ limits are used to assign samples into their classes.
- It can be seen that
  - Mouse 7 has $p < 0.05$ in Class A (diabetic), representing 1/30 or 0.033 of Class A samples.
  - For Class B, mice 33, 45, 53 and 66 have $p < 0.05$ representing 4/41 or 0.097 or about twice the expected number. However, these four samples are likely to have been misdiagnosed, so could be viewed as outliers that are not members of the population of non-diabetic mice.

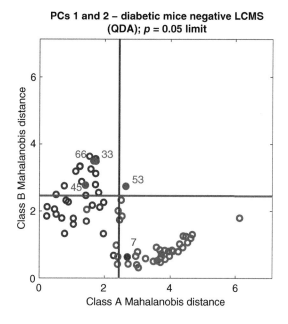

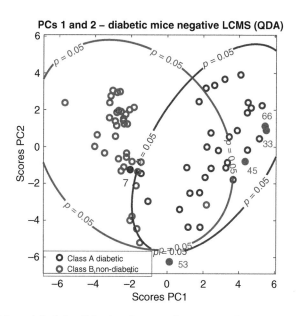

FIGURE 7.30   QDA model of the diabetic mice negative LCMS using 2 PCs.

- There are 16 ambiguous mouse samples from class A (diabetic) and 5 for class B (non-diabetic).
- Hence, over 20 of the diabetic mice appear ambiguous or misclassified, and 6 of the non-diabetic mice.
- This is best shown in the contour plot, where the non-diabetic mice (red class B) appear to be distributed over a very large area.

- The reason for this large area is that mice numbers 33, 45 and 66 appear to have been misdiagnosed as non-diabetic, and distort the shape of class B, expanding its size.
- A solution is to remove these 3 mice, resulting in a new 68×32 matrix, consisting of 30 diabetic mice and now 38 non-diabetic mice (we could have changed the diagnosis of these samples but as they are rather suspect the best is to remove them).
- To correctly analyse this new dataset, after row scaling, standardise each column again and then perform PCA.
- The new figures are presented in Figure 7.31.
- We can see that the areas representing each group are now more compact. There are 3 out of 68 (=4.4%) of the mice outside $p = 0.05$ limits, which is excellent as 5% of 68 is 3.4, given that the model is only approximate.
- There are also very few ambiguous samples.

Different numbers of ambiguous and misclassified samples can be predicted by changing the number of PCs in the model or the p value of the decision threshold, remembering these are only predictions.

If desired this preliminary analysis can be used to sift the samples, and remove any ambiguous samples, retaining only those that are unambiguously in each group, and then use these to determine which are the most significant metabolites or potential markers. We leave this often important stage to the experimentalists.

One advantage of one-class classifiers is that they can be used for any number of groups. We will exemplify using case study 5, the NMR of maize, obtained at four different temperatures during the growth cycle.

- For this illustration, the data are log scaled (to the base 10) as described in Section 4.6.4 of Chapter 4 and then PCA is performed on the centred log scaled dataset. No standardisation is performed.
- The plots of the scores of the first 2 PCs together with the $p = 0.05$ QDA boundaries (using the Mahalanobis distance calculated using the population standard deviation and χ^2-distribution for the critical value) for all four temperatures are presented in Figure 7.32.
- For readers checking their calculations $x_{100,17} = 1.93$ after log scaling representing sample 100 (harvested at 16 °C) and the 17th variable (isoleucine).
- The scores of the first two PCs for this sample are $t_{100} = [0.739 \ -0.120]$.
- From the QDA plot is can be seen that the highest temperature group is quite distinct, most of the 16 °C samples can be distinguished from those harvested at lower temperatures, and in the first two PCs the two lowest temperature groups are almost indistinguishable from each other.
- In Figure 7.33, we illustrate the class distance plots between each pair of classes (class A = 20 °C, class B = 16 °C, class C = 13 °C and class D = 8.5 °C) using a 2 PC model. Several conclusions can be drawn, for brevity, we do not discuss all these in detail.

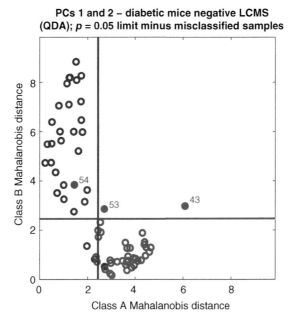

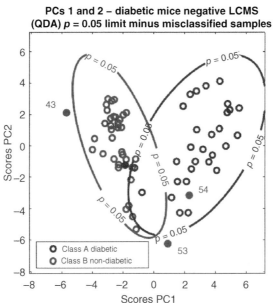

FIGURE 7.31   QDA model of the diabetic mice negative LCMS using 2 PCs, minus 3 misclassified samples.

- To check, for sample 21 the distances to the four classes are $\boldsymbol{d} = [0.83\ 4.76\ 4.17\ 4.08]$ making it well within the $p = 0.05\ \chi^2$ critical limit of 2.45 for class A (20 °C) but outside the critical limit for the other three classes. Although of course a normal distribution is only an approximation and the statistical conditions are unlikely to be exactly obeyed, it still provides a good preliminary picture.

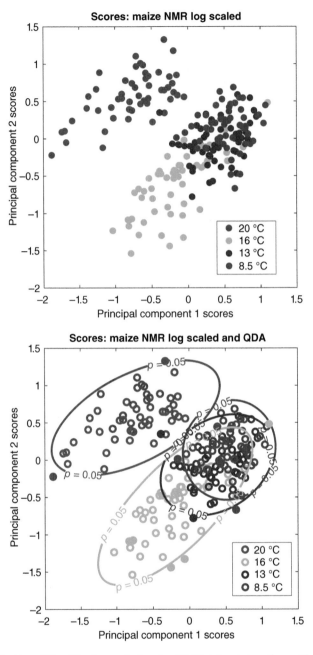

FIGURE 7.32   First two PCs of the log-scaled maize NMR dataset, together with the $p = 0.05$ QDA boundaries.

- The pattern using only two PCs misses a great deal of information and it is usually important to include a few more PCs in the model. A 10 PC model is illustrated in Figure 7.34, showing excellent separation.
- However, as discussed in Section 7.7.2, this can lead to over-fitting. The temptation to force complete separation between groups, which can be done by many methods and can lead to overinterpretation of differences.

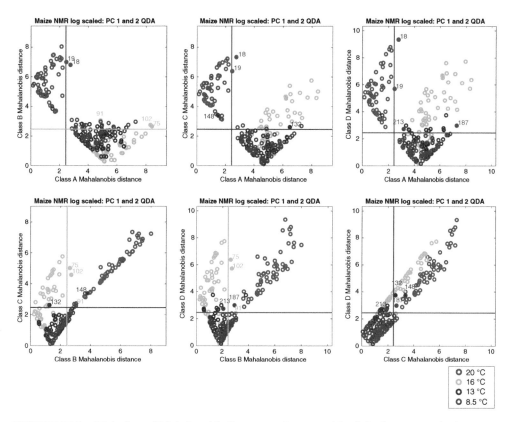

FIGURE 7.33 QDA plots of Mahalanobis distance to the centroids of the four groups in the case study of Figure 7.32; misclassified samples from the modelled groups are marked.

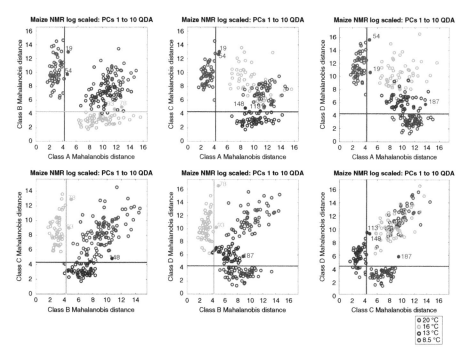

FIGURE 7.34 As Figure 7.33 but using a 10 PC model; misclassified samples from the modelled groups are marked.

- This exploratory approach, though, can identify outliers and also can help in determining which samples are most likely to be characteristic of different groups.

In some cases, it is unrealistic to force full separation between groups, but classification techniques can still be used to select which samples are most appropriate for further investigation. In Figure 7.35 we show the QDA boundaries for the scores of the first 2 PCs of the NMR diabetes study (case study 4). Of course, the separation could be improved by variable selection, using more PCs, etc. However, for brevity, we notice that there are several samples that are unique to the extreme classes of Normal Glucose Tolerance and Type 2 Diabetes. We could select these extreme samples, which are large enough groups, and look at the metabolic differences just in these samples. In this case PR methods can be employed not to determine the provenance of any individual sample or directly to determine marker compounds, but to select samples for further investigation.

7.5.2   SIMCA

SIMCA (Soft Independent Modelling of Class Analogy) is one of the first methods to be explicitly developed for the chemometrics literature, and can be considered as a one-class classifier, related to QDA.

There is a large literature on SIMCA developed over nearly 50 years, and it would be impossible to summarise all aspects comprehensively in this text.

7.5.2.1   Disjoint PCA

The main distinction between SIMCA and most other classification algorithms is disjoint PCA. Traditionally we perform PCA on all samples, usually representing several

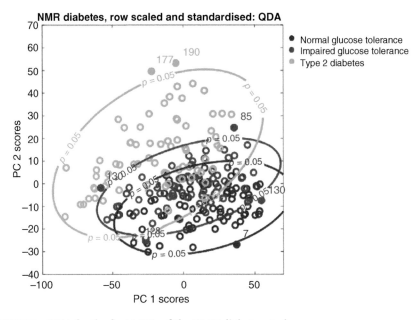

FIGURE 7.35   QDA for the first 2 PCs of the NMR diabetes study.

groups, at the same time. Then we can perform PR individually on each class separately, using the PC scores. However, a conceptual weakness is that classes are not treated independently. So, for example, if we were provided with samples from two classes, for example, two species, and used PCA and then QDA to develop a classification model, and then we were provided new data on a third species, which we had not considered originally if we had performed PCA on both of the original species together, we would have to reform the model again to introduce the third group. If however PCA is performed separately on each class, there would be no need to change the original models, if a new group is introduced at a later stage.

Performing PCA on all data together is called conjoint PCA. Performing it separately on each class is called disjoint PCA. The main differences are illustrated symbolically in Figure 7.36, where disjoint PCA is performed on the blue group. The axis of variation for the first PC for both conjoint and disjoint models is shown. In some occasions, as illustrated, a disadvantage of disjoint PCA is that in some cases the variation between groups is lost or reduced, as in the first PC of Figure 7.36.

In contrast, in most cases, the distance between the space defined by the disjoint PCs is much greater for samples not part of group being modelled than the group being modelled. Figure 7.37 illustrates this symbolically, the samples of red group are very far from the first PC of the blue group. These distances are called prediction errors, and are a second indicator that is usually necessary when defining SIMCA models. The red group samples are all far away from the first disjoint blue PC, and so would be predicted badly.

Hence, SIMCA measures two types of variation or error, the first is projection within the space of the disjoint model and the second distance from this model. We will describe this in more detail below.

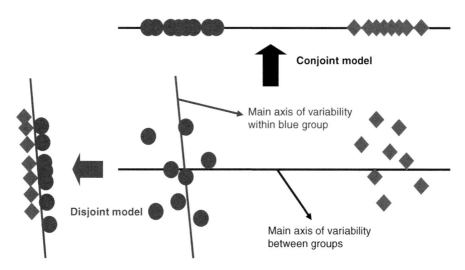

FIGURE 7.36   Difference between conjoint and disjoint PCA, the blue and red group representing two classes. The first PC is shown; for the conjoint model the variability between both groups is retained, but lost for the disjoint model of the blue group.

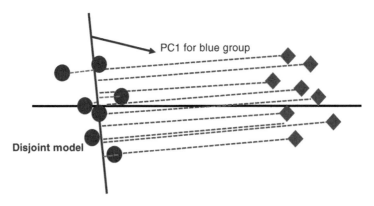

FIGURE 7.37   Illustration of prediction error for a 1 PC disjoint model.

We will illustrate the calculations of disjoint PCA for case study 10 (negative ion LCMS), involving a 71×34 matrix of which the samples 1–30 are from class A (diabetic mice) and class B (non-diabetic). Note that there are some misdiagnosed non-diabetic mice, but for the purpose of illustrating our methods, we will not remove these, and leave this as an exercise for readers that are interested.

- We will illustrate numerically how to form a class model for the 30 samples of class A.
- The first step is to row scale all the data to a total of 1, so if you are checking the data, $x_{7,12} = 0.0182$.
- The next step is to standardise just the samples belonging to class A by first calculating the column means and standard deviations. For variable 12, we find $\bar{x}_{12\{A\}} = 0.0150$ and $s_{12\{A\}} = 0.00364$ (using the population sd). These differ from the overall values of $\bar{x}_{12} = 0.0179$ and $s_{12} = 0.00549$.
- We then standardise the samples of class A so that ${}^{std}x_{ij\{A\}} = \left(x_{ij} - \bar{x}_{j\{A\}} \right) / s_{j\{A\}}$ or in the case of sample 7, variable 12, ${}^{std}x_{7,12\{A\}} = (0.0182 - 0.0150)/0.00364 = 0.9004$. Note that it is different from the value obtained for the overall standardised data, for which ${}^{std}x_{7,12} = 0.0578$.
- We also have to transform the data from class B using the mean and standard deviation of class A, so that, for example, for sample 40, ${}^{std}x_{40,12\{A\}} = 0.1601$; although we will not use these for creating the model we still may want to predict their provenance.
- The next step is to perform PCA on the new standardised class A, forming the model $X_A = T_A P_A + E_A$ as usual. To check we find $t_{7,2\{A\}} = 1.350$.
- For samples not members of class A, we can predict the scores, by $\hat{T}_{B\{A\}} = X_B P'_A$ where $X_{B\{A\}}$ represent the 41 samples in class B (non-diabetic controls) using class A model. and the '^' or hat symbol means predicted.
- Hence, we find $\hat{t}_{40,2\{A\}} = 1.824$. This allows us to see how well samples not part of the original class A fit into the model.
- If done correctly we should find the estimated values for the scores for class A equal the original computed values, that is, $\hat{T}_{A\{A\}} = T_A$.

The same can be done for a class B model. In this case, class B is standardised first and then class A transformed using the mean and standard deviation of class B. To check $t_{40,2\{B\}} = -0.843$ and $t_{7,2\{B\}} = 0.0353$.

The disjoint PC plots of the scores of the first two PCs for each class together with the conjoint plot (standardising the entire dataset) are illustrated in Figure 7.38.

- Ellipses are drawn at a Mahalanobis distance of 2.45 from the centroids. This corresponds to a p value of 0.05 using the χ^2-distribution with 2 degrees of freedom. Of course, the data are not perfectly modelled by a normal distribution but the limits are indicative and provide guidance.
- For the disjoint class A model scores, class B samples have been projected into the class A area.

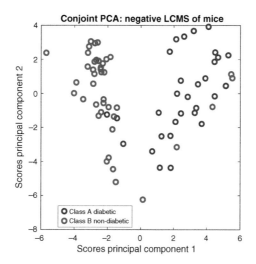

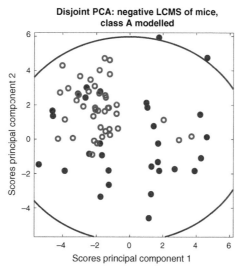

FIGURE 7.38 Conjoint and disjoint PC scores plots of the negative LCMS of the mice, standardising the modelled class with the modelled class represented by closed circles; cut-off limits at a Mahalanobis distance of 2.45 from the centroids are indicated.

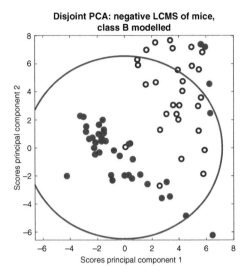

FIGURE 7.38 (Continued)

- For class A, 2 out of 30 samples are outside the $d = 2.45$ boundary, suggesting that although the samples are not closely approximated by a normal distribution, this model is still reasonable.
- For the disjoint class B model, we find that several of the class A (diabetic) samples are outside the region of class B. This is probably because the misdiagnosed samples influence the model, and of course, an alternative is to reform the model again without these outlier samples.
- As a further illustration, 3 misdiagnosed samples are removed and the class B model is reformulated and shown in Figure 7.39. The distribution of class B samples over the region defined is much more even, suggesting a more suitable model. However, for conciseness, we will only follow the calculations using the full 71 sample dataset below.

7.5.2.2   D- and Q-statistics

For conjoint one-class models such as QDA as described above, we only needed to calculate the Mahalanobis distance from the centre of the class model.

For SIMCA we need to calculate two statistics:

- The first is the D-statistic which is also called the Score Distance or Hotelling statistic.
- It is the squared Mahalanobis distance of a sample from the centre of the distribution in the disjoint PC space, defined by $D_{A,i} = \left(\hat{t}_{i\{A\}} - \overline{t}_{\{A\}}\right) S_{t\{A\}}^{-1} \left(\hat{t}_{i\{A\}} - \overline{t}_{\{A\}}\right)'$

where $S_{t\{A\}}$ is the variance–covariance matrix of the scores of the disjoint PC

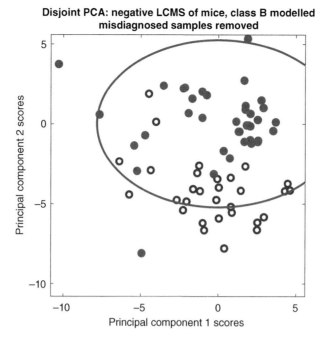

FIGURE 7.39 Removing 3 misdiagnosed samples and reforming disjoint model; cut-off limits at a Mahalanobis distance of 2.45 from the centroids are indicated.

model for class A, $\hat{t}_{i\{A\}}$ is the estimated scores (as a row vector) for sample i using a class A disjoint model (for samples of class A it is the same as the scores calculated using the model), and $\bar{t}_{\{A\}}$ the mean of the scores for class A using the disjoint model.

- To follow this calculation for a 2 component class A model and sample 20

  - $t_{20\{A\}} = \hat{t}_{20\{A\}} = \begin{bmatrix} -1.240 & -2.663 \end{bmatrix}$ and $S_{tA} = \begin{bmatrix} 8.459 & 0 \\ 0 & 5.926 \end{bmatrix}$ which is diagonal because the scores are orthogonal

  - $\bar{t}_{\{A\}} = \begin{bmatrix} 0 & 0 \end{bmatrix}$ as class A has been centred

  - so if we consider sample 20 $t_{20\{A\}} = \hat{t}_{20\{A\}} = \begin{bmatrix} -1.240 & -2.663 \end{bmatrix}$

  - so $D_{A,20} = \begin{bmatrix} -1.240 & -2.663 \end{bmatrix} \begin{bmatrix} 8.459 & 0 \\ 0 & 5.926 \end{bmatrix}^{-1} \begin{bmatrix} -1.240 \\ -2.663 \end{bmatrix} = 1.378$

- Note that it is traditional to quote the square of the Mahalanobis distance.
- For samples not part of the original class A training set, we use the loadings and mean of class A to estimate the scores using a class A model and calculate the relevant distance.
- The second is the Q-statistic, which is also called the SPE (Square Prediction Error) or orthogonal distance.

- This is defined by the error (or residual) between the predicted values of X using the disjoint PC model and the observed values; note that the squared value is usually calculated.
- This can be calculated by

$$e_{A,i} = \left(x_i - \hat{x}_{i\{A\}}\right)\left(x_i - \hat{x}_{i\{A\}}\right)' = \sum_{j=1}^{J}\left(x_{ij} - \hat{x}_{ij\{A\}}\right)^2 = \sum_{j=1}^{J} e_{A,ij}^2$$

- where $\hat{X}_A = T_A P_A$ or the predicted values of X (suitably pre-processed) using the PC model.
- For a 2 PC class A disjoint model we know for sample 20

  $t_{20\{A\}} = \hat{t}_{20\{A\}} = \begin{bmatrix} -1.240 & -2.663 \end{bmatrix}$ as described above.
- If you follow the calculations, check that the 32×32 matrix P_A is orthonormal. The value $p_{5,13} = -0.0897$.
- The estimate for sample 20, using the disjoint class A model is

  $\hat{x}_{20\{A\}} = \begin{bmatrix} -0.775 & 0.512 & 0.572 & ...... \end{bmatrix}$ (only the first 3 values are presented)
- and the pre-processed value is $x_{20} = \begin{bmatrix} 0.027 & -0.468 & -0.228 & ...... \end{bmatrix}$
- so $e_{A,20} = \begin{bmatrix} 0.772 & -0.979 & -0.800 & ...... \end{bmatrix}$
- and so $Q_{A,20} = \sum_{j=1}^{32} e_{A,20,j}^2 = 18.765$ where there are $J = 32$ variables on this occasion.

As we will see these two types of error or distance can be used to see how well a sample fits into a disjoint class model. We will present the results of the calculation in the next section.

7.5.2.3   Limits and Decisions

There is a great deal of discussion in the literature as to how to decide whether a sample is a member of a predefined group, based on the values of the Q- and D-statistics. We will not have room to describe all possibilities and just indicate one approach. To outline every proposed criterion in the literature would require a whole lengthy chapter or even a book in its own right. This section primarily outlines the main principles.

Usually, a cut-off value of D and Q can be defined. It can of course just be defined numerically or graphically, but also a statistical value is often used. SIMCA has the advantages that it is a statistically based method for classification.

- For the D-statistic, this can either by a limiting value of χ^2 or T^2. If we use a p value of 0.05, we find for two degrees of freedom this is 5.99 and for 10 degrees of freedom this is 18.31 using χ^2. We recommend χ^2 rather than T^2 as discussed in Section 5.3, unless the number of samples is substantially more than the number of variables.
- For the Q-statistic, this is a bit more complex as it is not so easy to define the number of degrees of freedom. One approach (of many) is as follows.

- Scale the Q-statistic by dividing by the mean of the Q values as follows. $^{new}Q_{A,i} = Q_{A,i} / \left[mean\left(\sqrt{Q} \right) \right]^2$, this is so the Q values are on a comparable scale for different calculations.
- Then calculate the limiting value of χ^2 for $J - A$ degrees of freedom where J is the number of variables and A *is* the number of PCs. For 32 variables and 2 PCs, this equals 30 degrees of freedom, and for $p = 0.05$ this is 43.773.
- Divide this by $J - A$ to give a limit of 1.459 for a two PC model and 1.542 for a 10 PC model, in this case.
- Anything with a scaled Q value above this is considered to be too far out of the relevant reduced PC space for the model.
- For a two PC class A disjoint model, the mean value of Q_A for a two PC model as defined above is 4.030, so the scaled value for sample 20 is $18.765/4.030^2 = 1.155$, below the limit.

The values of the D-statistic and scaled Q-statistic, together with the limits as described above are presented both for a 2 PC and 10 PC model in Figure 7.40 and Figure 7.41. Note that these limits are only advisory and just approximate to a statistically based decision criterion. However, they can be used to determine class membership based on both statistical indicators.

- For the two PC models and D-statistic, it is worth comparing to Figure 7.38 with the boundaries corresponding to the cut-off limits.
- The number of samples outside the cut-off limits is the same both in the 2 PC D-statistic graphs of Figure 7.40 and the contour plots of Figure 7.38.
- It can be seen that the Q-statistic cut-off is reasonably sensible.
- The Q-statistic tends to be more discriminatory to the D-statistic.
- As more components are employed in the model, apparent discrimination is better, and for cases where the number of variables exceeds the number of samples, perfect discrimination could be achieved, but as we will see in Section 7.7.2 this could be a consequence of over-fitting and provide an over-optimistic picture of the data.
- In many situations, it is important to be able to estimate the most appropriate number of PCs to be included in the model, which could differ for each class. This will be discussed later.
- As both Q- and D-statistics can be used to estimate which samples are outside the decision limits, there are numerous rules as to how to combine predictions or whether one rule or the other is primary. The simplest approach is to reject samples using each group-specific model that are outside decision limits using either of the criteria. Of course, this can still lead to ambiguous samples or outliers, as each class is modelled separately, so samples that are rejected as outside a class using a class A model could also be outside a class using a group B model.

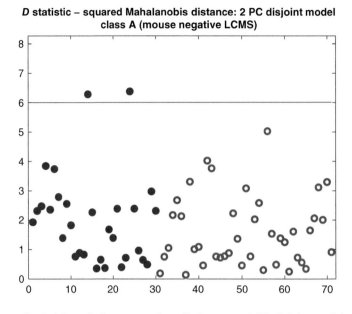

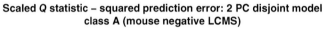

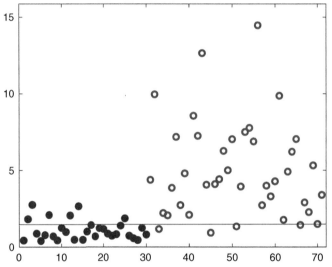

FIGURE 7.40   *D*- and *Q*-statistics plus critical limits for disjoint two PC models; the modelled group is indicated by closed circles.

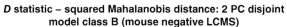

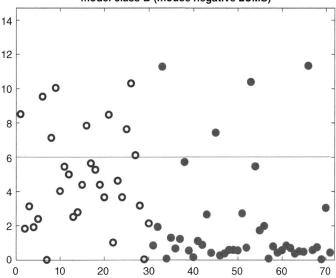

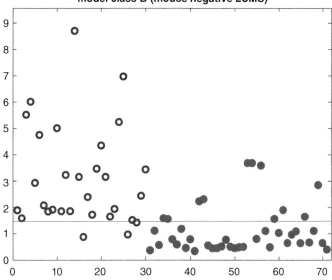

FIGURE 7.40 (Continued)

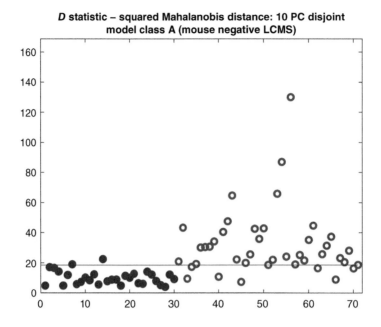

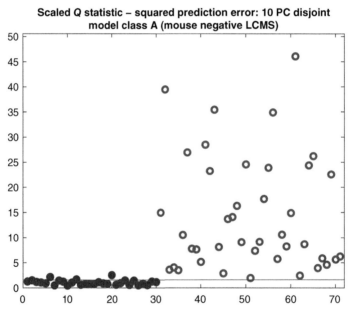

FIGURE 7.41   As Figure 7.40 but 10 PC model.

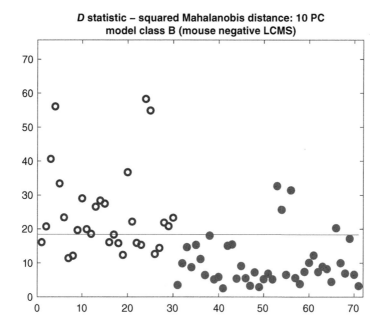

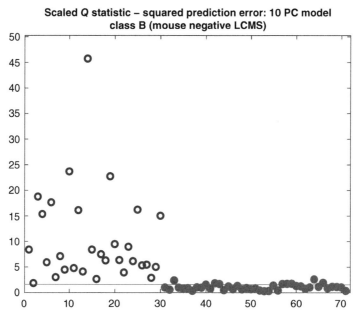

FIGURE 7.41 (Continued)

The main aim of this section has been to illustrate the steps in SIMCA modelling, which has spawned a vast literature that continually evolves, but involves the same main steps, namely disjoint PCA, QDA, calculating Q and D statistics and establishing a decision rule.

7.6   MULTICLASS CLASSIFIERS

In all the examples of Section 7.4, there have been only two classes. In many studies in metabolomics, there are more than two defined groups in the data. When using one class classifiers as of Section 7.5 such as QDA and SIMCA, we can increase the number of groups at any time, but for the classifiers of Section 7.4, we need extensions to cope with more than two classes.

The aim is to form a boundary between each class as illustrated in Figure 7.42. Samples are classified according to which side of the boundary they lie. We will look at the extensions of LDA and PLSDA.

A small simulated case study is introduced in Table 7.9, consisting of 30 samples, 10 of which originate from each of three classes A, B and C, and characterised by 2 variables.

7.6.1   LDA as a Multiclass Classifier

The extension here is quite simple. The Mahalanobis distance is calculated to each class centroid and samples assigned to the class whose centre it is closest to.

We calculate the Mahalanobis distances to the three class centroids in Table 7.9. We will not detail the numerical calculations in full, as the principle of LDA has already been illustrated numerically in Section 7.4.1. This is illustrated in Figure 7.43. The Mahalanobis distances to the centroids of each class are calculated and the four misclassified samples are marked. The figure and table should be compared. They are all very close to the class borders, and different models using different criteria may come to different conclusions.

We will illustrate with a real example, case study 5, using log10 scaled data, as discussed previously in Section 4.6.4.

- The scores of the first two PCs are presented in Figure 7.44 together with the boundaries as determined using LDA.

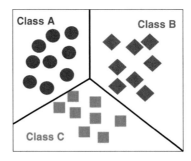

FIGURE 7.42   Multiclass classifiers.

TABLE 7.9 Three class two variable simulated case study, together with Mahalanobis distances to the centroids of each class, and misclassified samples marked.

Sample	x_1	x_2	d_A	d_B	d_C
A1	81	87	0.458	0.877	1.705
A2	94	86	0.342	1.206	1.458
A3	105	79	0.425	1.532	1.026
A4	77	60	0.806	1.312	1.055
A5	45	38	1.928	1.884	1.921
A6	73	79	0.414	0.699	1.588
A7	82	56	0.960	1.548	0.893
A8	109	106	1.237	1.784	2.153
A9	141	121	2.119	2.753	2.731
A10	79	71	0.371	1.011	1.223
B1	49	82	1.075	0.052	2.210
B2	40	69	1.267	0.539	2.141
B3	26	45	1.996	1.589	2.337
B4	81	103	1.131	1.181	2.326
B5	44	56	1.387	1.095	1.885
B6	77	92	0.716	0.838	1.961
B7	31	63	1.540	0.839	2.287
B8	45	103	1.699	0.960	2.946
B9	43	110	1.976	1.278	3.228
B10	36	90	1.547	0.518	2.728
C1	123	45	1.837	2.743	0.608
C2	134	70	1.299	2.404	0.644
C3	90	27	2.249	2.792	1.401
C4	84	50	1.222	1.802	0.859
C5	189	91	2.590	3.660	2.185
C6	102	55	1.132	1.985	0.378
C7	86	53	1.096	1.729	0.789
C8	118	60	1.200	2.210	0.102
C9	161	82	1.873	2.971	1.436
C10	78	44	1.477	1.938	1.080

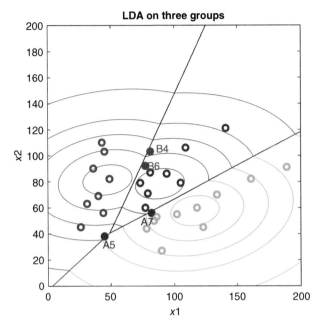

FIGURE 7.43   LDA on the three groups of Table 7.9, misclassified samples indicated by closed circles.

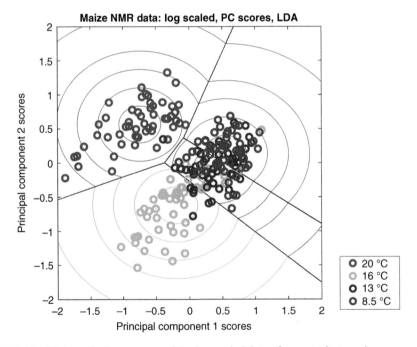

FIGURE 7.44   LDA on the four groups of the log-scaled data of case study 5, maize grown at four temperatures.

- It can be seen that the maize harvested at 20 °C is very distinct and well-classified, whereas the two lowest temperatures are hard to distinguish.
- The numbers and identities of the misclassified samples for the 2 PC model are presented in Table 7.10.
- There are 60 out of 216 misclassified. However only 1 sample is misclassified from group A (20 °C), 15 from group B (16 °C), 26 from group C (13 °C) and 18 from group D (8.5 °C).
- We can improve the model by using more PCs. The boundaries cannot be visualised graphically, but the number and nature of the misclassified samples can be obtained.
- In Table 7.10, we also present the results of LDA for a 5 PC and 10 PC model.
- For the 5 PC model there are now 25 misclassified samples of which 1 is from group A, 5 from group B, 7 from group C and 12 from group D.
- For the 10 PC model, there remain just 7 misclassified samples, of which 1 is from group B, 1 from group C and 5 from group D.
- This suggests that the samples harvested at lower temperatures are better distinguished using later PCs.
- Some samples are consistently misclassified using all models and should be examined as possible outliers.
- Although the LDA models appear very promising when 10 PCs are used, there is a possibility of over-fitting, that is, using too many PCs than can be justified. This will be discussed in greater detail later in Section 7.7.2 on validation. There will be some cases where we are certain that there can be distinction and so just need as good a separation as possible, but other cases, more commonly, where we cannot be sure the analytical or biological data is adequate to discriminate between different groups of samples.

This section has shown how LDA can be extended to any number of groups.

7.6.2 PLSDA as a Multiclass Classifier

There are numerous ways PLS can be used for multiclass models.

The most important issue for any use of PLSDA when modelling more than two classes, is to understand how PLS has been implemented, as different approaches can result in very different answers. For brevity, we will illustrate in detail just one approach, which we recommend, whilst mentioning two alternatives.

7.6.2.1 One Versus All

Probably the commonest, and most reliable, approach is one versus all. In this situation a PLSDA model is obtained between members of one class (the 'in group') and all other classes (the 'out group'). This is done for all classes. So if there are three classes, A, B and C, there are three models, namely

TABLE 7.10 Misclassified samples using LDA and the log scaled maize data, and different numbers of PCs in the model.

2 PCs	39(A→D)	58(B→C)	59(B→C)	63(B→C)	68(B→C)	74(B→C)	77(B→C)	80(B→C)	83(B→C)	85(B→D)
	91(B→D)	100(B→D)	103(B→C)	104(B→C)	106(B→C)	107(B→C)	109(C→D)	112(C→D)	113(C→D)	114(C→B)
	116(C→D)	117(C→B)	119(C→D)	123(C→B)	125(C→D)	127(C→D)	128(C→B)	129(C→B)	131(C→D)	132(C→B)
	135(C→B)	136(C→D)	139(C→D)	140(C→D)	143(C→D)	145(C→D)	146(C→D)	148(C→A)	151(C→D)	152(C→D)
	155(C→D)	161(C→D)	166(D→C)	168(D→C)	174(D→C)	181(D→C)	183(D→C)	184(D→C)	186(D→C)	187(D→C)
	189(D→C)	190(D→C)	192(D→C)	196(D→C)	198(D→C)	202(D→C)	210(D→C)	213(D→C)	214(D→C)	216(D→C)
5 PCs	39(A→C)	63(B→C)	87(B→C)	91(B→D)	93(B→C)	107(B→D)	113(C→D)	124(C→D)	125(C→D)	133(C→D)
	142(C→D)	157(C→D)	158(C→D)	163(D→C)	169(D→C)	181(D→C)	184(D→C)	187(D→C)	190(D→C)	193(D→C)
	199(D→C)	202(D→C)	205(D→C)	208(D→C)	214(D→C)					
10 PCs	63(B→C)	115(C→B)	163(D→C)	178(D→C)	184(D→C)	202(D→C)	208(D→C)			

- Class A versus Classes B and C
- Class B versus Classes A and C
- Class C versus Classes A and B

In our terminology, we denote the c value of the 'in group' as $+1$ and the 'out group' as -1. Hence, each classification model will be different.

We will illustrate this initially by the case study of Table 7.9.

- The first step is model Class A versus the rest.
- We assign $c = +1$ to the 10 samples A1 to A10 and $c = -1$ to the 20 samples B1 to C10.
- We recommend to weight mean centre the data. When performing a class A model we calculate $\bar{x}_A = \begin{bmatrix} 88.6 & 78.3 \end{bmatrix}$ and $\bar{x}_{not\,A} = \begin{bmatrix} 81.85 & 69.5 \end{bmatrix}$ so that the weighted mean $\bar{\bar{x}}_A = \begin{bmatrix} 85.225 & 73.9 \end{bmatrix}$
- We then correct the values, for example, $x_5 = \begin{bmatrix} 45 & 38 \end{bmatrix}$ before correction but becomes $\begin{bmatrix} -40.225 & -35.9 \end{bmatrix}$ after.
- Then, perform PLSDA keeping in this case both components. For sample A5, this becomes $t_{A,A5} = \begin{bmatrix} -52.97 & -15.51 \end{bmatrix}$.
- We can now plot the PLSDA scores for both PLS components and also predict c for all samples.
- This model can be repeated for all three classes. The scores plots, together with the $c = 0$ border are presented in Figure 7.45.
- We make several observations.
- If you look closely, you can see that the three figures are simply rotations of each other. That is because the data are fully modelled by 2 PLS components. Below we see that this is not so if the first two components do not model the data fully.
- The model for Class A does not look very promising. This is because class A is in the centre of the two classes, and it is not possible to find a good dividing line between class A and the others.
- However, Classes B and C are at the extremes and modelled very well.
- The resultant estimated values of c and their interpretation are presented in Table 7.11 using a full two-component model.
- The estimated values of c using all three models (and 2 PLS components) are listed.
- From these we can decide which side of the borderline a sample lies, for example 'B' means it has a value of $\hat{c}_B > 0$ and 'not B' a value of $\hat{c}_B < 0$; there are of course other possible decision rules.
- We then can combine these estimates, if for example, a sample has $\hat{c}_A < 0$, $\hat{c}_B > 0$ and $\hat{c}_C < 0$ we assign it to class B; there will be some ambiguous samples. There are 5 ambiguous and 1 misclassified in class A (60%), 6 ambiguous in class B (60%), and 3 ambiguous in class C (30%) or a total of 15 out of 30.

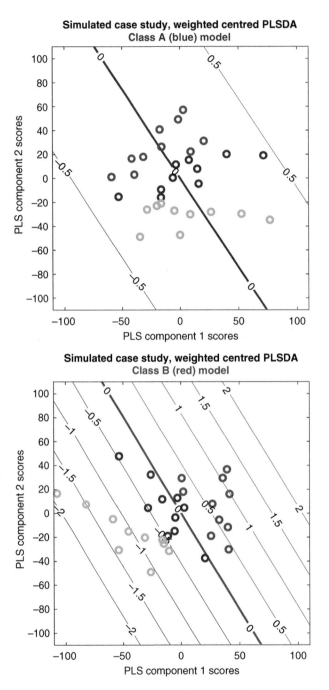

FIGURE 7.45 PLSDA one versus all models for the three classes of simulated data.

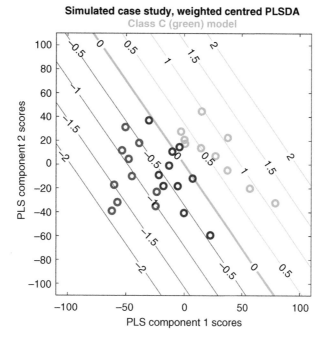

Simulated case study, weighted centred PLSDA
Class C (green) model

FIGURE 7.45 (Continued)

- However, a better criterion might be to take the highest predicted value of c over all three class models. This leaves just 4 misclassified samples all from Class A, and the rest correctly classified, or an overall rate of $26/30 = 87\%$ correctly classified.

- An alternate approach, which can be checked from Table 7.11 would be just to model classes B and C; any sample with a positive estimated value for either of the classes is assigned to its appropriate class, and for neither into class A.

- Of course, there are lots of possible decision rules, and it is important to be able to know which is being used as different conclusions could be drawn.

We will extend the illustration of this method using a real example, that of the NMR of maize harvested at four different temperatures, using the log-scaled data. For brevity, we will not illustrate the calculations for every step.

- We log scale the data first using logarithms to the base 10.

- PLSDA is performed separately using each of the four temperatures as an 'in group'.

- The two component PLSDA models are illustrated in Figure 7.46 for all four models.

- Unlike the simulated case study of Table 7.9, the PLSDA scores plots appear quite different for each of the models. This is because there are 34 variables, so the scores plots represent a favourable projection in the variable space, which is different according to each model, rather than just a rotation of the original variables.

TABLE 7.11 Results of t2 component PLSDA model on simulated data: see text for detail, misclassified samples are marked.

Sample	$\hat{C}_A$	$\hat{C}_B$	$\hat{C}_C$	Model A	Model B	Model C	Decision $c > 0$	Decision maximum c
A1	0.095	0.0927	−0.6302	A	B	not C	A or B	A
A2	0.106	−0.1686	−0.3808	A	not B	not C	A	A
A3	0.0679	−0.4933	−0.0228	A	not B	not C	A	A
A4	−0.1191	−0.2866	−0.0547	not A	not B	not C	None	C
A5	−0.335	−0.0550	−0.0844	not A	not B	not C	None	B
A6	0.0217	0.1085	−0.5779	A	B	not C	A or B	B
A7	−0.1427	−0.4479	0.1276	not A	not B	C	C	C
A8	0.2820	−0.1141	−0.5982	A	not B	not C	A	A
A9	0.4439	−0.4634	−0.4013	A	not B	not C	A	A
A10	−0.0314	−0.139	−0.2827	not A	not B	not C	None	A
B1	0.0102	0.6103	−1.0659	A	B	not C	A or B	B
B2	−0.1031	0.5608	−0.9115	not A	B	not C	B	B
B3	−0.3085	0.4201	−0.5812	not A	B	not C	B	B
B4	0.2184	0.362	−1.0123	A	B	not C	A or B	B
B5	−0.1976	0.2667	−0.5316	not A	B	not C	B	B
B6	0.1278	0.2521	−0.819	A	B	not C	A or B	B
B7	−0.1624	0.6291	−0.9244	not A	B	not C	B	B
B8	0.1664	1.039	−1.6368	A	B	not C	A or B	B
B9	0.2175	1.1945	−1.8387	A	B	not C	A or B	B
B10	0.0531	0.9895	−1.4825	A	B	not C	A or B	B
C1	−0.1684	−1.4041	1.1015	not A	not B	C	C	C
C2	0.0403	−1.1902	0.6953	A	not B	C	A or C	C
C3	−0.3549	−1.0865	0.9590	not A	not B	C	C	C
C4	−0.1861	−0.5865	0.3056	not A	not B	C	C	C
C5	0.2818	−1.8711	1.1479	A	not B	C	A or C	C
C6	−0.1216	−0.8409	0.4984	not A	not B	C	C	C
C7	−0.1601	−0.5736	0.2686	not A	not B	C	C	C
C8	−0.0599	−1.0576	0.6565	not A	not B	C	C	C
C9	0.1719	−1.496	0.8771	A	not B	C	A or C	C
C10	−0.2411	−0.5747	0.3448	not A	not B	C	C	C

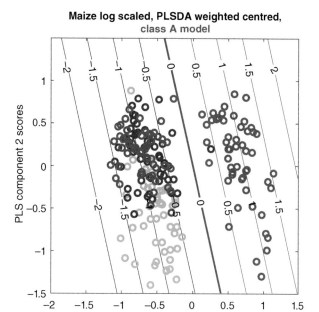

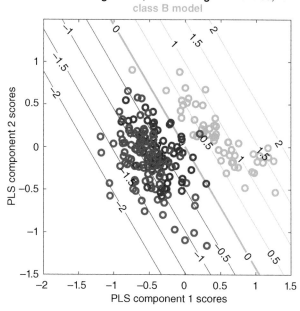

FIGURE 7.46   PLSDA models one versus all, using the four different temperatures and log-scaled data from the maize NMR case study 5 with predicted values of c indicated in the contour levels.

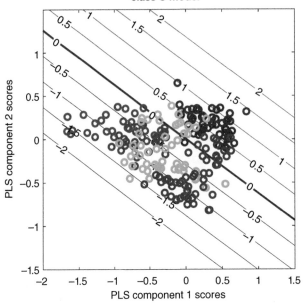

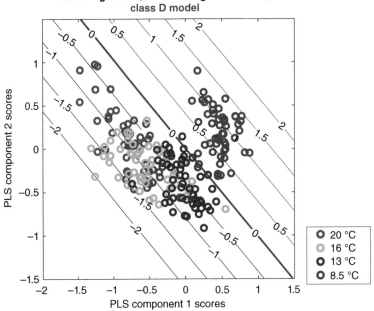

FIGURE 7.46   (Continued)

- Apart from class C (harvesting at 13 °C), the other three groups appear well separated when using 2 component models.

- For readers that are following these calculations numerically, we will not present the full calculations, but you can check the predicted values of c for sample 100, which are $[-1.22\ 0.78\ -0.66\ -0.52]$ and as $\hat{c}_B = 0.78$ we correctly assign it to Class B (samples harvested at 16 °C).

- We can illustrate the predicted values of c using a 2 PC model in Figure 7.47. The coloured symbols refer to the model rather than the origin of the samples, so each sample is represented by four symbols, representing the estimated values of c using the four models. Each quarter relates to one of the four classes.

- The misclassified samples using a 2 PLS component model and the highest estimated value of c as the criterion, as listed in Table 7.12. Ten samples are identified. Most are also identified in the 5 PC QDA model of Table 7.10 for the same data. Although there are some differences, comparing models can be used to identify obvious outliers.

- Of course, the number of PLS components could be increased if required.

In some situations in metabolomics, the classes are related sequentially, so this relationship could be taken into account and the classes not just treated as unrelated discrete groups. For example, we could assign a value of $c = +1.5, +0.5, -0.5$ and -1.5 to each of four groups. In our multiclass case studies, 2 (malaria), 3 (human diabetes) and 5 (maize) there is a relationship between each of the classes. This is discussed in Sections 7.6.3 and 9.3.5. However, in many examples in classical chemometrics, the classes have no specific relationship, for example, we may be trying to distinguish esters from acids from amides, or archaeological pottery from three sites, and Class

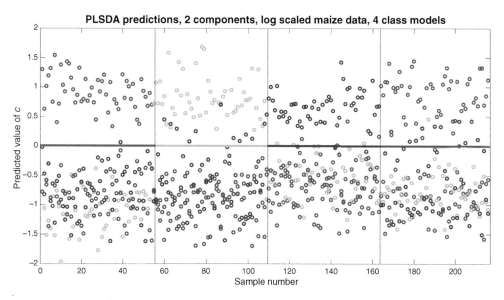

FIGURE 7.47 Predicted values of c using the four class PLSDA models. Coloured symbols refer to models rather than origin of samples, so each sample has four predicted values of c. Each quarter relates to a different harvesting temperature and the model with the most positive value of c corresponds to the predicted class.

TABLE 7.12   Misclassified samples using one versus all PLSDA model and the log-scaled maize data using two PLS components.

39(A→C)   63(B→C)   163(D→C) 169(D→C) 184(D→C) 187(D→C) 187(D→B) 190(D→C)

199(D→C) 202(D→C) 208(D→C)

A is equally connected to Class B and it is to Class C. This situation could occur in metabolomics, for example, if we were to study three genotypes, and most PLSDA applications are based on discrete and unrelated classes.

Potential biomarkers can also be identified using PLSDA loadings, but as the four classes are related, it makes little sense to, for example, identify biomarkers from Classes B and C (in this case). However, they could be identified from models based on the extreme groups A and D.

We can, however calculate the loadings where they can be interpreted. For the 20 °C (Class A) model, we see that the scores of PLS component 1 is a good separator with $t_1 = 0$ separating Class A from the rest of the samples (remember we use weighted centring to obtain these results) and so the loadings will give a clue as to which metabolites are most characteristic of high-temperature growth conditions. The loadings are presented in Figure 7.48. We can see, for example, alanine and glutamate appear characteristic of this temperature. The PLSDA loadings should be compared to the PCA loadings calculated for the standardised data in Figure 4.10, which would lead to similar conclusions in this case. Obviously, there is no single best approach, but using a variety of methods for visualisation can aid interpretation.

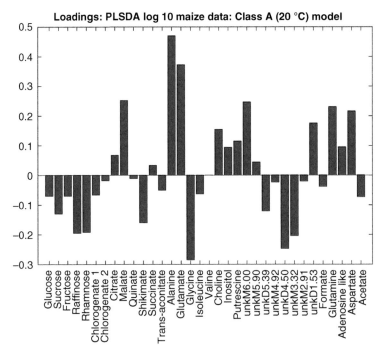

FIGURE 7.48   PLSDA model of log-scaled data of the maize using a Class A (20 °C) model.

In Chapters 9 and 10 we will discuss further approaches for biomarker identification. It is important to appreciate that chemometrics is mainly an exploratory step and points the way to more definitive experiments later. In this case study, the models for Classes B, C and D would not provide a definitive answer using loadings, although this could be improved via OPLS as discussed above. As the possibilities are large we will not explore all options in this chapter.

7.6.2.2   One Versus One

An alternative to one versus all is to model one versus one. This selects only samples from two classes and forms a two-class model. The samples left out are predicted using a two-class model. All samples are then assigned to one of these two classes. The model is repeated for every pairwise comparison. For example, if there are three classes, there will be three models, namely Class A versus B, Class A versus C and Class B versus C. Ideally, a sample from Class A will be assigned to Class A in the first two models and either class in the final model. The majority vote would be Class A, and it would then be assigned to its correct class.

The majority vote rule is effective but has two drawbacks. First, there is the possibility of ties, and second if there are several classes there are many binary comparisons. An alternative is to use a decision tree.

Two possible decision trees for a four-class problem are illustrated in Figure 7.49. The problem is that these will come to different conclusions under certain circumstances. Table 7.13 shows a situation where the two decision trees come to different conclusions, using each of the decision trees. The majority vote also comes to an ambiguous answer, so neither approaches are ideal. Decision trees can reduce the number of comparisons as only 3 rather than 6 comparisons are required for each sample. When there are only a few classes, this advantage is probably not very important but for many classes this can be time-saving. If there are N classes, it is only necessary to perform $N-1$ comparisons using a decision tree, but there are $N!/(2!(N-2)!)$ possible pairwise comparisons in total. For seven classes, this reduces the number of comparisons from 21 to 6.

However, the one versus one approach is sometimes used as an alternative to one versus all, and if using PLSDA multiclass classification models, it is always essential to understand how the predictions have been obtained.

7.6.2.3   PLS2DA

A final approach is PLS2. In this case, the c block can consist of several columns. If there are 4 variables, the c block will consist of a matrix of dimensions $I \times 4$, each column corresponding to a variable and each row a sample. We can denote $c = +1$ for members of each specified group and $c = -1$ otherwise. PLS2 calibration can be expressed in a similar way to the more usual PLS1 except that C is a matrix rather than a vector, so the main equations, using the NIPALS algorithm, become as follows:

$$X = TP + E$$

$$C = TQ + F$$

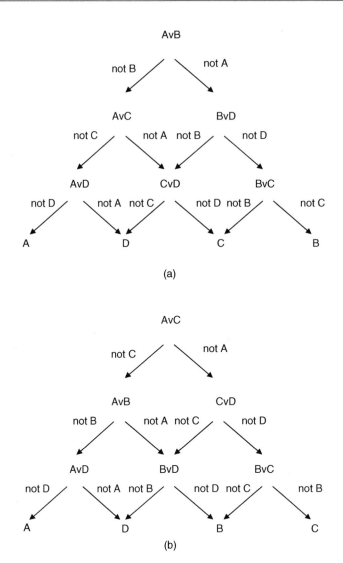

FIGURE 7.49   Two four-class decision trees.

We will not explore the PLS2 algorithm in detail in this text, but often the predictions are misleading. One problem relates to column centring, it is not possible to weight centre the X matrix, because each class has different centroids. There are other limitations to PLS2, but the literature on PLS is vast and stretches over 50 years, so it is impracticable to convey a detailed discussion of all approaches within the limitations of this text.

A modification of PLS2DA has been recently reported that is more successful. In this case, the X matrix is calibrated to an $I \times N$ matrix C using PLS2, where each column consists of 1s and 0s and corresponds to one of N classifiers. PCA is performed on the predictions of C which are then modelled using a variety of possible approaches, but most often QDA. We will not describe this, and other, modifications in depth in this text but suggest that as if employed one step in a classification strategy PLS2 can be more helpful.

TABLE 7.13   Assignment of sample using the decision trees of Figure 7.49; the result of all six binary tests is given in the first line, and then the three binary tests from the decision trees.

Binary tests		AvB	AvC	AvD	BvC	BvD	CvD
Result		A	C	A	B	D	C
Decision tree	(a)						
	Step 1	not B					
	Step 2		not A				
	Step 3						not D
	Decision						**C**
	(b)						
	Step 1			not A			
	Step 2						not D
	Step 3				not C		
	Decision						**B**

The main advice is to be aware of PLS2, if PLSDA is being used for classification where there are more than two groups, to check whether PLS2DA has been used. This text recommends using PLSDA one versus all, which we have illustrated in depth above, although other elaborations can provide good results but it is important to check how the method has been implemented carefully first.

7.6.3   Multilevel PLSDA

In some cases, classes can be numerically related. As an example, in case study 4, there are three related classes, Normal Glucose Tolerance, Impaired Glucose Tolerance and Type 2 Diabetes. Instead of treating these as independent groups, and creating three two-class PLSDA models, we could set $c = +1$ for NGT, $c = 0$ for IGT and $c = -1$ for diabetic donors. PLSDA models can then be formed between the experimental data X and this new c vector, instead of treating each class as separate and unrelated.

Although theoretically multilevel PLSDA could be used as a potential classifier, in practice it can be hard to successfully define criteria for the intermediate classes, so we will not illustrate this approach in depth in the context of a classification. However, it is useful for detecting variables that change monotonically over the different classes, and as such is a good approach for identifying which variables or markers are likely to be significant if there are several related groups and the aim is to see which metabolites follow a trend over these groups. Therefore, this method is discussed in detail in Section 9.3.5 of Chapter 9 in the context of variable selection rather than in this chapter.

This approach avoids using PLS2DA or one versus all/one versus one PLSDA methods but is only useful if the groups are related in a quantitative way.

7.7 VALIDATION, OPTIMISATION AND PERFORMANCE INDICATORS

A very important issue is how to determine how well a specific method performs. This allows us to validate models (to avoid over-fitting) and to optimise models (usually involving determining how many Principal Components or PLS Components are suitable for a specific model).

7.7.1 Classification Performance

It is important to establish a numerical indicator of how well a method has performed. Normally, this is based on the number or proportion of samples correctly classified, the CCR. These are often organised in the form of a table, often called a confusion matrix or contingency table.

7.7.1.1 Two Classes

Many problems consist of only two classes or categories. Case study 10, of the diabetic and non-diabetic mice is of such a case. Confusion matrices are most commonly used when there are two-class models such as LDA or PLSDA. Each model will produce slightly different results and it is often useful to compare them. They can also be called contingency tables although the latter terminology has a wider usage.

We will look in detail at the two PC LDA model in Table 7.3.

- 3 out of 30 of the diabetic and 5 out of 41 of the non-diabetic mice are misclassified using this model.
- Remember different confusion matrices could be obtained via different models.
- The data can be tabulated in a confusion matrix as in Table 7.14.
- The diagonal numbers represent the correctly classified samples, and the off-diagonal numbers the incorrectly classified samples.

A lot of literature defines one group as positive and one as negative. For example, we may be testing for a disease, the positive group may be diseased and the negative controls or normal. However, equivalent approaches can also be used if there are two (or more classes) none of which are considered specifically as positive or negative.

There are several common ways in which we can summarise these numbers, many used in the literature.

TABLE 7.14 Confusion matrix for the 2PC model of Table 7.3.

		Actual	
		Diabetic (Class A)	Non-diabetic (Class B)
Predicted	Diabetic (Class A)	27	5
	Non-diabetic (Class B)	3	36

Correctly Classified Rate and Related Indices

- The *CCR* (Correctly Classified Rate) or in percentage terms *%CC* (Percent Correctly Classified) can be calculated, as an overall statistic, equalling $(27 + 36)/71 = 0.8873$ or 88.73% in our example. This is also called the overall Accuracy (*ACC*).

- The *CCR* can be calculated for each class separately, being $27/30 = 0.9$ for Class A and $36/41 = 0.8780$ for Class B or $\%CC_A = 90.0\%$ and $\%CC_B = 87.80\%$. Note that in this case, the overall *%CC* differs slightly from the mean *%CC* for each class because the class sizes are different.

- The Misclassification Rate (*MCR*) or Error Rate (*ERR*) is $1 - CCR$ or 0.1127 for the overall dataset, 0.1 for Class A and 0.1220 for Class B. This can of course also be expressed as percentages.

- The Recall (*REC*) is the same as the *CCR*. In traditional analysis of confusion matrices, it is defined just for the positive group, but when dealing with classification models, it can be calculated for each class separately, or alternatively as an overall measure.

- The Precision (*PRE*) is the proportion of samples predicted to be a member of a given class that are correct. Over all the classes, it is the same as the *CCR*, but differs from the *CCR* for each individual class. For Class A, it equals $27/32 = 0.8438$ and for Class B, it equals $36/39 = 0.9231$.

- In our case, the *CCR* is the proportion of correctly assigned samples in each column, and the *PRE* is the proportion correctly assigned in each row. Note the rows and columns are sometimes swapped around in some definitions of a confusion matrix, so be careful what package you use.

- For each class, *PRE* answers how well the test performs whereas *CCR* how well a class is predicted.

- The *F1*-score is often employed in machine learning. For each class it is given by $F1 = (2 \times PRE \times CCR)/(PRE + CCR) = 0.8710$ for Class A and 0.900 for Class B. It is the harmonic mean of the correctly classified rate and the precision.

- The *F1*-score for the overall dataset is the same as the *CCR* and *PRE* since $F1_{overall} = (2 \times PRE_{overall} \times CCR_{overall})/(PRE_{overall} + CCR_{overall}) = (2 \times CCR^2_{overall})/(2 \times CCR_{overall}) = CCR_{overall}$.

- It is also possible to calculate the *F0.5*-score and the *F2*-score defined by $F0.5 = (1.25 \times PRE \times CCR)/(0.25(PRE + CCR))$ and $F2 = (5 \times PRE \times CCR)/(4(PRE + CCR))$ but we will not discuss these further.

- The Matthews correlation coefficient is often used as an alternative to determine the quality of the models and is defined by

$$MCC = \left(C_a C_b - M_a M_b\right) / \sqrt{\left[\left(C_a + M_b\right)\left(C_b + M_a\right)\left(C_b + M_b\right)\left(C_a + M_a\right)\right]}$$

where C_a is the number correctly classified from Class A and M_a is the number misclassified from this class. In other words, it is the difference between the product of the diagonal minus off-diagonal terms divided by square root of the product of the sums of the rows and columns. In our case, it equals $(27 \times 36 - 3 \times 5)/\sqrt{[(27+5) \times (36+3) \times (36+5) \times (27+3)]} = 0.7724$ and is regarded as a good overall measure of the quality of a binary classification model. For a binary classification procedure, it varies between +1 and −1.

Positives and Negatives and Related Indices

Often, one group is considered 'positive' and the other 'negative'. In our case, we will call the positive samples those diagnosed as diabetic and the negative samples non-diabetic. Some people like to view samples in this way. We can re-express the confusion matrix in terms of P (positives) and N (negatives) as in Table 7.15.

- TP = True Positives, TN = True Negatives, FP = False Positives, FN = False Negatives, as presented in the table.
- FPs are also sometimes called Type 1 errors and FNs are called Type 2 errors.
- There are a number of common statistics, which frequently appear in the literature.
- The sensitivity or recall or TPR (true positive rate) is given by $SEN = TPR = TP/(TP + FN)$ or in our case $27/30 = 0.90$. Note this equals CCR_+ and is simply a different terminology.
- The specificity or selectivity or TNR (true negative rate) is given by $SPE = TNR = TN/(TN + FP)$ or in our case $37/41 = 0.8780$, the same as CCR_-.
- We can define FPR (false positive rate) $= 1 - TNR = 0.1220$ and FNR (false negative rate) $= 1 - TPR = 0.1$ in our case, as the corresponding MCR.
- Some people like to use likelihood ratios. They are defined by $LR_+ = TPR/FPR$ or in our case 7.38 and $LR- = TNR/FNR$ or in our case 8.78. These are often used in medical and forensic diagnosis. A positive likelihood ratio of 9 would imply that if a test suggested a donor is positive for a disease, for example, whereas the test is not necessarily perfect it is 9 times more likely they have the disease than not.

Depending on the literature, different terms can be used, so it is worth checking what papers in certain areas cite.

Bayesian Statistics

In this textbook, we will not explore Bayesian methods in depth; however, certain schools of thought do use Bayesian approaches. They are not common in metabolomics.

TABLE 7.15   Confusion matrix re-expressed as positives and negatives.

		Actual	
		Diabetic	Non-diabetic
Predicted	Diabetic	TP (27)	FP (5)
	Non-diabetic	FN (3)	TN (36)

The principle is that the relative number of samples in each group does not necessarily represent the probability that an unknown sample is a member of a group. As an example, we may analyse 50 samples from donors with a given disease and 100 controls (no disease). Our confusion matrices will have a summation of 50 for Class A (diseased) and 100 for Class B (controls). Does this mean that the chances of a donor having this disease are 50/150 = 0.33. Probably it does not. This probability (0.33) can be considered a prior probability. In the general population, the probability may be much lower, whereas if a sample has already been diagnosed as likely to have the disease and we just need an analysis to confirm or reinforce a clinical decision, it could be much higher. Hence, the prior probabilities can be adjusted according to what is known about the origins of the samples. In terms of confusion matrices, this can be done by changing the proportions of each column to reflect prior probabilities.

A Bayesian would argue that the analytical test and chemometrics are used to refine the probability of an outcome or class membership, to obtain a posterior probability. Let us say a doctor only sends samples to a laboratory that they are 80% sure are diseased, and then the analytical tests suggest this is correct, this may increase the prior probability of 0.8 to for example a posterior probability of 0.93. This could be quite crucial if the outcome is to recommend a patient for a serious operation.

Several methods such as LDA can quite easily be modified to include Bayesian terms if necessary, or the posterior probabilities can be calculated changing the confusion matrix so that the relative summation of each column is proportional to the prior probability of an outcome. Sometimes several tests can be performed, the result of each test refining the probability, treating the result of the first test as a prior probability for the second test to give a posterior probability, which in turn is treated as a prior probability for the third test and so on.

In metabolomics, we rarely need to make serious clinical or forensic decisions and are more interested, for example, which markers are most likely to be important as a precursor of a disease. Probabilities in the form of p values can be calculated, but it is unlikely the distribution of metabolites obey completely the underlying assumptions, so metabolomics could be regarded more as an exploratory science requiring further experiments at a later stage rather than a definitive study leading to significant clinical or forensic consequences. However, if several tests are performed successively, it is possible to use Bayesian methods to refine the p valuc.

In most cases the overall %CCR is taken as a good indicator of quality of a model and we will use this below, but readers should be aware of Bayesian criteria are sometimes cited as alternatives.

7.7.1.2 Multiclasses

When there are more than two classes, the confusion matrices become slightly more complicated, and cannot be expressed in terms of positives and negatives.

However, most of the concepts above can be extended when there are more than two classes. A confusion matrix for the 2 PC model of Table 7.10 is presented in Table 7.16.

TABLE 7.16   Confusion matrix for 2 PC model of Table 7.10.

			Actual			
			20 °C	16 °C	13 °C	8.5 °C
			A	B	C	D
Predicted	20 °C	A	53	0	1	0
	16 °C	B	0	39	6	0
	13 °C	C	0	12	28	18
	8.5 °C	D	1	3	19	36

PRE, *CCR* and the *F*1-score can be calculated for individual classes or the overall data, just as for a binary classifier. As in this example, the overall $\%CCR = 100 \times (53 + 39 + 28 + 36)/216 = 100 \times 156/216 = 72.22\%$. If more PCs are included in the model, the $\%CCR$ increases, for example, if 10 PCs are included it becomes 96.76%. Also, samples harvested at different temperatures have different $\%CCRs$ as can be calculated from the table.

The definition of the Matthews correlation coefficient is a little more complicated if there are more than two classes. However, one way of calculating this is

$$MCC = \frac{cs - \sum_k p_k t_k}{\sqrt{\left(s^2 - \sum_k p_k^2\right)\left(s^2 - \sum_k t_k^2\right)}}$$

where

- *c* is the total number of samples correctly predicted, or 156 in the example using a two PC model.
- *s* is the overall number of samples, or 216 in our example.
- There are *k* classes.
- p_k is the sum of row *k* or the number of samples predicted to be in class *k* equalling 54, 45, 58 and 59 for the four classes in our example.
- t_k is the sum of column *k* or the number of actual samples in class *k*, equalling 54 for each column in our example (this will not always be equal for each column in many practical examples).

In our case,

$$MCC = \frac{156 \times 216 - 54 \times 54 - 54 \times 45 - 54 \times 58 - 54 \times 59}{\sqrt{\left(216^2 - 54^2 - 45^2 - 58^2 - 59^2\right)\left(216^2 - 54^2 - 54^2 - 54^2 - 54^2\right)}} = 0.646$$

This can be used as an overall indicator of the quality of the model.

Most chemometricians use the overall $\%CCR$ as an indicator of the quality of the model but sometimes dive deeper and any of the indices above can be computed.

7.7.1.3 One-Class Models

For one-class models, such as SIMCA, it can be a bit more complicated as to how to express the success or otherwise of a model.

The simplest approach is to produce individual confusion matrices for each model. As an example, we will use the QDA 2-component model of 38 samples (excluding the three misdiagnosed samples) illustrated in Figure 7.31.

- There are two separate confusion matrices, for class A and class B models, as presented in Table 7.17.
- The QDA class boundaries are set at $p = 0.05$, so we would expect around 5% of the 'in group' if an approximate multinormal model is obeyed, to be outside the boundaries, for example, classed as 'not A' or 'not B'. If a sample is 'not A' this does not necessarily imply it is a member of class B.

TABLE 7.17 Confusion matrices and contingency table for the 38 samples modelled by two PCs and QDA as presented in Figure 7.31.

Model Class A confusion matrix

		Class A model	
		Actual	
		A	not A
Predicted	A	29(96.67%)	4(10.52%)
	not A	1(3.33%)	34(89.47%)

Model Class B confusion matrix

		Class B model	
		Actual	
		not B	B
Predicted	not B	27(90%)	3(7.89%)
	B	3(10%)	35(92.10%)

Joint model contingency table

		Actual	
		A	B
Predicted	A	27(90%)	1(2.63%)
	B	1(3.33%)	32(84.21%)
	AB	2(6.67%)	3(7.89%)
	Neither	0(0%)	2(5.26%)

- These values are 3.33% (Class A – diabetic) model and 7.89% (Class B – non-diabetic) groups which are reasonably close to 5% suggesting that our models and data are not too far away from a normal distribution. The proportions in the 'out group' are unrelated to how well the model performs.
- We can also combine both models into a single contingency table. All confusion matrices can also be called contingency tables, but the reverse is not so. A confusion matrix must be square, whereas a contingency table can have more outcomes than there are original categories.
- The contingency table is presented in Table 7.17. When there are only two categories, there are only four possible outcomes, but if there are more than two categories, the contingency table becomes much more complex and impracticable.
- The contingency table can be checked either by calculations which for brevity we will not detail here, or graphical inspection of Figure 7.31. As an example, there are two samples from Class B that are predicted to be members of neither class (using the $p = 0.05$ criterion), 43 and 53, and 1 sample that is misclassified as class A, 54, as annotated in the figure.
- Of course, using different criteria and algorithms, the confusion matrices and contingency tables will differ.
- Usually, for the purpose of model performance, we can calculate the %CCR. For the class A model, this is 96.67% and for the class B model 92.10%. Note that each model has a separate overall %CCR. However, different authors might use other criteria.

7.7.2   Validation

Validation plays a crucial role in classical chemometrics and PR. It is especially important for multivariate data, as it is easy to over-fit models. It is therefore often vital to see whether a classification model is over-optimistic or realistic.

The normal way of doing this is to perform a model on a selection of samples, usually called the training set or calibration set, and then check how well it does on an independent selection of samples, not part of the training set, usually called the test set or prediction set.

There are huge debates as to how to set up a test set. What proportion of samples should be included? What if there are outliers either in the training or test set? Is a single test set representative? Some authors iteratively reform the test set by selecting a different portion of samples many times over, typically 100 times, taking advantage of modern computer power.

Every model can be validated, but to illustrate the principle we will illustrate with case study 1 (LCMS of pre-arthritic and control donors) using LDA after first reducing the data using PCA on the standardised data and retaining the scores.

- The first step is to divide the data into training and test sets.
- The full dataset consists of a matrix of size 49×81, of which the first 19 samples are controls and the last 30 are pre-arthritic.

- 33 out of 49 of the samples (67%) are selected for the training set and 16 for the test set as presented in Table 7.18. Typically training sets consist of 2/3 to 3/4 of the samples. Sometimes the test set is proportionately balanced according to the relative number of samples in each class in the overall (auto-predictive) dataset, or in other cases, each class is equally represented – there is no general rule. In our case the ratio of a number of samples in the training set for class A: class B is 2, and for the test set 2.16. These cannot be exactly equal in this example, but are very close.

- The first step is to prepare the training set samples. The data have to be standardised. For those that are following the calculations, before standardisation $^{train}x_{9,32} = 4.232 \times 10^5$. For the 33 training set samples $^{train}s_{32} = 4.284 \times 10^5$, $^{train}\bar{x}_{32} = 6.946 \times 10^5$ so the standardised value $^{train}x_{9,32} = -0.633$. Note that this differs from the value obtained when standardised over all 49 samples, which would include the test set in the calculation of mean and standard deviation and is −0.560.

- Performing PCA on the training set results in a scores matrix T. To check, $t_{9,2} = -1.356$.

- The test set scores can be obtained by first transforming the test set using the mean and standard deviation of the training set so that $^{test,train}x = \left(^{test}x - ^{train}\bar{x} \right) / ^{train}s$, so $^{test}x_{3,32}$ is transformed from 4.198×10^5 to −0.642.

- Note we will use the following notation in this section. The training set samples for Class A are numbered from 1 to 13 and for Class B from 14 to 33; the test set for Class A is separately numbered from 1 to 6 and for Class B from 7 to 16. This is to have a sequential list of samples for both training and test set. Hence, test set sample 3 corresponds to original sample 16 and corresponds to class A (controls). Obviously, other numbering systems are possible.

- We can now predict the scores for the test set by $^{test}\hat{T} = ^{test}X \, ^{train}P'$ where X represents the pre-processed test set data as described above. In our case $^{test}\hat{t}_{3,2} = -0.022$.

- The next step is to calculate the Mahalanobis distance to the centroids of each class using LDA and a varying number of PCs.

- For a two PC model, the variance–covariance matrix for the training set is given by $S = \begin{bmatrix} 12.293 & 0 \\ 0 & 9.882 \end{bmatrix}$ and the means of the two classes for the training set are $\bar{x}_A = \begin{bmatrix} 0.692 & -0.611 \end{bmatrix}$ and $\bar{x}_B = \begin{bmatrix} -0.450 & 0.397 \end{bmatrix}$.

TABLE 7.18 Training and test set samples from case study 1.

	Training set	Test set
Controls (Class A)	13 samples (1–13)	6 samples (14–19)
Pre-arthritic (Class B)	20 samples (20–39)	10 samples (40–49)

- This gives distances, using the Mahalanobis distance model, for the training set to classes A and B of, $d_{A7} = 3.993$ and $d_{B7} = 4.441$, correctly placing sample 7 into Class A. For the test set $^{test}d_{A3} = 0.866$ and $^{test}d_{B3} = 0.536$; the third test set sample is sample 16 overall which should be a member of Class A, but is incorrectly assigned to Class B using this model.

We can look at performance of models when we change the number of PCs.

- The differences between the distances $d_A - d_B$ can be calculated for both training and test set samples and for different numbers of PCs in the model.
- The results are presented graphically in Figure 7.50.
- A negative bar means the sample is closer to Class A, and a positive bar to Class B. Samples whose provenance is Class A are presented in blue, and Class B in red.

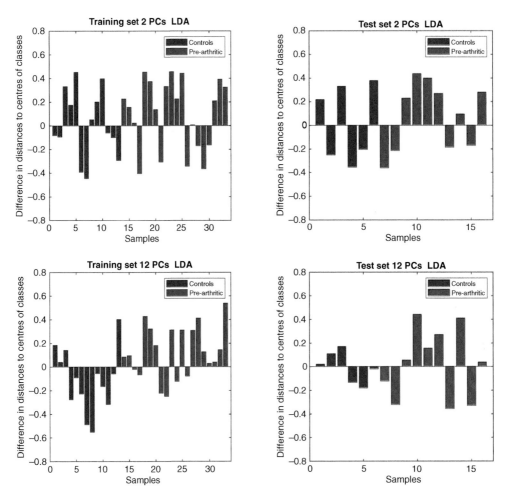

FIGURE 7.50   Training and test set difference in Mahalanobis distances from centroids using LDA for the case study 1 the pre-arthritis dataset as described in the text.

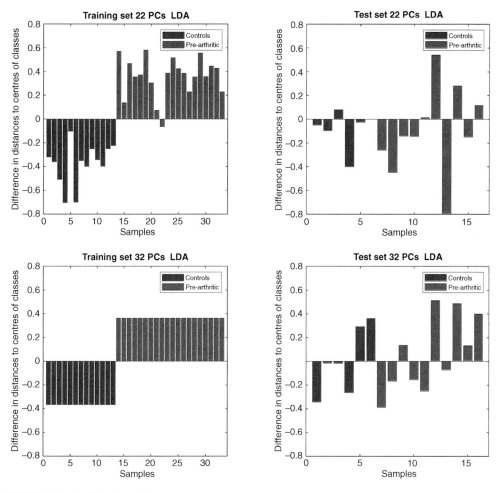

FIGURE 7.50 (Continued)

- When only two PCs are used in the model both training and test sets are poorly predicted.
- As the number of PCs are increased, the training set appears much better modelling whereas there is not much difference in the test set using the number of PCs presented.
- As there are 33 samples in the training set, the training set predictions for 32 PCs appear perfect. However, the test set suggests otherwise.

We can express the performance numerically as in Table 7.19 and see in the cases visualised, the overall %CCR for the test set is the same for all models, and around 50% suggesting none of the models have any discrimination.

This at first appears rather bleak, suggesting that the data and methods are not valuable. However, if we then decide to run all models using 2–32 PCs, we can obtain a finer structure. We can calculate the *%CCR* for both the training and test sets as the number of PCs are changed, as illustrated in Figure 7.51.

TABLE 7.19   Training and test set CCR for LDA and case study 1 using different numbers of PCs in the model.

	Training		Test		Training	Test
No. of PCs	CCR_A	CCR_B	CCR_A	CCR_B	%CCR	%CCR
2	7	14	3	6	63.64	56.25
12	9	14	3	6	69.70	56.25
22	13	19	5	4	96.97	56.25
32	13	20	4	5	100.00	56.25

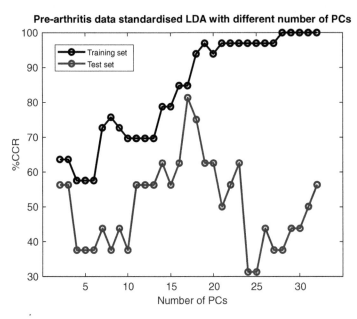

FIGURE 7.51   Changing the number of PCs for the LDA model of case study 1 the pre-arthritis dataset, showing the optimum model based on the test set is with 17 PCs.

- We see that a 17 PC model provides the best results.
- Only 3 out of 16 test set samples are misclassified, all from Class B, giving an overall %CCR of 81.25%.
- The difference $d_A - d_B$ is calculated for both training and test set and presented in Figure 7.52. This suggests we have found an acceptable model.

Although these calculations suggest that a 17 PC model is best, this critically depends on choice of test set, and a different conclusion may be obtained with a different test set so a common approach is to repeatedly generate different training/test splits to provide a consensus solution. Also, somewhat different conclusions might be obtained using different performance measures. However, it shows that the training set model will usually be over-optimistic as the number of PCs is varied. This does not mean that training set models are of no use. If we are sure that there are distinctions

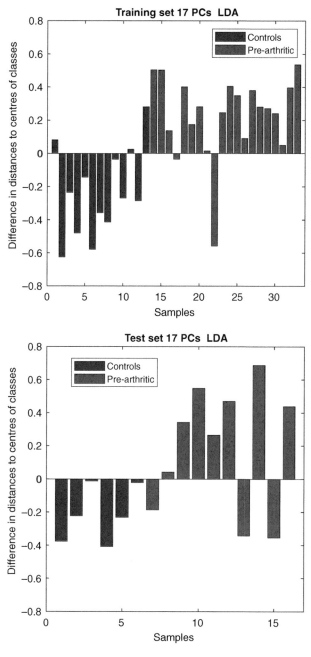

FIGURE 7.52 Training and test set for optimum number of PCs for case study 1 the pre-arthritis dataset LDA model.

between two or more groups and our main aim is exploratory, for example, to propose which metabolites are most likely to be markers and then go on to perform further experimental work or narrow our search down, it may be perfectly good to run training set models.

In most of traditional statistics, we are interested in hypothesis testing so validation allows us to provide an opinion as to whether two or more groups can be separated from their analytical evidence. In metabolomics, this is not always the main aim.

In this section, we have shown how to separate samples into training and test sets and demonstrated how to use this information to validate a specific model and case study. For every case study and method, the conclusions will be different, but the same principles of separating into training and test sets apply to all classification models.

7.7.3   Optimisation

Optimisation primarily involves estimating the number of components to retain in the model.

For PLS, we describe approaches to determine the optimum number of PLS components in Section 8.4 in the context of PLSR (PLS calibration). Similar approaches can be employed for PLSDA and we will not repeat these discussions in this chapter, but refer the reader to Chapter 8.

There is a very long historical interest in estimating the number of significant components for PCA models in chemometrics. The origins of chemometrics was primarily in analytical and physical chemistry. Under such circumstances, an accurate estimate of the number of components often has a physical meaning, for example, how many reactants are in a reaction mixture or how many coeluting compounds are in an HPLC peak cluster, so a huge number of methods, probably up to 100 formally described approaches in the literature, have been developed.

In metabolomics, components do not usually have a physical interpretation, so although it is certainly important to retain a sensible number of components in a model, there often is not a sharp interpretable cut-off in number of components.

In many areas of metabolomics except when organisms are grown under very controlled conditions, for example, plant metabolomics, there will be many significant factors in addition to the effect we are interested in, influencing the metabolic signal. Hence, the optimum number of PCs for describing the entire variability in a series of samples may be too many to describe the variability we are interested in. Additional PCs might be useful to describe overall variability but could be viewed as error or noise when our interest is in one or two key factors, and are often describe as orthogonal variation.

One way is to vary the number of PCs in a training set model, to see how the performance of a chosen indicator such as the %CCR as the quality of test set prediction changes. We show this in Figure 7.51 where we find that 17 PCs are optimum for modelling the data of case study 1 (pre-arthritis identified by LCMS). The optimum number of PCs has been determined using the c block.

In this section, we describe how to determine the optimum number of PCs in the x block without reference to the c block. This not usually necessary in metabolomics and not generally recommended when there is a c block, but for completion, we describe the steps as it is common in many software packages that readers might encounter.

One of the most widespread and historically popular methods is cross-validation. There are many forms of cross-validation, but LOO (leave one out) is very common and often the default. Cross-validation this way is not always successful, and there will be a lot of variability besides the factor of interest, but we illustrate its calculation for case study 8 (environmental effects on *Daphnia* metabolism) where the method has some success.

The principal of LOO cross-validation is to leave each sample out at a time. Then PCA is performed on the remaining samples, and the left out sample's properties are predicted, as if the left out sample is a test set of size 1. This is repeated until every sample is removed once, so in effect generating I training sets of size $I-1$ and I test sets of size 1. The entire procedure (PCA) is then repeated for models with increasing numbers of components for each of the training sets. Once the correct number of components is reached, the cross-validated error, being the total error of the left-out samples, should increase, as additional components model noise which decreases the quality of predictions.

The detailed method is described below:

- The first step is to prepare the data. We standardise the 24×47 dataset, in this case. To check, after standardisation $x_{17,20} = -0.504$. Remember to use the population definition of the standard deviation fi repeating the calculations from this book.
- We can check first the auto-predictive (or training set) PCA on the entire dataset. As an example $t_{17,4} = -0.437$ and $p_{4,20} = 0.0249$.
- It is then usual to calculate the estimated values of X as gradually more PCs are included in the model, $\hat{X} = TP$; so for a 4 PC model $\hat{x}_{17,20} = -1.014$.
- We next calculate the error matrix $E = X - \hat{X}$ so check $e_{17,20} = 0.511$.

- Finally, for a k component model calculate the Sum of Square Error
$$SSE_k = \sum_{i=1}^{I}\sum_{j=1}^{J} e_{ij\{k\}}^2 \text{ or for four components } SSE_4 = 358.578.$$
- Most authors then divide by the number of degrees of freedom to calculate either
  - the Mean Sum of Square Error $MSSE_k = SSE_k/(I-1-k)$ or for four components $MSSE_4 = 358.578/(24-1-4) = 358.578/19 = 18.873$.
  - or the Root Mean Sum of Square Error $RMSSE_k = \sqrt{SSE_k/(I-1-k)}$ or for four components $RMSSE_k = 4.344$.
- We will use the former below, and not illustrate both numbers for brevity although the trends are the same.
- We can plot a graph of $RMSSE$ against the number of components. It should decrease or level off with number of components, although there can be some small increases when the number of components is large, due to the divisor including the term k, but this should be very small.
- We will illustrate this together with $RMSSCV$ below.

We now compare with the equivalent cross-validated error.

- The first step is to remove each sample, one by one.
- Starting with sample 1, we now form a matrix $X_{\{-1\}}$ which is an $I-1 \times J$ or in our case 23×47 matrix consisting of samples 2–24.
- To follow our calculations, recentre this matrix. There are various approaches to centring during cross-validation, which we do not have space to compare. So for example in this case $\bar{x}_{20,\{-1\}} = 0.0294$.
- We will run PCA on the centred dataset $X_{\{-1\}}$. We find $t_{17,4\{-1\}} = -0.542$, where '17' refers to the original 17th sample and is compared to -0.437 when the full 24×47 dataset is used in the model.
- What is important is the loadings vector for the reduced dataset, so for example $p_{4,20\{-1\}} = 0.0610$ compared to 0.0249 for the full dataset.
- In Figure 7.53, we compare loadings using all 24 samples and samples 2–24; there will be a similar relationship as each sample is removed.
- The next step is to predict x for the left out sample by $\hat{x}_{cv1} = \hat{x}_{1\{-1\}} = x_1 P'P + \bar{x}_{\{-1\}}$: note that $P'P$ is not a unit matrix (whereas PP' is) if we use the terminology of $X = TP + E$ as in this text and $\bar{x}_{\{-1\}}$ is the mean of samples 2–24 (after overall centring). This is the predicted value of the first sample after cross-validation.
- As an example, we find that if sample 1 is removed, the mean of the remaining 23 samples for variable 20 is given by $\bar{x}_{20\{-1\}} = 0.043$ (there are several ways

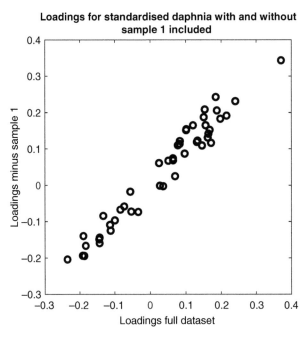

FIGURE 7.53   Loadings of PC 4 for case study 8 the Daphnia dataset, both including all samples, and excluding sample 1.

of centring the data, but they must be compatible – we will not compare these, but the calculation in this section involves first centring all 24 samples during standardisation, then recentring the remaining 23 samples). Hence, for a $k = 4$ component model the estimated value of sample 1, variable 20, using cross-validation is $\hat{x}_{cv1,20} = \hat{x}_{1,20\{-1\}} = -0.919$.

- This compares to $x_{1,20} = -0.675$ for the original standardised data, and $\hat{x}_{1,20} = -0.936$ for the full (24 sample) PC model using $k = 4$ components.

- We next calculate the error matrix $\boldsymbol{E}_{cv} = \boldsymbol{X} - \hat{\boldsymbol{X}}_{cv}$ where each row of $\hat{\boldsymbol{X}}_{cv}$ corresponds to the cross-validated estimate of x, so if you are following the calculations, the check $e_{cv17,20} = -0.491$ for a 4 component model.

- We can now calculate $SSECV_k = \sum\limits_{i=1}^{I}\sum\limits_{j=1}^{J} e^2_{cvij}$ or for four components $SSECV_4 = 689.302$.

- We can then calculate the Mean Sum of Square Error $MSSECV_k = SSECV_k/(I-1-k)$ or for 4 components $MSSECV_4 = 689.302/(24-1-4) = 689.302/19 = 36.279$. (We omit the root mean square error $RMSECV$ but identical conclusions can be drawn if using the mean sum of square error or its square root).

- This is substantially higher than $MSSE_4$ as calculated above.

The final step is to plot both $MSSE$ and $MSSECV$ again component number as in Figure 7.54. After 5 PCs, the value of MSSECV starts to increase, suggesting that additional PCs are adding noise and so an optimum is around 5 PCs.

It is essential to realise that there are numerous criteria for determining the optimum number of PCs and no one criterion is definitive. We have also chosen an example where the trend is clear using cross-validation. On the whole, this method is not

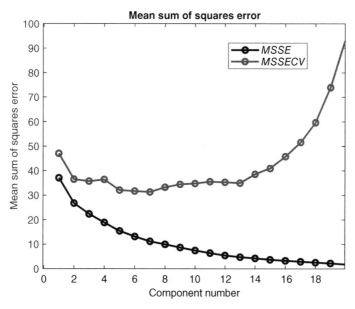

FIGURE 7.54 Mean sum of squares errors as a function of increasing number of components in the model for case study 8 (GCMS of Daphnia extracts).

always suitable for metabolomics investigations because PCs look at total variability and usually the aim is just to study the influence of one or a small number of factors. However, there is a large literature on this topic, and many readers will encounter cross-validation in multivariate software, so should be aware of the method. In many classical cases, such as in spectroscopy or chromatography, combined with PCA or factor analysis, it can be very effective, for example, to determine how many coeluting peaks are in a chromatographic cluster.

An alternative approach is simply to divide into training and test sets and gradually increase the number of PCs for the training set model, and see how well the test set is predicted. In Figure 7.51, we did this but saw how well the c block was predicted, this can be done instead just on the x block. If this is repeated by repeatedly generating different test/training set splits, an average prediction can be obtained rather analogous to cross-validation. Due to the huge literature in this area, we will not describe all the multitude of variations on this theme here.

Whereas the topic of how many significant components should be retained in a PC model is very widely reported, it is less important in metabolomics so we will restrict the discussion of this section and refer interested readers to extensive classical literature.

CHAPTER 8

Multivariate Calibration

8.1 INTRODUCTION

For around 20 years, multivariate calibration was one of the main foci of classical chemometrics. Indeed, books published on chemometrics in the 1970s and 1980s hardly mentioned pattern recognition. This was because the early growth points were primarily in quantitative analytical and physical chemistry, and a major focal point was to determine the concentration of components in a mixture, either an extract, for example, from the food or petrochemical industry, or laboratory-generated such as in pharmaceutical reaction monitoring or process control. Chemometrics was developed as an aid to the interpretation of the spectra of mixtures, especially using NIR (near-infrared) but also UV/Vis and related approaches.

In metabolomics this need is less widespread. The majority of work commonly called metabolomics involves biomarker discovery in which case selecting the most promising markers (Chapters 9 and 10) and Pattern Recognition (Chapter 7) are the most important applications. However, a few metabolomic studies do involve calibration, for example, from the chemical or spectroscopic profile, can we predict biological activity?

In this book, case studies 3 (measurement of triglycerides in children's blood) and 9 (bioactivity of herbal medicine) involve calibrating analytical measurements to an external parameter of direct interest.

Although multivariate calibration in itself could be the subject of an entire book, and has a prominent role in chemometrics, to provide a balanced text in metabolomics, we will restrict this chapter just to describing PLS (partial least squares) and its use in calibration. PLS for discrimination has already been described in Chapter 7 reflecting its relative importance to the published metabolomic literature.

Data Analysis and Chemometrics for Metabolomics, First Edition. Richard G. Brereton.
© 2024 John Wiley & Sons Ltd. Published 2024 by John Wiley & Sons Ltd.
Companion website: www.wiley.com/go/Brereton/ChemometricsforMetabolomics

8.2   PARTIAL LEAST SQUARES REGRESSION

PLSR (partial least squares regression) is one of the best-established chemometrics methods. Although its use is not as widespread in metabolomics compared to PLSDA, it still has a role and has been developed over many decades, and is well known.

As an illustration, we will use case study 9, the bioactivity of herbal medicine, which can be arranged into a 78×1120 data matrix X consisting of 78 samples and 1120 single wavelength HPLC data points, and a 78×1 vector c of bioactivity, as described in Chapter 3. The aim is to predict c from X.

In this section, we will describe PLSR using auto-predictive models (that is the entire dataset), whereas in Section 8.3 we will show how it can be divided into training (sometimes called calibration) and test (sometimes called prediction) sets, rather like classification models (Section 7.7.2).

We will restrict our discussion to the NIPALS algorithm and PLS1 (one variable in the c block).

PLS using the NIPALS algorithm forms two sets of equations as follows:

$$X = TP + E$$

$$c = Tq + f$$

where

- X is an $I \times J$ (in our case 78×1120) data matrix,
- T an $I \times A$ ($78 \times A$) scores matrix,
- P an $A \times J$ ($A \times 1120$) 'x' loadings matrix,
- c an $I \times 1$ (78×1) vector of calibrants,
- q an $A \times 1$ 'c' loadings vector,
- E and f represent residuals (often called errors although strictly speaking they are statistically different),
- and there are A PLS components in the model.
- In addition a weights matrix W is obtained of dimensions $J \times A$ ($1120 \times A$), which has some analogies to the loadings matrix,
- and for a model of A components a regression vector b of dimensions $J \times 1$.
- For the training or auto-prediction data, $Tq = Xb$.
- For any new data, not part of the original dataset, we can predict $\hat{c} = xb$.

More detail is provided in Section 7.4.2.1 of Chapter 7.

It is normal to centre both blocks prior to PLSR. We will initially describe the calculation of the first PLS component.

- The first step is to centre both x and c blocks, so, for example, the value $x_{20,291} = 794.59$ before column centring, and 24.83 after; the corresponding value $c_{20} = 365.47$ and after column centring -130.95. The chromatographic intensity at data point 291 corresponds to an elution time of 14.81 min.

- After PLS to check your calculations for a 1 PLS component model if you are following the methods
  - $t_{20,1} = 1.232 \times 10^3$,
  - $p_{1,291} = 0.157$
  - $w_{291,1} = 0.099$
  - $b_{291,1} = 0.0053$
  - and $q_1 = 0.0534$.
- The centred value of $c_{20} = -130.95$ and its predicted value is $\hat{c} = 65.81$.
- As these are centred to convert to the original values, since $\bar{c} = 496.42$, after correction for the mean, $c_{20} = 365.47$ and its predicted value is $\hat{c}_{20} = 562.23$. The residual (or error) $e_{20} = 365.47 - 562.23 = -196.76$.
- Note that in PLS we usually calculate c block rather than x block errors, although it would be possible (but unusual) to use the x block. This is in contrast to PCA where we have to use the x block error (Section 7.7.3).
- The overall error (strictly the residual) of the 1 PLS component calibration model can be measured in various ways:

  - RSS: residual sum of squares or $RSS = \sum_{i=1}^{1} e_i^2 = \sum_{i=1}^{78} e_i^2 = 2.437 \times 10^6$ in this case.

  - $RMSE$: root mean square error or $RMSE = \sqrt{RSS/(I - A - 1)} = \sqrt{RSS / 76} = 179.06$

    in this case, where $A = 1$ for a 1-component model (some authors do not make this small adjustment of including A in the divisor). The divisor $I - A - 1$ equals the number of degrees of freedom.

  - $\%RMSE$ is the percentage $RMSE$ given by $100 \times \left(RMSE / \bar{c} \right) = 36.07\%$ and is sometimes quoted.

  - R^2 which is the correlation coefficient between observed and predicted and equals 0.317 for this model, which is not terribly encouraging.

Figure 8.1 shows the mean intensity over all 78 chromatograms.

The results for a 1-component model can be presented graphically in Figure 8.2, as follows:

- In the graph of predicted against observed; there is quite a lot of scatter, as reflected in the value of R^2.
- The loadings, weights and regression coefficients, presented as a graph against elution time. These show which peaks are most influential for the first PLS component.
- The loadings and weights, although they emphasise the same main peaks, have different relative intensities.
- When there is only one component, the regression coefficients are proportional to the weights.

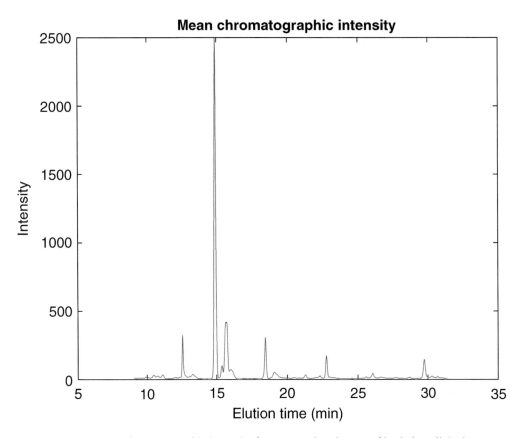

FIGURE 8.1   Mean chromatographic intensity for case study 9 (HPLC of herbal medicine) over all 78 chromatograms.

- It can be shown that $\boldsymbol{b} = \boldsymbol{w}\,q$ when there is only 1 component. So, for example for sample 8, $b_8 = 1.174 \times 10^{-4} = w_8\,q = 0.0220 \times 0.0534$.

We now increase the number of PLS components to 5.

- Comparable graphs are presented in Figure 8.3.
- Note that the loadings and weights involve only to component 5, whereas the predictions and regression coefficients include all 5 components in the model.
- The predictions are now much improved.
  - $RSS = 3.509 \times 10^5$.
  - $RMSE = 69.81$.
  - $\%RMSE = 16.32\%$
  - $R^2 = 0.902$.
- Note that the regression coefficients are no longer linearly related to the weights.
- All peaks in the chromatograms have some influence by the time 5 PLS components are used.

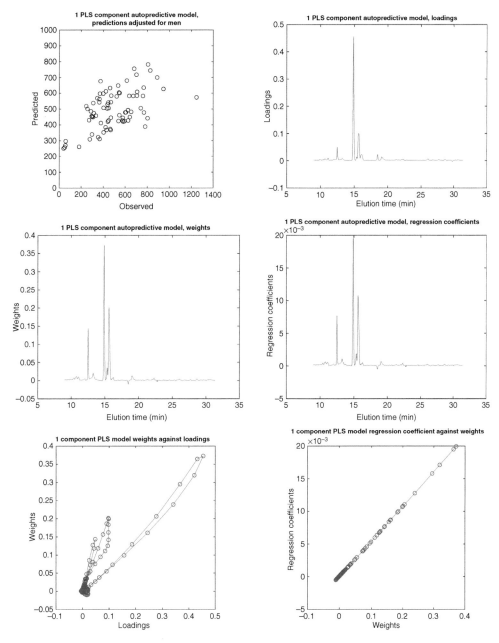

FIGURE 8.2 1 component PLS model for case study 9, HPLC of herbal medicine.

We can visualise the performance as the number of PLS components is increased using the RMSE as presented in Figure 8.4. The trends are as expected. Sometimes these plots are used to estimate the number of significant components in a model, although as we will see later methods such as cross-validation are considered more useful.

In these graphs, we see that there is a small kink between 9 and 10 components, and some will use this to recommend retaining 9 components in the model. Sometimes these trends are clearer if the graphs are on logarithmic axes.

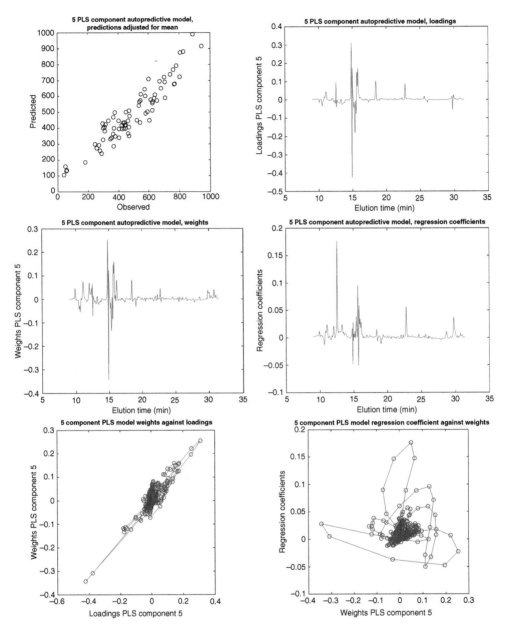

FIGURE 8.3   5 component PLS model for dataset of Figure 8.2.

8.3   TRAINING AND TEST SETS

In most studies involving multivariate calibration, data are split into training and test sets. This concept has been introduced previously (Section 7.7.2). When talking about calibration, many authors prefer to use the concept of calibration (rather than training) and prediction (rather than test) data. The former terminology arises primarily from the machine learning literature, whereas the latter is primarily from

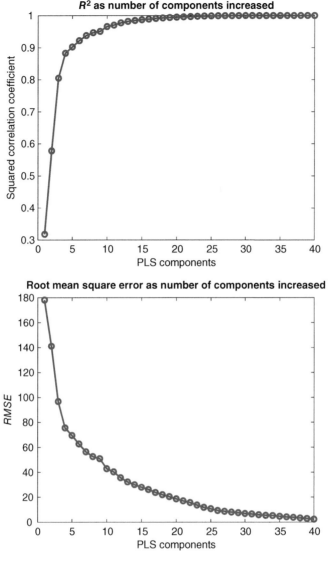

FIGURE 8.4 PLS performance for auto-predictive model of case study 9 as the number of PLS components is increased.

the analytical chemistry literature. To be consistent in this book, we will use the former terminology but denote errors using the latter as is usual in most chemometrics articles and books.

Usually between 1/3 and 1/4 of the data are assigned to the test set, the remaining to the training set. The training set is used to develop a model, and the test set is to determine independently how well the model performs. In Section 8.2, we see that the more the components in the model the better it apparently performs when all the data are used (auto-prediction = assessing model using combined training and test sets). However, later components may overfit the data, and provide over-optimistic predictions; if too many components are used in a model, predictions on new samples (of unknown properties) may be seriously in error because later components are a consequence of noise. So, we need to see how well a model predicts the properties of unknowns.

We will divide the samples of case study 9 (bioactivity of herbal medicine) into

- Training set: samples 1–50 ($I_{train} = 50$)
- Test set: samples 51–78 ($I_{test} = 28$).

We will initially illustrate the calculations using a 5 PLS component model.

- The first step is to mean centre the training set. The mean $^{train}\bar{c} = 558.46$ and for the 291th variable $^{train}\bar{x}_{291} = 865.84$.
- Hence, the mean centred values for the 20th sample are $c_{20} = 365.47 - 558.46 = -192.99$, and for $x_{20,291} = 794.59 - 865.84 = -71.26$. Note these differ from the centred auto-predictive model. because the overall means differ from the training set means.
- Performing PLSR on the training set gives for the first PLS component (except for b which is for all components)
  - $t_{20,1} = 645.15$
  - $p_{1,291} = 0.170$
  - $w_{291,1} = 0.0700$
  - $b_{291,5} = 0.0106$ for the full 5 component model
  - and $q_1 = 0.0570$.
- We also can calculate for the training set using 5 components where 'C' denotes calibration
  - $RSSC = 2.165 \times 10^5$.
  - $RMSEC = 70.14$.
  - $\%RMSEC = 12.56\%$
  - $R^2 = 0.904$.
- $RMSEC$ is usually calculated by $RMSEC = \sqrt{RSSC / (I_{train} - A - 1)}$. Note the divisor equals the number of degrees of freedom $I-A-1$.
- We can now estimate the value using 5 components by $\hat{c} = {}^{cent,\ train}X_{train}\, b + {}^{train}\bar{c}$. The estimated value of $\hat{c}_{train,20} = 595.23$, so the error $e_{train,20} = 365.47 - 595.23 = -229.76$ (using a 5 component PLS training set model).
- For the test set, we first need to adjust by the mean of the training set. The adjusted value of the 10th test set sample (sample 60 overall, as the first 50 samples are from the training set) is $^{test}c_{10} - {}^{train}\bar{c} = 278.24 - 558.46 = -280.22$.

- Then simply calculate $\hat{c} = ^{cent,train}X_{test}\,b + ^{train}\overline{c}$ (the '*cent,train*' superscript means that the test set matrix is adjusted by the centre for the training set).
- So the estimated value of the 10th test set sample (the 60th overall) $\hat{c}_{60} = \hat{c}_{test,10} = 354.75$, and the error $e_{test,10} = 278.24 - 354.74 = -76.51$.
- We can now calculate for the test set where P denotes the prediction.
  - $RSSP = 1.812 \times 10^5$.
  - $RMSEP = 80.44$.
  - $\%RMSEP = 20.85\%$
  - $R^2 = 0.821$.
- $RMSEP$ is usually calculated by $RMSEP = \sqrt{RSSP / I_{test}}$; note the denominator is not usually adjusted for the number of components for the test set, in contrast to the training set.
- Some authors call the test set squared correlation coefficient Q^2 rather than R^2.
- Note that R^2 and $RMSEP$ are worse than for the training set. $RSSP$ is not comparable because there are 50 samples in the training set but only 28 in the test set. $\%RMSEP$ is comparable and is calculated relative to the mean value of c for the test set (385.63) rather than the training set (558.46) which differ by quite a lot due to choice of samples (this is because there are a small number of outliers we have not removed). R^2 of course is a comparable indicator for both training and test sets.
- The test set model for 5 PLS components seems slightly but not substantially worse than the training set model.

The predicted versus observed estimates for both training and test sets are illustrated in Figure 8.5.

We can compare the change in error as the number of components is increased. This often gives a good idea of how many PLS components are optimum for the model. The error decreases for training set models as the number of components is increased (although if $RSSC$ is divided by $I - A - 1$ rather than I to give $RMSEC$ there can sometimes be an increase if the number of components is very large – but this technical issue is beyond this text), whereas for test sets, it usually increases after the optimum number of components is found. If we visualise the predictions for a 10 component model as in Figure 8.6, we see there is a substantial difference between the training and test set.

We can plot the change in various measures of errors as in Figure 8.7.

- For the first 4 PLS components, there is not much difference in performance for the training and test set models.
- Slight differences relate to the choice of samples in the two subsets of samples.
- After 4 components, there is a marked divergence in the performance of the training and test sets. This suggests that 5 or more components overfit the data, and so an optimum model would involve four PLS components.

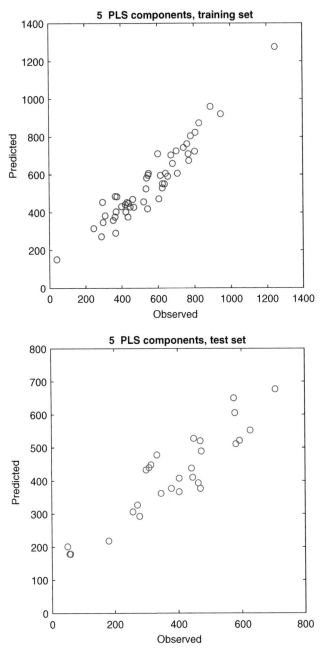

FIGURE 8.5   Training and test set predictions for the 5 component model of the data for case study 9 (HPLC of herbal medicine to measure bioactivity).

Some modern software involves iterative approaches of selecting training and test sets, so that the *RMSE* and their curves are the means of different ways of splitting the data; however, most software only allows the user a single split.

Test sets are not always used as a way of determining the optimum number of components, but can just be used as an independent way of validating a model, the

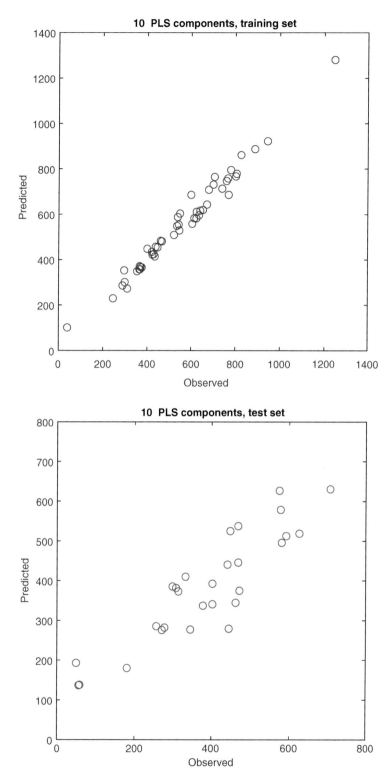

FIGURE 8.6   As Figure 8.5 but with a 10-component model.

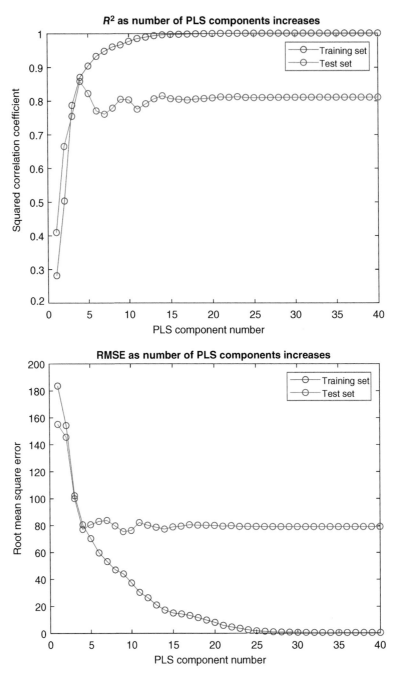

FIGURE 8.7 *RMSE* and R^2 as the number of PLS components are increased for training and test sets and the case study in this chapter.

number of components being chosen internally from the training set, using methods discussed in Section 8.4.

8.4   OPTIMISATION: NUMBER OF PLS COMPONENTS

There are a huge number of ways of estimating the number of significant components in a PLS model. For the first years of the niche chemometrics literature, an enormous literature was developed, as quantitative calibration was important for the correct estimation of properties and concentrations. It would be impossible to summarise this literature in detail, and many software packages use different favoured approaches. We illustrate just one common approach.

One approach is as outlined in Section 8.3, to see how many components are needed to optimise the model on a test set. It is sometimes possible to divide the data into three splits:

- Training set,
- Test set to determine the number of significant components,
- Validation set which is a completely separate set of samples external to the training and test set.

The test set could be resampled many times, perhaps 100 times, each time obtaining a different optimum number of components and a consensus or average used for the model. With modern computing power, this can be done quite easily. There are a large number of such iterative approaches.

However, in the early days of chemometrics computing power was limited, and such computationally intense approaches were impracticable on early micros. Methods developed during these early days were incorporated into the first software and are still commonplace.

Many of the early methods were based on cross-validation. LOO (Leave One Out) cross-validation has been described in the context of PCA in Section 7.7.3. In this chapter, we will illustrate its application for PLS. The method below could be used for PLSDA (Chapter 7) as well as PLSR but we will describe its application in the context of PLSR.

Cross-validation involves the following:

- Leaving out one or more samples from a dataset,
- Forming a model with the remaining samples,
- Then estimating the value of the left-out sample(s) viewing them as a test set of 1 (or a small number of samples),
- Then iterating again to leave out all the samples once.
- Then obtaining predictions for the left-out samples and comparing them to the observed values, to calculate the *RMSECV* (root mean square error of cross-validation) and other comparable statistics for the estimated values of the left-out samples.

- Usually, the training set (minus the left-out samples) is not re-centred or re-standardised although there are various different approaches.
- This is done for models with different numbers of PLS components.

If there are I samples in a training set, each sample is left out once.

As we have described numerically the calculations for a training/test set split, and they are fairly similar in the case of LOO, we will only provide a summary using case study 9 (HPLC of herbal medicine). The auto-predictive model is on the entire dataset (not the training set) in this case. A small difference from the calculations in Section 8.3 is that commonly data are not recentred each time a sample is removed as only one is removed so there is not much difference in means, although some authors and software developers might choose to do this.

- Using a 5 component model, the estimated value for sample 20 is
  - ${}^{cv}\hat{c}_{20} = 499.28$ for cross-validation
  - ${}^{all}\hat{c}_{20} = 468.72$ for an overall auto-predictive model
  - compared to the observed value $c = 365.47$
- If we use a 20-component model we find
  - ${}^{cv}\hat{c}_{20} = 461.95$ for cross-validation
  - ${}^{all}\hat{c}_{20} = 393.42$ for an overall auto-predictive model
- Increasing from 5- to 30-component model reduces the cross-validated error for sample 20 from $e_{cv,20} = -133.82$ to -96.48 but the auto-predictive error from $e_{cal,20}$ -103.24 to -27.95 for sample 20 which is a much more dramatic reduction.

- $RMSE$ is calculated by $RMSE = \sqrt{\sum_i^I e_{all,i}^2 / (I - A - 1)}$, and $RMSECV$ is calculated by $RMSECV = \sqrt{\sum_i^I e_{cv,i}^2 / I}$. Note the different divisor.
- As an example, using 5 PLS components, $RMSE = 69.81$ whereas $RMSECV = 87.13$; as we increase the model to 20 components, $RMSE = 18.65$ but $RMSECV = 77.28$. Hence, $RMSE$ (auto-prediction) has decreased a great deal whereas $RMSECV$ only by a small percentage, suggesting not much advantage in the additional components.

We can best visualise the trends graphically by calculating R^2, $RMSE$ and $RMSECV$ as in Figure 8.8. We can come to similar conclusions as in Section 8.3. Alternatively, some would perform the cross-validation on the training set rather than the overall dataset, and then use the optimum number of PCs obtained from the training set to predict the test set, rather than perform cross-validation on the entire dataset. There are a large number of variations of this theme. Cross-validation is often more

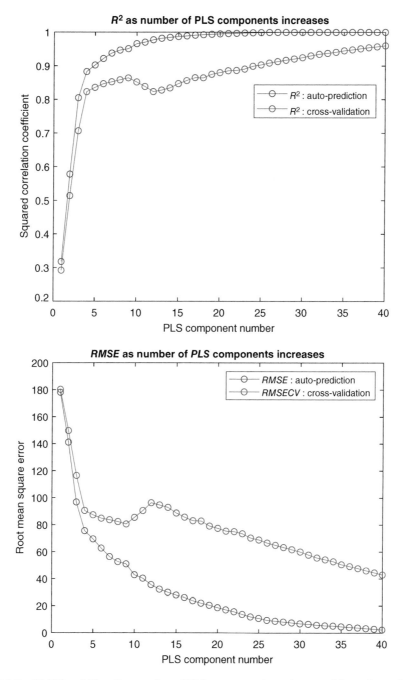

FIGURE 8.8   *RMSE* and *R²* as the number of PLS components are increased for auto-predictive and cross-validated models and the case study in this chapter.

successful when used on PLSR than on PCA when the latter is used for classification as discussed in Chapter 7. However, cross-validation is part of almost all classical chemometrics software and most readers of this book are likely to have come across this method.

CHAPTER 9

Selecting the Most Significant Variables and Markers

9.1 INTRODUCTION

The most important task of most metabolomics laboratories is to provide an indicator of which variables correspond to likely markers. In many studies, there are two or more groups of samples, for example, a control group and a diseased or treated group, and the aim is to guide investigators as to which metabolites are most significant in characterising the difference between these groups.

Much of metabolomics is exploratory, and then often if of sufficient interest, a company may take on further tests and exploration. The majority of published academic projects, for example, are preliminary, but play a valuable role in narrowing down the search for markers. Some variables or metabolites can then be proposed for investigation in the next steps. Not many such metabolites will eventually make the final stage, for example, in clinical diagnosis, which could be many years, or even decades, ahead of the first experiments. But biomarker discovery is still a major business and an essential first step to generate leads.

In earlier chapters, we have already discussed some approaches for variable selection, so where appropriate will only briefly remind the reader referring back to the previous chapters but will also detail additional methods in this chapter.

9.2 UNIVARIATE APPROACHES

There are numerous univariate approaches available.

Simple classical approaches are to use the t, z, F or χ^2 statistics to determine a p value as described in Section 5.4. It is quite common to cite t values as to whether

Data Analysis and Chemometrics for Metabolomics, First Edition. Richard G. Brereton.
© 2024 John Wiley & Sons Ltd. Published 2024 by John Wiley & Sons Ltd.
Companion website: www.wiley.com/go/Brereton/ChemometricsforMetabolomics

individual metabolites are significant markers to discriminate between groups of samples. Whether these are meaningful depends on the variables falling approximately into a normal distribution for each class. There are several ways this can be checked if desired. A plot in which p values are plotted against rank should be approximately linear, except for extremes, which may indicate outliers, which is a simple graphical approach and is illustrated in Figure 7.28. For best practice such rank plots or more formal statistical tests of normality should be done for each variable to check they approximately obey normality; however, in reality standard parametric tests such as the t-test will identify extreme differences and certainly permit ranking of variables as potential markers via p values.

As not all distributions are well represented by a normal distribution alternatives are non-parametric tests, taking into account whether the intensity of a variable (usually representing a metabolite) is greater or lower in one class than another. If there are 20 samples in Class A and 20 in Class B, and all 20 estimated concentrations of a potential metabolite are higher in all samples from Class A, then it is likely to be a good marker for Class A.

The Mann–Whitney U test is a common non-parametric univariate test used frequently in metabolomics, also called the Wilcoxon rank sum test. We will initially illustrate the numerical calculation with the simulated 20×5 dataset of Table 5.1 in Chapter 5. The first 10 samples are members of Class A and the second 10 of Class B.

- The test is performed individually on each variable.
- The first step is to calculate U. There are two ways of doing this. A simple approach is to calculate S_{ij} for all pairs of samples i and j in two classes A and B (which can have different sizes I_A and I_B) so that
  - $S_{ij} = +1$ if $x_{Ai} > x_{Bj}$
  - $S_{ij} = 0$ if $x_{Ai} < x_{Bj}$
  - For equal values $S_{ij} = 0.5$ but this only happens sometimes in metabolomics, mainly if there are missing values which are replaced by an estimate that is the same for each of the missing values as discussed in Section 4.7.1.
- A second approach is to calculate U based on the ranks of each variable. As these two methods result in identical outcome, we will only describe one approach in detail for brevity.
- So if we take the case study, for variable 1, comparing sample $i = 3$ (sample C) in Class A and $j = 5$ (sample O) in Class B (numbering samples from 1 to 10 in each class separately), $x_{A3} = 36.062$ and $x_{B5} = 10.182$, therefore $S_{35} = +1$.
- We can then calculate $U = \sum_{a=1}^{I_a} \sum_{b=1}^{I_b} S_{ab}$ which in our case for variable 1 equals 89.
- The maximum value of U is $I_A I_B = 10 \times 10 = 100$ in our case, and the minimum value is 0. A value of 89 suggests that this variable is overwhelming of higher intensity in Class A samples and likely to be a good marker for Class A.

- Some authors calculate two U values U_A and U_B. The former is as defined above, whereas the latter can be defined by $U_B = I_A I_B - U_A$ or in our case equals $100 - 89 = 11$. However, they both result in the same p value and conclusions. We will follow the calculations just with a single U below.
- The next stage depends on sample size, but if there are a reasonable number of samples, we can calculate
  - the expected mean $m = I_A I_B / 2$ or in our case $(10 \times 10)/2 = 50$
  - and the expected standard deviation $s = \sqrt{\left(I_A I_B \left(I_A + I_B + 1 \right) \right) / 12}$ or in our case $s = \sqrt{\left(10 \times 10 \times 21 \right) / 12} = 13.23$
- We can then calculate the number of standard deviations U is from the mean, which in our case equals $(89 - 50)/13.23$ or 2.95.
- This can be approximately modelled by a z- or normal-distribution (for simplicity we will not discuss other approaches). The two-sided p value of a measurement 2.95 standard deviations from the mean is $p = 0.0032$, which suggests this variable is a very significant marker.
- The p values for all five variables are presented in Table 9.1.
- This can be compared to Table 5.7 of Chapter 5. The general trend is the same and the variables with the lowest p values, 1, 3 and 4, can be identified in both cases. The p values using the Mann–Whitney test are usually not so extreme as those obtained from parametric tests, and in some ways probably more realistic, but for the purpose of identifying potentially discriminatory variables, both approaches work well in this case.

We also illustrate this using case study 1, the pre-arthritic and control donors, studied using LCMS, using standardised data as in Section 5.4, although standardising makes no difference in the results of the Mann–Whitney test.

- There are 19 samples in Class A and 30 in Class B, making a maximum U value of 570 and a mean of 285.
- For readers that like to perform these calculations, variable 1 is Methionine, the U value is 256 so its corresponding p value is 0.552, so not likely to be a discriminatory variable. Note that the calculated p value in this case is somewhat lower than using the t-test ($p = 0.948$), as discussed in Section 5.4.

TABLE 9.1 U and p values using the Mann–Whitney U test for the simulated dataset of Table 5.1.

	1	2	3	4	5
U	89	61	77	0	48
p (mw test)	0.0032	0.4057	0.0413	0.0002	0.8798

- We can present the p values for all 81 variables in Figure 9.1. This should be compared to Figure 5.25. Although high p values differ somewhat between the two figures, low p values (significant markers) are fairly similar suggesting both approaches approximately identify the same markers.
- The six variables with $p < 0.05$ are presented in Table 9.2.
  - Note variable 19 has a half-integer value of U. This is due to the way missing values were replaced (by 0) although other methods of estimating missing variables may provide a different result.
  - Note variables 10 and 77 have the same p values. This is because the U values are both 96 away from the mean of 285.
- We can compare the p values using the t-distribution of Table 5.8.
- Variables 10, 19, 24 and 45 are all identified using both criteria. As $p < 0.05$ is an arbitrary cut-off and depends on many different assumptions, none of which will be exactly obeyed, we do not expect perfect agreement.

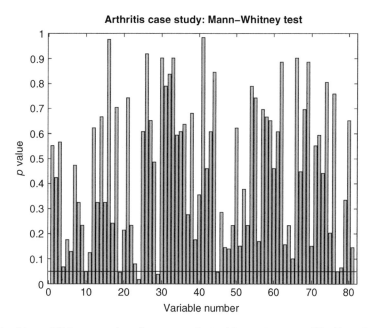

FIGURE 9.1   Mann–Whitney p values for case study 1 with $p = 0.05$ cut-off indicated.

TABLE 9.2   Five variables from the arthritis case study 1 with p values < 0.05 using the Mann–Whitney U statistic.

Variable	10	19	24	29	45	77
Metabolite	Hypoxanthine	Lyso-PC (16:0)	Lyso-PC (14:0)	Glutarylcarnitine	Oleic acid	B-Hydroxypalmitic
U	189	187.5	169	386	382	381
p	0.049	0.045	0.017	0.038	0.046	0.049

- It is important to remember that out of 81 variables, we would anticipate around 4 would have a $p < 0.05$ even when there is no significant difference between the two groups. This does not mean the 6 variables we have identified are not potential markers, but this evidence alone would not be sufficient to be very confident, but it could be regarded as an important an exploratory step to focus further research on a shortlist of candidate metabolites. A marker should have a low p value, but having a low p value does not conversely imply it is definitely a marker. If the same metabolite was identified using two independent tests, this would be good evidence it is a marker.

9.3   LOADINGS, WEIGHTS AND VIP SCORES

Many chemometricians prefer to use multivariate indices. The reason is that metabolites may interact, that is, the concentrations are not always independent of each other. In practice, multivariate approaches often come to similar conclusions to univariate ones, but nevertheless are widespread and certainly provide complementary assessments of significance.

9.3.1   Principal Component Loadings

We have already discussed the use of PC (physical component) loadings in Chapter 4. In this chapter, we will briefly discuss under what circumstances they can be used for determining which variables or metabolites are most likely to be markers.

If PC1 (or another early PC) shows good separation between groups, its loadings can be used to provide indicators as to which metabolites (or variables) are potential markers. In some situations, the difference between two or more groups is the main distinguishing feature of a dataset.

- For case study 10, using the negative ion LCMS, the scores of PC1 after row scaling to a constant total of 1 and standardising, show good separation between the two groups as in Figure 9.2(a), with four probably misdiagnosed samples included (the calculations could be repeated removing these four samples, but for brevity we just present one calculation, as the majority are grouped correctly).
- We can see that the scores of PC1 for Class A (diabetic) are positive and for Class B negative in this case (remember that different algorithms may reverse the signs but this will be consistent for both scores and loadings).
- Hence, loadings that are positive represent possible markers for Class A and negative for Class B.
- We can choose an essentially arbitrary cut-off, in this case, $p = \pm 0.2$ (where p denotes for loadings and not probability), to select those markers that are likely to be significant. Fourteen variables are identified and tabulated in Table 9.3 together with the loadings on PC1. 7 are potential markers for Class A (diabetic) and 7 for Class B (non-diabetic).

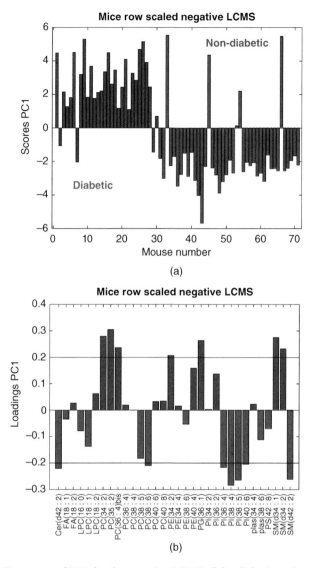

FIGURE 9.2   (a) Top scores of PC1 for the negative LCMS of the diabetic and non-diabetic mice after row scaling and standardising and (b) bottom the corresponding loadings with $p = \pm 0.2$ cut-off indicated.

For some case studies, for example, in animal or plant metabolism, the main factor of interest can often be identified using PCA (Principal Components Analysis). This is because conditions can be controlled, for example, mice of identical genetic stock can be kept in identical environments, with the same food supply, temperature, bedding, etc. Although there will sometimes be small individual differences, in most cases, these will be relatively minor. However, when studying human metabolomics, especially over a long period, for example, disease progression, this is not easy, there will always be significant differences in diet, genetic profiling, personal habits, environment, health profiles, etc. and even difficulties finding a suitably balanced and homogeneous number of donors. Hence, the factor that is most of interest will not necessarily be the most important factor differentiating individuals, and PCA will not be an adequately

TABLE 9.3   Variables from case study 10 (negative ion MS of diabetic and non-diabetic mice) that have loadings of PC 1 > abs(0.2), significant variables for diabetic mice blue and non-diabetic red, using PCA.

Variable	7	8	9	16	20	30	31
metabolite	PC (34 : 2)	PC (35 : 2)	PC (36 : 4)bis	PE (34 : 2)	PG (36 : 1)	SM (d34 : 1)	SM (d34 : 2)
p(loadings)	0.2797	0.3053	0.2368	0.2071	0.2636	0.2751	0.2322

Variable	1	13	23	24	25	26	32
metabolite	Cer (d42 : 2)	PC (38 : 6)	PI (36 : 4)	PI (38 : 4)	PI (38 : 5)	PI (40 : 6)	SM (d42 : 2)
p(loadings)	−0.2205	−0.2096	−0.2163	−0.2834	−0.2638	−0.2045	−0.261

powerful tool as the main source of variability will be represented by other factors, often not systematically studied.

If in contrast, we look at the PC scores of the pre-arthritis dataset (standardised case study 1) we see no particular relationship to class membership (see Figure 9.3) and as such could not meaningfully relate the PC scores to whether a specific metabolite is a marker. To do this, we need to use PLSDA (Partial Least Squares Discriminant Analysis) as a tool.

9.3.2   PLSDA Loadings and Weights

An alternative approach is to employ PLSDA.

We will initially exemplify this with the negative LCMS mouse dataset (case study 10).

- Figure 9.4(a) presents the scores of the first PLSDA component using weighted centring and the method described in Section 7.4.2.2. As this type of calculation has been discussed in detail previously, we will not describe the steps numerically again in this chapter. There is in practice very little difference between the PCA and PLS (Partial Least Squares) scores in this case.
- Figure 9.4(b) illustrates the corresponding PLS loadings, again a very similar pattern to the PCA loadings.
- Using our arbitrary cut-off of $p = \pm 0.2$ (where p stands for PLS loadings) we now find there are 13 significant metabolites listed in Table 9.4, all of which are the same as for PCA (Figure 9.2) except for variable 26, which is very close to the cut-off.
- As the two groups are well separated using PC1 because this is the main factor in differentiating the groups, using PLS makes very little difference to the solution. PLS can be viewed as a rotation of the original data, and if the rotation is optimal for separating into groups because this is the main factor of variability, there is little advantage in using PLS.

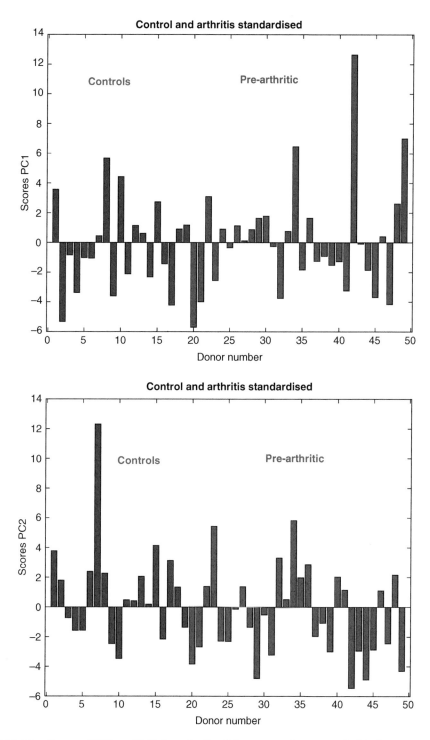

FIGURE 9.3   Scores of PC1 and PC2 (standardised) for case study 1 pre-arthritis dataset.

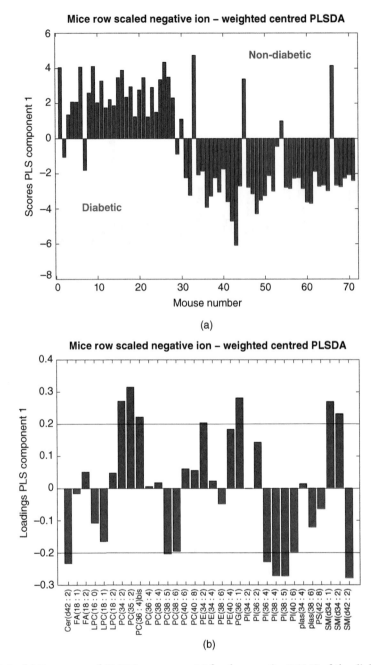

FIGURE 9.4   (a) Top scores of PLSDA component 1 for the negative LCMS of the diabetic and non-diabetic mice after row scaling and standardising and (b) bottom the corresponding PLS loadings with $p = \pm 0.2$ cut-off indicated.

TABLE 9.4 Variables from case study 10 (negative ion MS of diabetic and non-diabetic mice) that have loadings and weights of PLS component 1 > abs(0.2), significant variables for diabetic mice blue and non-diabetic red, using PLSDA.

Loadings

Diabetic

Variable	7	8	9	16	20	30	31
Metabolite	PC(34:2)	PC(35:2)	PC(36:4)bis	PE(34:2)	PG(36:1)	SM(d34:1)	SM(d34:2)
p(loadings)	0.2708	0.3144	0.2216	0.2035	0.2808	0.2686	0.2312

Non-diabetic

Variable	1	12	23	24	25	32
Metabolite	Cer(d42:2)	PC(38:5)	PI(36:4)	PI(38:4)	PI(38:5)	SM(d42:2)
p(loadings)	−0.2327	−0.2033	−0.2286	−0.2719	−0.2724	−0.2792

Weights

Diabetic

Variable	7	8	19	20	30	31
Metabolite	PC(34:2)	PC(35:2)	PE(40:4)	PG(36:1)	SM(d34:1)	SM(d34:2)
w(weights)	0.2308	0.323	0.2191	0.2788	0.2285	0.2029

Non-diabetic

Variable	1	12	23	24	25	32
Metabolite	Cer(d42:2)	PC(38:5)	PI(36:4)	PI(38:4)	PI(38:5)	SM(d42:2)
w(weights)	−0.2075	−0.2062	−0.2410	−0.2282	−0.2727	−0.3056

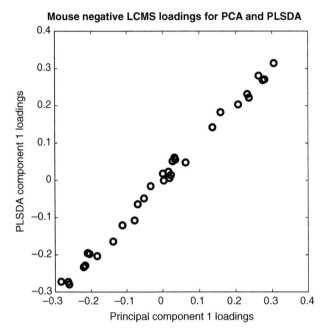

FIGURE 9.5   Loadings for the mouse negative LCMS data of case study 10 using PCA and PLSDA.

- In Figure 9.5, we present a graph of the PLSDA versus PCA loadings of the first component and see there is very little difference, so either could have been used in this case.
- Many people prefer to use the PLS weights rather than loadings. These often only differ a little but have better theoretical properties, in that they are orthogonal to each other. The weights for metabolites with a value > ±0.2 are presented in Table 9.4 and can be compared to the metabolites obtained using loadings. There is no difference for Class B (non-diabetic) and only a small difference for Class A (diabetic). As the choice of cut-off is arbitrary, this difference is not very important (Figure 9.6).

In contrast, for the case of pre-arthritis (case study 1), PLS makes a major difference compared to PCA. This is because the PLS solution is forced to emphasise the difference between the predefined two groups, even though this is not the major factor differentiating the donors.

- Figure 7.16 (Chapter 7) shows the scores plot of the first 2 PCs of the standardised data using both PCA (centred) and PLSDA (weighted centred). Whereas there is no real separation in the PC scores plot, there is a promising separation in the scores of the first 2 PLS components.
- Figure 9.7 shows the corresponding scores, loadings and weights of the first PLSDA component, calculated as described in Section 7.4.2.2 of Chapter 7.

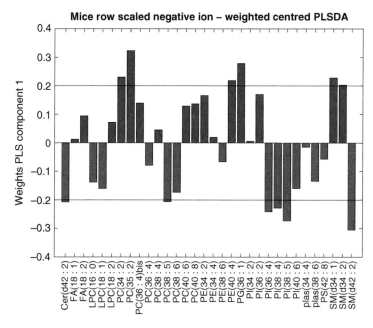

FIGURE 9.6   PLSDA weights of the first component corresponding to the loadings in Figure 9.4.

- The cut-off of $p = \pm 0.2$ and $w = \pm 0.2$ are illustrated. It is important to understand this is an arbitrary cut-off and has no formal statistical basis, but still can be used to sift out the most significant markers.

- The metabolites that fall outside the cut-off criteria are listed in Table 9.5. Less metabolites are identified using weights rather than loadings, with only 2 as markers for Class A (controls) and 4 for Class B (pre-arthritic) compared to 10 and 6, respectively, but it is important to appreciate the cut-off is arbitrary. As the sum of squares of the weights is 1, the more variables there are, the less are likely to have $w > 0.2$.

- However, of the 6 markers identified using weights, 5 are also found using loadings, and these have the highest absolute values of loadings, so there is reasonable agreement as to the likely most significant metabolites.

- For reference, we can plot the loadings of PLSDA component 1 versus the loadings of PC1 in Figure 9.8 and see they are very different, unlike for the mouse data of Figure 9.5, primarily because in human data there are many other factors that influence variability that are important.

9.3.3   VIP Scores

Sometimes more than one PLS component is needed to characterise the provenance of a sample using an A component model. In such case, the loadings or weights of one component are insufficient.

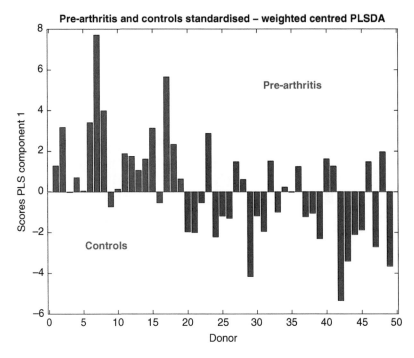

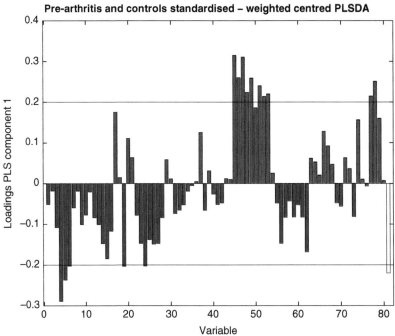

FIGURE 9.7　Scores, loadings and weights of PLSDA component 1 of case study 1 (pre-arthritis) data; loadings and weights are coloured by sign.

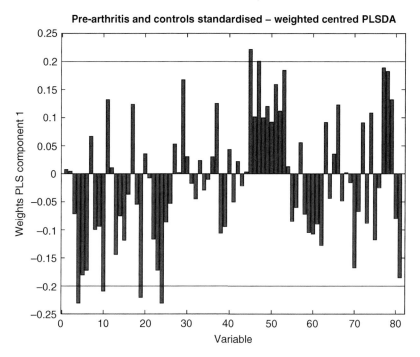

FIGURE 9.7 (Continued)

VIP (variable importance in projection) scores are a way of obtaining the average importance over several components.

We will illustrate the calculation numerically with the simulated 20×5 dataset (X matrix) of Table 5.1 in Chapter 5.

- The first step is to perform PLSDA centring the X matrix, labelling samples 1–10 (A–J) with $c = +1$ and samples 11–20 (K–T) with $c = -1$, and retain 5 non-zero PLS components.
- To check we find $t_{73} = 5.255$ (the score of object G and component 3), $p_{42} = 0.345$ (the loading of component 4 and variable 2) and $w_{24} = 0.363$ (the weight of variable 2 and component 4), remember to ensure the loadings for each component are row vectors, whereas the weights are column vectors, in the notation in this book; some authors transpose the loadings.
- To check your calculations, you should check also that $\sum_{j=1}^{J} w_{ja}^2 = 1$, that is, the sum of weights for each component $a = 1$ (this is not so for the loadings using the NIPALS algorithm).
- The VIP score for variable j for an A component model is defined by

$$VIP_j = \sqrt{J \sum_{a=1}^{A} w_{ja}^2 q_a^2 t_a' t_a \Big/ \sum_{a=1}^{A} q_a^2 t_a' t_a} = \sqrt{J \sum_{a=1}^{A} w_{ja}^2 q_a^2 \sum_{i=1}^{I} t_{ia}^2 \Big/ \left(\sum_{a=1}^{A} q_a^2 \sum_{i=1}^{I} t_{ia}^2 \right)}$$

- where the data matrix has dimensions $I \times J$ (in our case 20×5)
- t, w and q are as defined by PLS (see Section 7.4.2)

TABLE 9.5 Variables from case study 1 (control and pre-arthritic donors) that have loadings and weights of PLS component 1 > abs(0.2), significant variables for controls blue and pre-arthritic red, using PLSDA.

Loadings

Class A (controls)	Metabolites	Loadings	Class B (pre-arthritic)	Metabolites	Loadings
45	Oleic acid	0.3144	4	Kynurenine	−0.2894
46	Linolenic acid	0.2594	5	Tryptophan	−0.2368
47	Cis-gondoic acid	0.3104	6	Phenylalanine	−0.2025
48	Docosahexaenoic acid	0.2232	19	Lyso-PC(16 : 0)	−0.2034
49	Docosapentaenoic acid	0.2584	24	Lyso-PC(14 : 0)	−0.2027
51	Palmitic acid	0.2398	81	3-Indolelactic acid	−0.2197
52	Cis-9-palmitoleic acid	0.2134			
53	Stearic acid	0.2198			
77	B-Hydroxypalmitic acid	0.2140			
78	Beta-hydroxylauric acid	0.2504			

Weights

Class A (controls)	Metabolites	Weights	Class B (pre-arthritic)	Metabolites	Weights
45	Oleic acid	0.2216	4	Kynurenine	−0.2306
47	Cis-gondoic acid	0.2007	10	Hypoxanthine	−0.2092
			19	Lyso-PC(16 : 0)	−0.2204
			24	Lyso-PC(14 : 0)	−0.2308

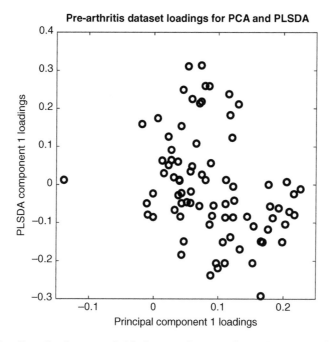

FIGURE 9.8   Loadings for the pre-arthritis dataset of case study 1 using PCA and PLSDA.

- in our case $\boldsymbol{q}' = [0.0476\ 0.0113\ 0.0093\ 0.0047\ 0.0005]$
- For each of 5 components $\sum\limits_{i=1}^{20} t_{ia}^2 = 10^3 \times \left[7.798\ 1.070\ 0.618\ 1.051\ 2.354\right]$
  $(a = 1-5)$; note that the sum of squares of the PLS scores does not monotonically decrease unlike for PCA.
- So for variable 1, $\boldsymbol{w}_1' = [0.373\ -0.559\ 0.015\ 0.713\ 0.198]$
- Hence, for variable 1 using a 5 component model,

$$VIP_1 = \sqrt{5 \times (2.461 + 0.043 + 0.000 + 0.012 + 0.000)/17.905}$$
$$= \sqrt{5 \times 2.516/17.905} = 0.838$$

- Note that the contributions for components 3 and 5 are small but if presented to more significant figures not zero.
- The values of *VIP* using 5 PCs are $\boldsymbol{VIP} = [0.838\ 0.181\ 0.740\ 1.922\ 0.146]$ for all five variables.
- Note that the mean sum of squares $\sum\limits_{j=1}^{J} VIP_j^2/J = 1$ as can be verified and VIP scores will always be positive.
- If a different number of components are used in the model, the VIP scores of course will differ, for example, if there were only 2 PLS components, $\boldsymbol{VIP} = [0.838\ 0.154\ 0.742\ 1.926\ 0.118]$; although the difference is not great it is because the larger components contribute to most of the discrimination in this case.

- Often a VIP score of 1 is set as a cut-off, variables with scores higher than this are considered significant. Rather like the arbitrary cut-off of 0.2 we used for loadings and weights, this has no specific statistical basis, but is for guidance. This is indicated in Figure 9.9 for a 5 component model. Variable 4 is seen to be significant by this criterion, whereas 2 and 5 have little discriminatory power.

We can calculate VIP scores for case studies 1 (pre-arthritis) and 10 (mice using the negative LCMS data) to compare with one component PLSDA and PCA criteria.

- The VIP scores using a weighted centred 10 PLS component model, after first scaling and standardising, for case study 10 negative LCMS mouse data are presented in Figure 9.10 together with a cut-off of VIP = 1.
- 15 metabolites are identified with VIP scores above this cut-off and listed in Table 9.6. Most are the same as those identified using other approaches.
- The group for which each metabolite is a marker is not identified using VIP scores, but can be obtained using numerous complementary approaches, for example, the signs of the loadings or weights using the first PLS component where this is discriminatory.
- 15 metabolites are almost half the total number identified (32) and suggests this cut-off is somewhat low. However, there is no specific statistical justification for using this cut-off, and we will discuss how p (probability) values can be calculated in Section 9.3.4. Nevertheless, VIP scores are useful multivariate indicators when several components are needed to form a model and can definitely be used for ranking variables.

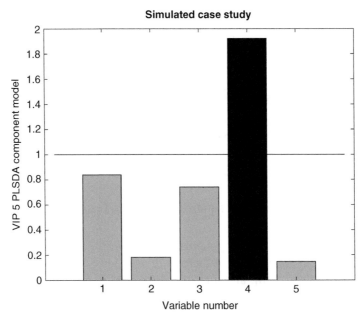

FIGURE 9.9   VIP values for a 5 PLSDA component model for the simulated case study of Table 5.1, with a significance cut-off of 1 indicated.

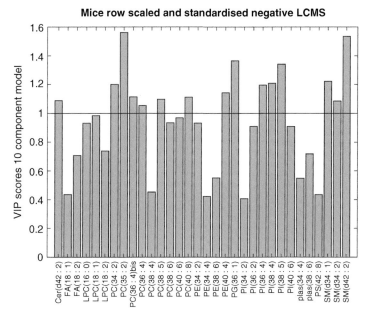

FIGURE 9.10 VIP scores for a 10 PLS component model for the mouse negative LCMS dataset of case study 10 with a limit of 1 indicated.

TABLE 9.6 Variables from case study 10 (negative ion MS of diabetic and non-diabetic mice) that have VIP scores > 1, using a 10 PLS component model.

Variable	1	7	8	9	10	12	15	19
Metabolite	Cer (d42 : 2)	PC (34 : 2)	PC (35 : 2)	PC (36 : 4) bis	PC (36 : 4)	PC (38 : 5)	PC (40 : 8)	PE (40 : 4)
VIP scores	1.0883	1.2001	1.5609	1.1141	1.0542	1.097	1.1114	1.1426

Variable	20	23	24	25	30	31	32
Metabolite	PG (36 : 1)	PI (36 : 4)	PI (38 : 4)	PI (38 : 5)	SM (d34 : 1)	SM (d34 : 2)	SM (d42 : 2)
VIP scores	1.3662	1.1959	1.2084	1.342	1.222	1.0853	1.5349

- The corresponding VIP scores for a 10-component PLSDA model for the pre-arthritic dataset of case study 1, after standardising and weighted centring, are illustrated in Figure 9.11. For brevity, we will not list all the metabolites whose VIP score is greater than 1, but note that 33 metabolites out of 81 are identified by this criterion, which is almost certainly too many. The suggestion would be to increase the VIP cut-off to filter out only the most significant metabolites. VIP scores can also be used to rank the metabolites in order of influence in a model consisting of several PLS components.

VIP scores are commonly employed in chemometrics for determining the relative significance of variables.

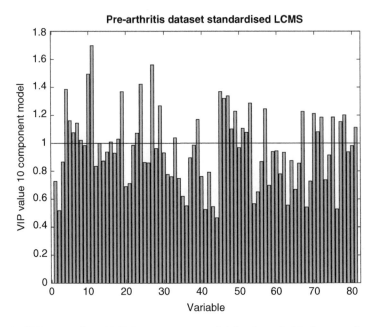

FIGURE 9.11   VIP scores for a 10 PLS component model for the arthritis dataset of case study 1 with a limit of 1 indicated.

TABLE 9.7   Metabolites with p values less than 0.1 using VIP values for a 10-component PLSDA model for the arthritis case study.

$p < 0.05$				
Variable	10	11	24	27
Metabolite	Hypoxanthine	Niacinamide	Lyso-PC(14 : 0)	Isovalerylcarnitine
p values	0.0192	0.0014	0.0459	0.0117

$0.05 < p < 0.1$				
Variable	4	19	45	46	47
Metabolite	Kynurenine	Lyso-PC(16 : 0)	Oleic acid	Linolenic acid	Cis-gondoic acid
p values	0.0641	0.0714	0.0596	0.0863	0.0794

9.3.4   P Values

It is not common to attach a *p* value to loadings, weights or VIP values, but there are some empirical ways this can be done. As always *p* values should be viewed as indicative rather than having a precise statistical meaning, but they can nevertheless be used as a general indicator of which metabolites or variables are likely to correspond to potentially significant markers.

A simple approach is via the jack-knife, which is similar to cross-validation discussed in Sections 7.7.3 and 8.4 but instead of removing a sample or variable and calculating performance or errors on the sample/variable removed, it is calculated on the remaining data. We also are primarily interested in the effect of removing individual variables rather than samples. For calculating the p value of a variable, if there are 20 variables in a dataset, each variable is removed individually and the VIP scores are calculated on the remaining 19 variables. We can then see where the VIP score of the left-out variable (using the full dataset) would fit into the distribution formed by the remaining 19 variables.

We will illustrate this with the pre-arthritis data (case study 1). We need sufficient variables to be able to be able to adequately define a distribution of VIP scores, so the method would not be suitable for the simulated case study (five variables). As case study 1 consists of 81 variables it is suitable for illustrating the method.

- The first step is to standardise the data so for example $x_{17,20} = -0.992$.
- The next step, in our implementation of PLSDA, is to weight mean centre each column, as there are 19 samples in Class A (controls) and 30 in Class B (pre-arthritic) so that $x_{17,20}$ becomes -1.002. Note that there are other implementations of PLSDA that involve slightly different methods of column transformation.
- We label the 19 Class A samples as $c = +1$ and the 30 Class B samples as $c = -1$.
- We then calculate the VIP values for the full 49×81 matrix, choosing 10 PLS components in this case (it can be any number up to 49). As an example $VIP_{20} = 0.690$ (for Carnitine, the 20th variable).
- Then each variable is removed one by one. If we removed variable 40, we then perform PLSDA on the 49×80 matrix consisting of variables 1–39 and 41–81, and find $VIP_{20(\sim40)} = 0.699$ being the VIP value for variable 20 when variable 40 has been removed. There will be a distribution of 80 VIP scores for the background distribution as each of the 81 variables is removed one by one.
- We now calculate the mean and (population) standard deviation for the VIP scores when each variable is removed. So, when variable 40 has been removed, $\bar{v}_{\sim40} = 0.964$ and $s_{\sim40} = 0.266$.
- The VIP score of variable 40 for the entire dataset is $VIP_{40} = 0.761$.
- We now calculate the number of standard deviations the VIP score of this variable is from the mean, of its distribution or $z_{40} = \left(VIP_{40} - \bar{v}_{\sim40}\right)/s_{\sim40} = \left(0.761 - 0.964\right)/0.266 = -0.763$.
- To obtain a p value, we use the one-sided normal (or z) distribution and find $1 - normcdf(-0.763) = 0.777$, which is the p value for variable 40 (Malic Acid) using 10 PLS components and VIP scores.
- In order to check that a normal distribution is a reasonable approximation, we could produce a rank graph, in which the p value for each variable is plotted against its rank, and we find it is nearly linear, justifying the use of a normal- or z-distribution, as in Figure 9.12.
- We can now present the p values for each metabolite as in Figure 9.13. If we set a limit of $p < 0.05$, we find there are only 4 metabolites below this limit, and a further 5 metabolites with $0.05 < p < 0.1$. These are listed in Table 9.7.
- Note that all six metabolites with an absolute value of the weight of <0.2 in the first PLS component also have p values using VIP of <0.1.

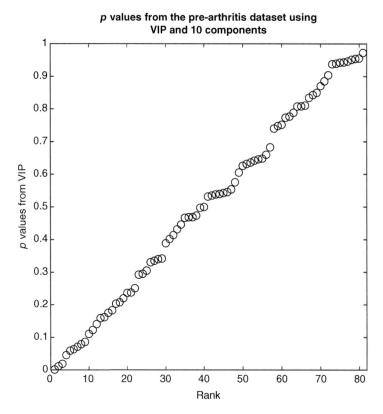

FIGURE 9.12   Graph of p values using a 10-component PLSDA model, against rank for case study 1 as discussed in the text.

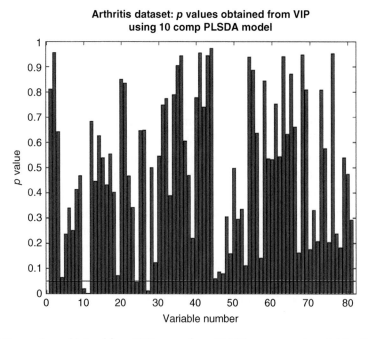

FIGURE 9.13   p values obtained from VIP scores for a 10 PLS component model for the pre-arthritis dataset of case study 1 with a limit of $p = 0.05$ indicated.

It is very important to understand throughout that if the data were normally distributed, we would expect around 5% (or 4) metabolites to have a p value of <0.05 and 10% (or 8) <0.1, and we find this is closely followed in the empirical model. This could at first suggest that there are no real discriminatory variables. However, in practice, variables with low p values are candidate markers and if the same metabolites are found to have low p values using independent tests, this is good evidence of their significance. In this text, we do not discuss how to combine many different tests, for example, by using Bayesian statistics, but this is a recommended next step should resources be available. In many preliminary projects, the work stops at this phase, and it would be up to the funder to then decide whether to invest in studying some of the potential markers further by performing new experiments. The p values can still be used to remove from consideration those metabolites that have high p values and focus on the remaining ones for further investigation. Any real marker should have a low p value, although the converse that a low p value implies a metabolite is a marker is not true.

This approach using the jack-knife could also be applied to loadings and weights, or using models with different numbers of PLS components, but for brevity, we just illustrate the method on one dataset. There need to be enough variables to obtain a reasonably well defined normal distribution for the method described above to be valid.

9.3.5 Multilevel PLSDA

In some cases when there are more than two groups, they are sequentially unrelated, for example, three genotypes, and the only practicable approach is to perform one versus all PLSDA, but in other cases, they are related sequentially. There are several examples in this text. Case studies 2 (malaria), 4 (human diabetes) and 5 (maize at different harvesting temperatures).

In this case, we can perform PLSDA at several levels. For the wheat, we could put $c = +1.5$ for class A (20 °C), $c = +0.5$ for class B (16 °C), $c = -0.5$ for class C (13 °C) and $c = -1.5$ for class D (8.5 °C). We will not discuss column centring in detail in this chapter, which becomes a bit complicated for unequal class sizes, but as there are equal (54) samples in each class, for this case study, we simply standardise X (the 216×34 data matrix) and the columns will be centred, as well as the 216×1 c vector. Note that we use standardised rather than log scaled data for convenience, as the former transformation automatically centres the columns. Obviously, there are many other possible approaches to pre-processing the data.

We briefly outline the main steps:

- To check your calculations, if you are following this text numerically, $x_{23,7} = 1.459 \times 10^3$ (involving sample 23 at 20 °C and the metabolite identified as chlorogenate 2. After standardisation (using the population standard deviation), this becomes $x_{23,7} = -1.129$. The corresponding $c_{23} = +1.5$ in our notation.
- The corresponding PLSDA score of the first component for sample 23 is $t_{23,1} = 3.126$ using a four-level model. The loading of variable 7 in the first PLS component is $p_{1,7} = -0.0349$ and the weight is $w_{7,1} = -0.0058$. The scores of PLS components 1 and 2 are presented in Figure 9.14.

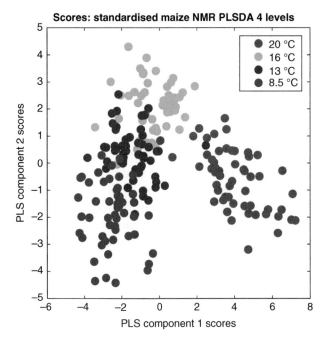

FIGURE 9.14   PLSDA scores of the first 2 components for the four-level PLSDA model for standardised case study 5 maize data.

- We can present the scores of PLS component 1 as a bar chart, with the corresponding loadings in Figure 9.15. We see that the largest loadings in magnitude correspond to malate, alanine and glutamine, all markers for high temperature. We could of course select those loadings whose magnitude is above a defined cut-off if preferred. We also see that the higher the harvesting temperature, the more positive the scores (remember some software will reverse the sign of the PCs).
- We will illustrate using a 5 component PLSDA model.
- There is no strict agreement about the decision criteria according to which class a sample is predicted to belong to, but as the data were calibrated to a vector with $c = +1.5, +0.5, -0.5$ and -1.5, a simple criterion might be that a sample belongs to Class A (20 °C) if $c > 1$, Class B (16 °C) if $0 < c < 1$, Class B (13 °C) if $-1 < c < 0$ and Class D (8.5 °C) if $c < 1$. This simple rule may not be appropriate if there are different class sizes.
- We find that using a five-component model $\hat{c}_7 = 1.636$. Note that if a 1 component model is used we would find $\hat{c}_7 = 2.422$. However, this suggests sample 7 is well placed in its correct class.
- The predicted values are presented in Figure 9.16. It is important to appreciate that this is an auto-predictive model and to see if it is useful the model should be validated by methods discussed in Section 7.7.2. We recommend all classification models are validated as good practice.

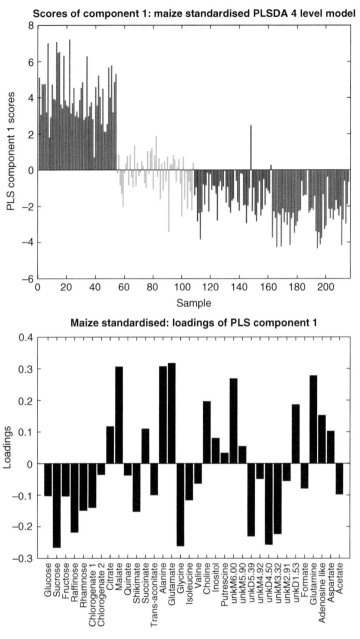

FIGURE 9.15   Scores and loadings of the first component of four-level PLSDA model of the standardised maize data.

- The VIP scores and p values as calculated above are also presented in Figure 9.16, for the 5 component model. If you are checking your calculations, we find that $VIP_{18} = 0.714$ corresponding to a p value of 0.685 (for valine) suggesting this is not likely to be a significant marker for temperature effects. The $VIP = 1$ and $p = 0.1$ cut-offs are indicated but for VIP are essentially arbitrary.

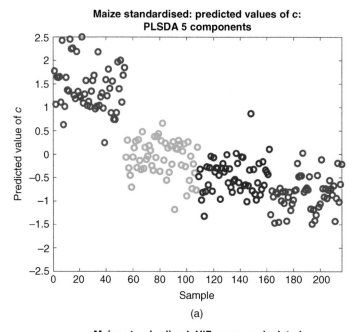

(a)

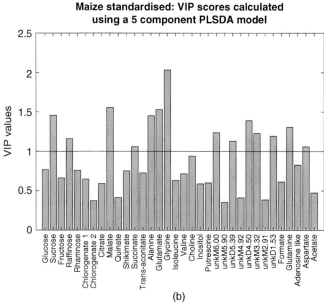

(b)

FIGURE 9.16   PLSDA at four levels for the standardised maize data, using a five-component model, (a) top predictions of c, (b) middle VIP values and (c) bottom p values calculated from VIP, with a VIP $= 1$ and $p = 0.1$ cut-off thresholds indicated.

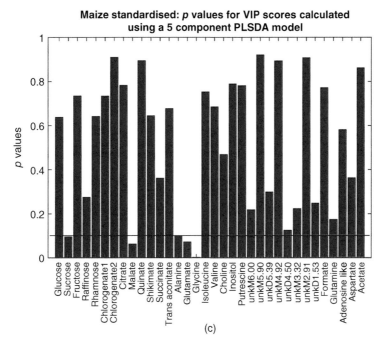

FIGURE 9.16   (Continued)

- The four variables with p values < 0.1 are presented in Table 9.8. These can be compared with other criteria such as the PC loadings (Figure 4.13). As these criteria are different we do not expect an exact correspondence, but they all have relatively large absolute values of loadings in PC1.
- Note that 4 out of 34 variables with $p < 0.1$ represent approximately 10% of the variables, so these could have been found just by chance, however, they are the variables that exhibit the most extreme variation over the temperature range, and if they also have low p values via an independent test are good candidates as markers.

As another example, we will briefly illustrate the results of PLSDA using three levels labelled by $c = 1$ (normal glucose tolerance), $c = 0$ (impaired glucose tolerance) and $c = -1$ for the human diabetes case study 3. Note that the numbers in each group differ but only very slightly and we have centred the columns of the X matrix, which will not

TABLE 9.8   Four metabolites with p value < 0.1 using VIP scores of the four-level PLSDA model on the standardised maize data.

Variable	2	9	15	16
Metabolite	Sucrose	Malate	Glutamate	Glycine
VIP scores	1.455	1.5521	1.5274	2.031
p values	0.0947	0.0623	0.071	0.0022

seriously affect the direction of the weights or loadings or magnitude of p values or VIP scores. We will row scale but not standardise the data in this case.

Figure 9.17 is of the mean intensity and the weights of the first PLS component using a three-level model, as a function of NMR chemical shift. The weights indicate which regions of the spectrum are most discriminatory, and the sign which are more intense when c is positive (normal glucose tolerance) and vice versa. It presents a straightforward visual way of identifying potential markers. Of course, we could do a full analysis, for example, with loadings, scores, VIP scores, etc. to gain complementary insights but for brevity illustrate how a three-level PLSDA model can be applied to this dataset.

Multilevel PLSDA models are not very successful in assigning samples to their groups as discussed in Section 7.6.3 of Chapter 7 because of the problems of assigning samples to intermediate groups but are useful in searching for potential markers whose concentrations vary monotonically over the groups. We will describe complementary approaches for this purpose using ANOVA in Chapter 10.

9.4   SELECTIVITY RATIOS

Selectivity ratios are an alternative to VIP scores.

We will illustrate this use case study 10 and negative ion LCMS. Usually, selectivity ratios are calculated as a consequence of PLSDA models, but they could be defined for other methods for multivariate regression, the key is the regression vector $\boldsymbol{b}$. We will describe this method in the context of PLSDA as this is the most common and widespread application.

- The data forms a 71×32 matrix with samples 1–30 (class A) being mice diagnosed as diabetic and samples 31–71 (class B) as non-diabetic.
- The data are first row scaled to 1 and then standardised down each column to give a new $\boldsymbol{X}$ matrix which we use below. To check, $x_{23,12} = -0.976$.
- A 71×1 $\boldsymbol{c}$ vector is defined, for which class A is labelled as +1 and class B as −1.
- The first step is to perform PLS. We will initially illustrate with a 10-component model. To check your calculations $t_{23,5} = -1.539$.
- We also obtain the $J \times 1$ $(=32 \times 1)$ regression vector $\boldsymbol{b}$. Note that this vector depends on the number of PLS components in the model. For our 10-component model in this example, for the 12th variable, we find $\boldsymbol{b}_{12} = 0.383$.

- The next step is to scale this vector to give a weights vector $\boldsymbol{w}_{tp} = \boldsymbol{b} / \sqrt{\sum_{i=1}^{I} b_i^2}$. In our case $\sqrt{\sum_{i=1}^{I} b_i^2} = 1.1478$ so for the 12th variable $w_{tp,12} = 0.334$. The subscript 'tp' stands for target projection and is a step in the calculation of selectivity ratios.

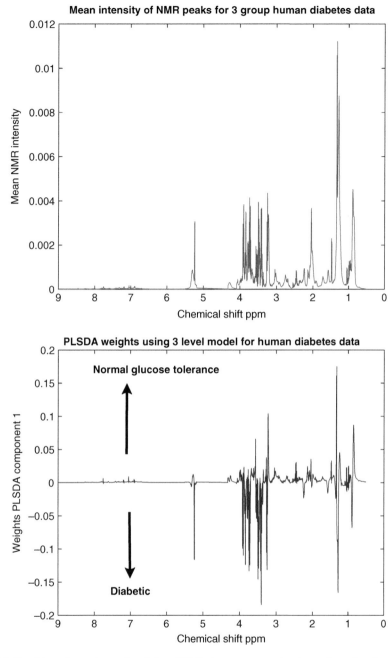

FIGURE 9.17 Mean intensity and weight of PLSDA component 1 using a three-level model for the human diabetes case study as described in the text.

- The next step is to calculate the target projection scores given by $t_{tp} = X w_{tp}$. Note that there is only one component in this model although 10 components were used to calculate the regression vector. In our case $t_{tp,23} = 0.6944$. The target projection scores correspond to variability than can be explain by the classifier c.
- The corresponding target projection loadings can be obtained by $p_{tp} = t'_{tp} X$ so $p_{tp,12} = -0.6540$.

We can now make predictions. The key is to be able to separate the data that is predicted using the classifier from the remainder, which can be viewed as an error (or within-group variability or orthogonal variance).

- $X_{tp} = t_{tp} p_{rp}$ is the predicted data matrix using just one component from target projections. It should include only variation that is correlated to the c value (or classifier). Other variation will not be correlated to classification and is equivalent (but not numerically exactly the same) as the orthogonal variation in OPLS (Orthogonal Partial Least Squares) as discussed in Section 7.4.2.3 of Chapter 7. Obviously, this depends on the number of PLS components used in the model and for proper interpretation requires a judicious choice of how many components are employed to estimate b.
- The error is given by $E_{tp} = X - X_{tp}$, which can also be called orthogonal or within group variability.
- In our case, we can calculate various sum of squares
  - The total sum of squares for the row-scaled standardised X

$$SSA = \sum_i^I \sum_j^J x_{ij}^2 = 2272 = 71 \times 32.$$

  - The sum of squares for X_{tp} $SSTP = \sum_i^I \sum_j^J x_{tp,ij}^2 = 430.12$

  - The sum of squares for E_{tp} $SSETP = \sum_i^I \sum_j^J e_{tp,ij}^2 = 1841.87$

- Note that $SSTP + SSETP = SSA$
- Although $SSTP$ is less than a quarter $SSETP$, this mainly suggests that the effect we are looking at (whether a mouse is diagnosed diabetic or not) is only a small factor in metabolic variability. This is not unexpected as there will be many other factors influencing the metabolic profile of the mice.

- The selectivity ratio is defined for each variable j as $sr_j = \sum_{i=1}^I x_{tp,ij}^2 / \sum_{i=1}^I e_{tp,ij}^2$. So for variable 12 (PC(36:4)) we find $sr_{12} = 18.610/52.390 = 0.355$. As this is considerably below 1, it is not a good discriminatory variable.
- The selectivity ratios for all 32 metabolites are presented in Figure 9.18.
- We can compare to Figure 9.10 which is of the 10-component VIP scores for the same dataset. We see that the metabolites with the highest VIP scores (PC(35 : 2)

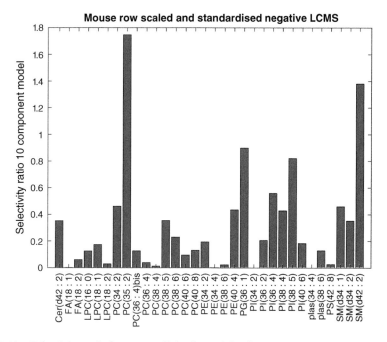

FIGURE 9.18   Selectivity ratio for mouse diabetic model using 10 PLSDA components.

and SM(d42 : 2)) are the same for selectivity ratios, but the latter emphasises the difference between the metabolites better.

- A selectivity ratio greater than 1 is likely to be an excellent marker. There is no strong guideline as to where to cut off the selection, however, using the mean ratio as a cut-off, selects 12 metabolites as possible markers.
- We can compare selectivity ratios to VIP scores as in Figure 9.19. Although most fall onto a monotonic curve, there are some variables with high VIP scores and low selectivity ratios that may not be good markers. VIP scores may select some false marker candidates because they may pick variables where the explained variance is dominantly orthogonal.

9.5   VOLCANO PLOTS

A common graphical method for visualising which variables (or metabolites) are significant to differentiate two groups of samples is a volcano plot. It combines a p value (e.g. using the t statistic) with a fold change. The latter is the ratio between the mean intensity of a variable i in Class A to that in Class B, $f_{iAB} = \bar{x}_{iA} / \bar{x}_{iB}$.

We will illustrate the method using the 71×146 case 10 dataset of the positive LCMS of the mice, of which the first 30 are known diabetic (Class A) and the remaining 41 diagnosed as non-diabetic (Class B).

- The first step is to row scale. This can be followed by standardising, to check the scores plot, which should be the same as Figure 4.17(a).

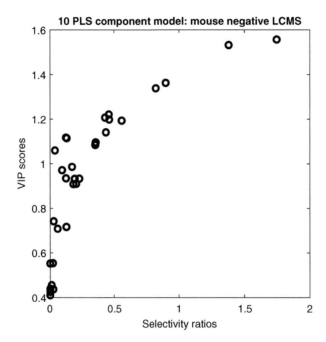

FIGURE 9.19   Comparison of VIP scores and selectivity ratio for case study 10 negative LCMS of diabetic and control mouse using 10 PLSDA component model.

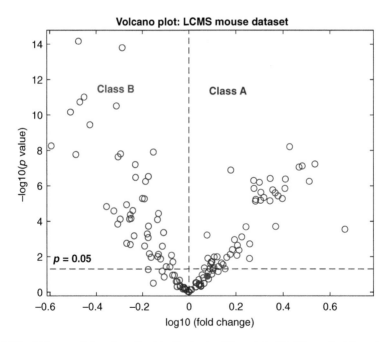

FIGURE 9.20   Volcano plot as described in the text, with metabolite 73 marked in red.

- As an example, the value of the row-scaled and standardised element $x_{20,73} = -0.567$, corresponding to mouse 20 (diabetic class A) and metabolite identified as PO(38:4). The corresponding score for sample 20 and PC 2 is $t_{20,2} = 0.680$ for readers that wish to check their calculations.
- The data though should not be standardised, but should be row-scaled if appropriate, to calculate the fold change; as well as row scaling (in this case) the same result would be obtained for p values using most methods whether standardised or not. The fold change cannot be meaningfully calculated using standardised or centred data. We standardised the data primarily to examine the scores plot and is a recommended first step when performing PCA in this case.
- Row scaling, but without standardising, gives a value of $x_{20,73} = 2.71 \times 10^{-1}$.
- For variable 73, using row-scaled data we find $\bar{x}_{73A} = 4.204 \times 10^{-4}$ and $\bar{x}_{73B} = 3.503 \times 10^{-4}$ so that the fold change $\bar{x}_{47A}/\bar{x}_{47B} = 1.200$, that is, the average intensity of metabolite 73 in Class A is around 83% of that in Class B.
- We can also calculate a p value for the difference between means as described in Section 5.4, using the t-distribution (other distributions and calculations of p values are possible, so for metabolite 73, $t = 1.524$ giving a p value of 0.132.
- It is now possible to calculate both the fold change and p values for all variables.
- In order to visualise these, logarithms are taken both of f and p.
  - We transform f to $\log(f)$, This means there is symmetry so that (using logs to the base 10)

    $\log(5) = 0.699$ and $\log(0.2) = \log(1/5) = -0.699$
  - and transform p to $-\log(p)$
- Hence, in our case for metabolite 73, $\log(f_{73}) = 0.079$ and $-\log(p_{73}) = 0.879$.
- We now can plot a graph of $-\log(p)$ versus $\log(f)$ for all metabolites (146 in this example) as in Figure 9.20. Metabolite number 73 is marked in red.
- The candidate markers will have relatively high values of $\log(f)$ and of $-\log(p)$. We can divide the graph according to the sign of f which tells us which class a variable is characteristic of, and if we want place a critical value of p, in the figure, we have marked $p = 0.05$ ($-\log(p) = 1.301$).
- In this dataset, there are some extreme p values, which present good evidence that there are likely to be markers present. Remember, volcano plots look at two types of evidence and so may be more definitive than using one indicator alone; although these types of evidence are not completely independent, but still have somewhat different basis.

Volcano plots are a simple univariate visual way of seeing which metabolites are likely to be markers. They are useful providing there are a reasonably large number of variables (at least 30 as a rule of thumb), and that there are two groups. There is, of course, no need to row scale or to use the t-test to obtain p values, and so such plots can be used in other situations; however, the application illustrated in this section is typical of its use in metabolomics and it is a common and simple visual approach to highlight which are potentially good candidate markers.

CHAPTER 10

Which Factors are Most Significant

10.1 INTRODUCTION

Many applied statisticians are concerned with the significance of factors in a model. Most traditional statistics involves univariate models, that is there is only one variable, but there may be several factors. So traditionally, a statistician might ask whether a factor has a significant effect on a response. A chemometrician may be more interested in which variables are significant classifiers, for example, for a disease or a genetic group or which have a significant effect on a process, such as the change in metabolic profile over time.

In practice, these two questions can come to identical conclusions, and if assuming an underlying normal distribution in the response, should result in identical p values.

- We can view a response as a variable, such as the concentration of a metabolite
- and a factor such as a classifier (or similar).

For a univariate response, we could ask:

- Can two or more groups of samples be discriminated adequately using this metabolite (a question phrased as a chemometrician might ask) or
- How significant is the classifier in describing variability in the concentration of the metabolite (a question phrased as a statistician might ask).

Data Analysis and Chemometrics for Metabolomics, First Edition. Richard G. Brereton.
© 2024 John Wiley & Sons Ltd. Published 2024 by John Wiley & Sons Ltd.
Companion website: www.wiley.com/go/Brereton/ChemometricsforMetabolomics

Much of the statistical literature is concerned with the significance of factors. This often allows the inclusion of additional terms in univariate models, for example, if x is the concentration of a metabolite, and c_1 the value of a classifier for one effect and c_2 another, we could write $\hat{x} = b_0 + b_1 c_1 + b_2 c_2$ as a two-factor model or include further quadratic terms and interaction terms. A chemometrician will probably form two separate classification models one with c_1 as a classifier and one with c_2. Even though the statistician's model may seem more sophisticated, especially with the inclusion of interaction terms, it is less easy to extend to multivariate models, although we will discuss the ASCA (ANOVA Simultaneous Component Analysis) method in Section 10.5 below.

Statisticians often use a method called ANOVA (Analysis of Variance) to determine the significance of factors in a model.

The principle of ANOVA is that two variances are compared, one involving variation of interest (sometimes called between group variance) and one involving uninteresting variation often called error (or sometimes residuals or orthogonal variance or within-group variance). The significance of the ratio of these two variances is then usually determined assuming underlying normality using hypothesis tests based on the F- or t-distribution (see Chapter 5). We will illustrate these calculations below.

As we will see, categorical regression (where the value of one variable is discrete) yields identical answers to a traditional ANOVA test if applied appropriately. The latter was developed around a hundred years ago, and is computationally relatively straightforward, involving comparing the means of different groups, but provides less information than regression, which is computationally less straightforward and would have been more time-consuming before the advent of computers. ANOVA is used for determining p values or whether specified factors are significant using both regression and the traditional ANOVA test, but the latter can be considered a shortcut, and we will compare both where appropriate. Whereas all problems studied using the ANOVA test can also be studied using regression, the converse is not always so.

The statistical literature on ANOVA and regression models is vast and in this text, we will only discuss some of the more common designs and approaches. We recommend as far as possible to try to stick to straightforward methods if feasible as it can be hard to interpret the more complex designs or organisation of data. Chapter 6 introduces the principles of experimental design.

10.2   TERMINOLOGY AND DEFINITIONS

Historically, statisticians have developed a terminology that is somewhat different to the terminology used by most chemometricians. We will describe various terminologies below.

- Most statistical texts use the terminology factor for the categorical variable and denote this by 'x' and response for the continuous variable and denote this by 'y'.
- Unfortunately, the opposite notation is more common in chemometrics, with the categorical variable denoted 'y' and the continuous variable by 'x'.

- This is because most statistics was developed with a univariate y (response) and often multivariate or multifactor x (factors), whereas the majority of traditional chemometrics problems, for example, those analysed using PLS1, involve a univariate y (e.g. a classifier) and a multivariate experimental dataset (x). Of course, both blocks can be either univariate or multivariate.
- In order not to swap notation around in different chapters of this book, we stick to 'c' for the categorical variable or factor and 'x' for the continuous variable or experimental data. When comparing to most software for regression or ANOVA it is important to appreciate the notation used in the book, which allows the reader to follow this text without getting confused by a changeover in x and y.
- For the examples in this chapter, variable x (or the dependent variable) will be univariate, except in Section 10.5, where we will extend to the multivariate case. Multivariate ANOVA can be quite complex and has several different forms of which we will describe ASCA which is popular in chemometrics. When the x block is multivariate, it is of course possible to treat each variable on its own, so retaining univariate x.
- The various notation and terminology are presented in Table 10.1.

So far in this text we have written mainly about classifiers, but in the ANOVA literature, authors normally use the term levels or groups or conditions. There is no fundamental difference. For example, if there are three classes in an experiment such as three phenotypes, these could also be called three groups or more commonly three levels. Hence, a three-level design or three-level ANOVA involves looking at the difference between three groups.

One unfortunate historical confusion is in the term 'level'. To many, obtaining data at three levels means at 3 related values, for example, at 3 temperatures, 10, 15 and 20 °C. Although this indeed can be considered a three-level design for the purpose of the ANOVA test, the 3 temperatures would be treated as unrelated, that is as three independent groups, so that for example data obtained at 20 °C would be considered as closely related to data from 10 °C as data from 15 °C. One way round is by regression, as we will see below, treating the factor as a quantitative variable, but this will provide different results to the ANOVA test. However, the term level is still generally used even if there is no quantitative relationship between the different categories.

TABLE 10.1   Differences in terminology and notation.

Common terminologies	Factor	Response
	Classifier	Experimental measurement
	Independent variable	Dependent variable
	Categorical variable	Continuous variable
	Treatment	
	Discrete variable	
Traditional statistics	x	y
Traditional chemometrics	y	x
This text	c	x

The number of factors and levels are important to guide how best to analyse the data and is a consequence of the design of the experiment (as discussed in Chapter 6).

- The term N way or N factor ANOVA is often used, where N equals the number of factors.
- One-way (or factor) ANOVA is when there is only one categorical variable, and two-way (or factor) when there are two categorical variables.
- It is rare to go beyond two-way ANOVA, and although factorial designs may consist of far more than two factors, they are usually treated in other ways such as by regression modelling (e.g. case study 8 – environmental factors on *Daphnia*).
- Each factor is also studied at a specified number of levels.
- Data for two-way ANOVA can often be presented as a table, with the levels of each factor on the sides.
- Methods using ANOVA are easiest applied to factorial designs, and the organisation of the data (or design) can be denoted by a product of N integers, the number of terms equalling the number of factors and the values of these terms equalling the number of levels K for each factor.
- A $K_1 \times K_2 = 2 \times 3$ factorial design is presented in Table 10.2. The design is called factorial because all combinations of levels are studied.
- We can envisage three-way data, for example, a $2 \times 4 \times 3$ design, which would consist of 24 unique combinations of levels. Normally, if more than two factors are considered, the number of levels and/or combinations are reduced as the number of experiments can be very high; fractional factorial designs as in Section 6.2.3.3 are common when there are two levels but several factors.
- There are a large number of ways the data can be organised, and special approaches have to be used according to the experimental arrangement. In this chapter, for reasons of length, we will restrict to common designs or arrangements of data. More specialist arrangements often require a quite detailed understanding of how to analyse the data and whole texts and software have been developed to analyse outcomes from for a wide variety of designs. However, most chemometricians will usually encounter primarily a few of the more straightforward designs.

In Table 10.2, we see that there are 6 combinations of conditions or levels.

- At each level, it is often normal to perform several repeat measurements, these are usually called replicates.
- It is important to recognise that there can be different types of replication, for example, we may have several donors and each donor may be sampled several times which in turn is analysed several times. All these levels of replication can be taken into account so each donor could be considered a factor, rather than a replicate. However, the method for analysis of such data can be complex, and we will not distinguish source of replication in this chapter.
- Ideally, the number of replicates at each level should be the same. For example, for our 2×3 design, we may decide to perform 5 replicates at each level, resulting in $6 \times 5 = 30$ measurements overall. Replication of some form is important

TABLE 10.2 Typical layout for 2×3 factorial design.

		Factor 2		
		Level 1	Level 2	Level 3
Factor 1	Level 1	X	X	X
	Level 2	X	X	X

to perform error analysis and determine the significance of a model or terms in a model.

- If the number of replicates at each level is different, the design is said to be unbalanced.
- For one-way (single-factor) data, this rarely is a problem but can cause difficulties if more than one factor is involved. It is best to use balanced data. In practical situations, there may be instrumental or sampling issues meaning that some samples will be destroyed or result in poor-quality analytical data. There are several ways of overcoming this.
  - Always record more replicates than necessary. For case study 8 (environmental toxicology of *Daphnia*), 5 replicates are obtained at each combination of levels. However, only 3 are retained at each unique combination, allowing for up to 2 out of 5 analyses to be discarded, the basis being we are very confident that there will be at least three out of five good-quality chromatograms corresponding to each unique combination of conditions.
  - Estimate the values of the missing data points as described in Section 4.7.1. If the number of missing data points is small, this may not be a major problem, especially if an approach such as kNN (*k* Nearest Neighbour) is used.
  - Use statistical approaches that take missing data into account. We will not describe these in detail here. For unbalanced multiway ANOVA Type I, Type II and Type III errors can be calculated under these circumstances and each may provide different answers when we want to estimate interactions. If necessary, readers are recommended to consult the more detailed statistical literature.
  - Forget about interactions (see below) and perform N one-way tests. These are reliable for unbalanced data.

Methods below assume that errors are normally distributed and that the variance of samples at each level is the same. In practice, these assumptions may not be fully obeyed, but reasonable deviations from these assumptions are usually tolerable. Remember that in metabolomics a main aim is often to suggest markers that are likely to be significant, but further experimentation is often necessary to confirm candidate metabolites, so p values should be viewed as indicative rather than used to make definitive final decisions. If a marker has a low p value using more than one independent test, it is a very promising candidate. Many preliminary scoping or academic studies are just aimed at suggesting some candidates, with sponsors deciding the next steps.

10.3   SINGLE FACTOR (ONE-WAY – ONE-FACTOR) ANOVA TEST AND REGRESSION

Interestingly, both regression and ANOVA are mathematically the same, and produce the same results in terms of p values and significance. Most in chemometrics prefer to talk about regression models, but many statisticians prefer to talk about ANOVA tests. We will compare the approaches. In common with all approaches, it is necessary to understand the concept of degrees of freedom, first introduced in Chapter 5.

10.3.1   Balanced Design at Two Levels

Initially, we will illustrate this with the first column (or variable) 40×12 dataset of Table 7.1 excluding sample B21.

This could be viewed as a one way, $K = 2$ level, dataset, with 20 replicate measurements at each level.

10.3.1.1   Degrees of Freedom

The first step is to determine the number of degrees of freedom. This concept was first introduced in Chapter 5. Usually, these are associated with a sum of squares as we will describe later.

- In the simulated dataset, there are $I = 40$ observations. The total sum of squares of the original response data is characterised by $\nu(T) = I = 40$ degrees of freedom.
- Usually, we are interested in variance around the overall mean, which has $\nu(M) = M = 1$ degree of freedom.
- This leaves $\nu(A) = I - M = I - 1 = 39$ degrees of freedom for the mean adjusted total. This is sometimes alternatively called the total sum of squares, but we will use 'A' in this text to distinguish from the uncentred data, with the total sum of squares referring to the uncentred data.
- In this simple example, the overall variance is then divided into two parts:
  - $\nu(R)$ is the degrees of freedom for regression or between group variance, and in this case $\nu(R) = K - 1 = 1$, where $K = 2$ is the number of levels or groups; this also equals the number of parameters in the regression model P as we will see.
  - $\nu(E)$ is the remaining number of degrees of freedom used for estimating the error or within group variance, in this case, $\nu(E) = I - M - (K - 1) = I - M - P = 40 - 1 - (2 - 1) = 38$
- We see that
  - $\nu(A) + \nu(M) = \nu(T)$
  - $\nu(R) + \nu(E) = \nu(A)$
- These can be represented by a degree of freedom tree as in Figure 10.1. There will be the same degree of freedom tree for each of the 12 variables.

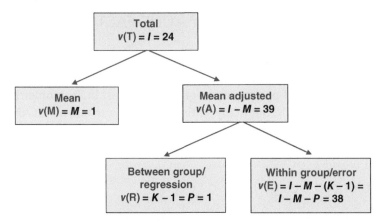

FIGURE 10.1   Degree of freedom tree for each variable one-way two-level simulated data with 20 replicates at each level.

- In this simple example we are interested in how the variance due to regression compares to the variance due to error or the between group variance compares to the within group variance.

We will use this to determine the mean of each source of variation.

10.3.1.2   ANOVA Test

We can illustrate the calculation of ANOVA on the first variable of the simulated dataset calling this x, so that $x_{12} = 44.49$ in our example.

- The total sum of squares (sometimes omitted from an ANOVA tables) is
$$SST = \sum_{i=1}^{I} x^2 = 130,856.$$

- The mean is $\bar{x} = \sum_{i=1}^{I} x_i \, / \, I = 53.44$ where I is the total number of samples. Hence, the total sum of squares for the mean is $SSM = I\,\bar{x} = 40 \times 53.44^2 = 114,221$.

- Hence, the total mean adjusted sum of squares $SSA = SST - SSM = \sum_{i=1}^{I}\left(x_i - \bar{x}\right)^2 = 130,856 - 114,221 = 16,334$ (to within integer rounding).

- The between group sum of squares is given by $SSR = \sum_{k=1}^{K} I_k \left(\bar{\bar{x}}_k - \bar{x}\right)^2$ where $\bar{\bar{x}}_k$ is the mean of data from level k and $I_k = 20$ in this case, is the number of samples at level k,
  - so $\bar{\bar{x}}_1 - \bar{x} = 8.768$ and $\bar{\bar{x}}_2 - \bar{x} = -8.768$ (since the data are balanced and centred)
  - and $SSR = 20 \times 2 \times 8.768^2 = 3075$
- This leaves the within group or error sum of squares to equal $SSE = SSA - SSR = 16,334 - 3075 = 13,559$ in this case. An equivalent equation is

$$SSE = \sum_{k=1}^{K} \sum_{i_k}^{I_k} \left(x_{i_k} - \overline{\overline{x}}_k \right)^2$$ where there are I_k samples at level k, our case $I_k = 20$ for both levels which yields the same answer.

- We now calculate the mean sum of squares by dividing the sum of squares for each term by its respective degree of freedom (which can be called a variance) so
  - $MSST = SST/\nu(T) = 3.271 \times 10^3$
  - $MSSM = SSM/\nu(M) = 1.308 \times 10^5$
  - $MSSA = SSA/\nu(A) = 426.52$
  - $MSSR = SSR/\nu(R) = 3.075 \times 10^3$
  - $MSSE = SSE/\nu(E) = 356.81$
- Although various different ratios can be calculated to determine the relative importance (or significance) of sources of variation, the usual question is whether the difference in means at each level is significant compared to the residual error, or alternatively phrased whether the between group variance is significant compared to the within group variance.
- This is done by calculating the ratio $MSSR/MSSE$, which is often called the F-ratio and in this case $F = 8.62$.
- This can be converted to a p value assuming an F-distribution (see Chapter 5), with $\nu_1 = 1$ (the number of degrees of freedom for SSR) and $\nu_2 = 38$ (the number of degrees of freedom for SSE), so that $p = 1 - Fcdf(F, 1, 38) = 0.0056$ and the difference in means is viewed as highly significant occurring only about 1 time in 200 by chance.

No further information is obtained using the ANOVA test.

Often the information is presented as an ANOVA table as in Table 10.3. Note that usually SST and SSM are not listed, although it is possible: an F value could be obtained to see whether the overall mean is significant or not, however, this is rarely done. Often SSA is called the Total Sum of Squares but in order to be consistent we will call it the Mean Adjusted Sum of Squares.

TABLE 10.3   ANOVA table for variable 1 of the simulated case study of Section 10.3.1.

Source of variation	Sum of squares	Degrees of freedom	Mean sum of squares	F	p
(Between group) SSR	3075	1	3075	8.619	0.0056
(Within group) SSE	13,559	38	357		
(Total mean adjusted) SSA	16,634	39			

The information can also be presented in a degree of freedom tree as in Figure 10.2, which is unique to variable 1.

- Using a similar approach we can calculate F and a p value for each of the 12 variables separately as in Table 10.4.
- We can also present these graphically as in Figure 10.3, with a critical value of $p = 0.05$ marked. This graph should be compared to Figure 7.2(c) where t rather than F has been used to obtain (identical) p values.
- We might note that variables 2, 9, 11 and 12 have a p value insufficient to reject the null hypothesis and therefore represent variables that have low discriminatory power.

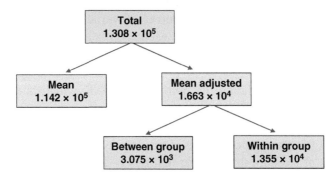

FIGURE 10.2   Degree of freedom tree for sums of squares for variable 1 in the simulated case study.

TABLE 10.4 F and p values for the 12 variables and the first 40 samples from Table 7.1 using ANOVA as described in the text.

Variable	F	p
1	8.6188	0.0056
2	1.4394	0.2377
3	11.3776	0.0017
4	15.2388	0.0004
5	5.5057	0.0243
6	11.0103	0.002
7	6.975	0.0119
8	20.8516	0.0001
9	2.4854	0.1232
10	11.6917	0.0015
11	1.3904	0.2457
12	0.034	0.8547

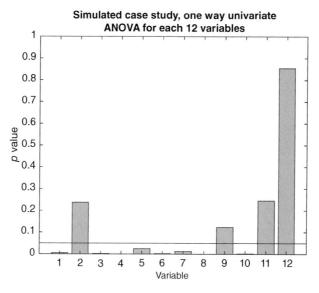

FIGURE 10.3 p values using one-way ANOVA for the 12 variables of the simulated case study with $p = 0.05$ limit marked.

10.3.1.3 Regression

An alternative is to use regression. As we will see this does come to identical conclusions as to whether a variable is a significant marker, but also provides additional information.

We will illustrate using the same variable 1, of Table 7.1 excluding sample B21, as above.

- We form a model between the response x and the classifier (or factor) c.
- We denote c with the values $c_i = +1$ if $i = 1$ to 20, and $c_i = -1$ if $i = 21$ to 40.
- The design is balanced in that there are 20 samples with $c = +1$ and 20 with $c = -1$.
- Usually, the data are column centred first so that after centring, for example, $x_{12,6} = -21.89$; as there are as many samples with $c = +1$ as -1, centring does not change the values of c in this example.
- The next step is to form a model between x and c. If we consider each variable at a time and call it x, $\hat{x} = bc$ or $x = bc + e$ in vector notation, or $x_i = bc_i + e_i$ where i is the sample number, providing both columns have been centred. The vector e is usually called the error, although strictly it is a residual.
- The key is to obtain an estimate b which can be done by simple regression.
  - If $x \approx bc$
  - then $xc'\left(cc'\right)^{-1} \approx bcc'\left(cc'\right)^{-1}$
  - so $\hat{b} = xc^{+} = x'c\,/\,c'c = \sum_{i=1}^{I} x_i c_i \,/\, \sum_{i=1}^{I} c_i^2$ where c^{+} is the pseudoinverse of c: this concept has been introduced in Section 4.3.4 and is discussed in most articles and texts on basic matrix algebra, and x_i is the value of sample i and variable 1 in this example.

- For variable 1, we then find $b = 350.73/40 = 8.768$; to simplify we will omit the $\wedge$ over the b below, to be compatible with most texts, but have to remember this is an approximation. The true underlying value (not measurable experimentally) is often denoted by β.
- Hence, for variable 1, $x = 8.768c + e$, so the estimate of the response for the first 20 samples is $\hat{x} = 8.768$.
- For each sample, we can calculate the residual (often also called the error, although strictly speaking these are different concepts). For example, $e_4 = x_4 - \hat{x}_4 = -7.68 - 8.76 = -16.45$. We plot the residuals in Figure 10.4.

We can now use ANOVA, again.

- In our case, we can ask whether the variance of the modelled data (SSR) is significant compared to the residual error (SSE). Many statisticians formulate a 'null hypothesis'. In our case, the null hypothesis is that there is no functional relationship between the variable or response (x) and the calibrant or classifier or factor (c). In formal analysis, it is always best to define this null hypothesis prior to further interpretation.
- A full analysis of sums of squares (for variable 1) can be obtained as follows:
  - The Total Sum of Squares prior to centring $SST = \sum_{i=1}^{I} {}^{uncentred} x_i^2 = 1.308 \times 10^5$.

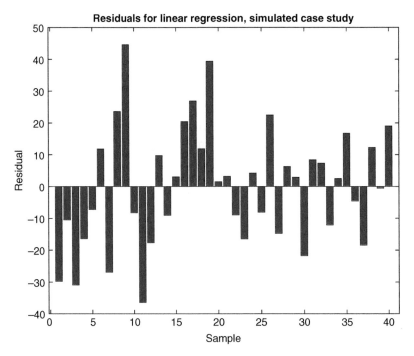

FIGURE 10.4   Residuals or errors for estimate of c for the 40 samples of the simulated case study, using variable 1 and linear regression.

- The Sum of Squares for the Mean $SSM = I \times {}^{uncentred}\bar{x}^2 = 1.142 \times 10^5$.

- The Mean Adjusted Sum of Squares $SSA = \sum_{i=1}^{I} x^2 = 1.663 \times 10^4$ (we will not use the right-hand subscript if centred in this example).

- The Sum of Squares for regression SSR, in our case $SSR = \sum_{i=1}^{I} \hat{x}^2 = 3.075 \times 10^3$.

- The Sum of Squares for the residuals (often called error) SSE, in our case $SSE = \sum_{i=1}^{I} e^2 = 1.355 \times 10^4$.

- These numbers are identical to those obtained using the ANOVA test but obtained using regression rather than by differences of means.
- The sum of square errors can also be presented in an ANOVA table just as for the ANOVA test, but instead of 'Within Group Sum of Squares' we have 'Error Sum of Squares' and instead of 'Between Group Sum of Squares' we have 'Regression Sum of Squares'. The p value of the model is 0.0056 as above.
- However, we can obtain additional information which is the p value of each factor in the model. In the case above, the p value of the single factor c is the same as that of the overall model, but when there are several factors, this will be different as described below.
- However, the F ratio and p value are identical using both methods for calculation in this example, so identical conclusions will be obtained about the significance of the factor.

The advantage of using regression is that more information is available, for example, the estimates of c and the errors (residuals) which could for example show if there are any outliers or which samples are well estimated using our model, and relate to methods described in Chapter 7. We could, for example, estimate the values of c for each variable separately so from the residuals see which sample is estimated well using each variable separately.

10.3.1.4 The t-test

In Section 5.4, we introduced the use of the t-test for assessing whether a variable is a significant marker.

If we follow the procedure in that section, we find that the t statistic for variable 1 is 2.936. The square of this equals the F value as obtained by the methods above. Some software prefers t rather than F values as the former has a sign whereas F is always positive.

If we calculate the p value using 38 degrees of freedom, we find that it is identical to the p value obtained by ANOVA.

Hence, for two-level, one-way (or one-factor) data, the ANOVA test is not strictly necessary and provides identical results to using the t-test. As we see later, ANOVA based methods are most useful when there are more than two levels. The t-test as usually employed will only compare two groups.

10.3.2   Unbalanced Design at Two Levels

An unbalanced design is one in which the number of samples at each level is different. For example, there may be 20 samples at level 1 but only 15 at level 2.

In some studies, it is easy to obtain balanced data, for example, in plant metabolomic studies where plants can be grown and harvested under controlled conditions. The same is so in many laboratory-based studies, an example being case study 8, the study of environmental conditions on metabolites extracted from *Daphnia*. However, in other cases, such as human clinical studies, this may be hard, as donors may drop out, or have defective samples or balanced sampling being very hard. Under such circumstances the data are unbalanced.

There is no fundamental problem with one-way ANOVA and unbalanced data.

As an example, let us return to the data of Table 7.1, but remove the last 10 samples, to give a 30×12 dataset, of which samples 1–20 belong to level 1 (or Class A) and 21–30 to level 2 (or Class B), and perform ANOVA on variable 1. Calculations are very similar to those using balanced data.

- The total sum of squares (sometimes omitted from an ANOVA table) is
  $$SST = \sum_{i=1}^{I} x^2 = 106,725.$$

- The mean is $\bar{x} = \sum_{i=1}^{I} x_i / I = 55.34$ where I is the total number of samples. Hence, the total sum of squares for the mean is $SSM = I\bar{x} = 30 \times 55.34^2 = 91,862$.

- Hence, the total mean adjusted sum of squares $SSA = SST - SSM = \sum_{i=1}^{I}(x_i - \bar{x})^2 = 106,725 - 91,862 = 14,862$.

- The between group sum of squares is given by $SSR = \sum_{k=1}^{K} I_k(\bar{\bar{x}}_k - \bar{x})^2$ where $\bar{\bar{x}}_k$ is the mean of data from level k; this time the number of samples at each level is different so $I_1 = 20$ and $I_2 = 10$,

- so $\bar{\bar{x}}_1 - \bar{x} = 6.869$ and $\bar{\bar{x}}_2 - \bar{x} = -13.739$ (note the mean for the second group is minus twice that of the first group, as the second group is half the size of the first group),

- and $SSR = 20 \times 6.869^2 + 10 \times 13.739^2 = 2831$.

- This leaves the within group sum of squares to equal $SSE = SSA - SSR = 14,862 - 2831 = 12,031$ in this case. An equivalent equation is $SSE = \sum_{k=1}^{K}\sum_{i_k}^{I_k}(x_{i_k} - \bar{\bar{x}}_k)^2$ where there are I_k samples at level k.

- The degrees of freedom $\nu(R) = 1$ as there are still two levels, and $\nu(E) = 30 - 1 - 1 = 28$.

- Hence, we calculate
  - $MSSR = SSR/\nu(R) = 2.831 \times 10^3$
  - $MSSE = SSE/\nu(E) = 429.68$.
- The F ratio is now $F = 6.59$, and the p value can be obtained by $p = 1 - Fcdf$ $(F, 1, 28) = 0.0159$.
- Notice that this is a higher p value than obtained for the balanced design. As there are less samples in Class B (or at level 2), compared to Class A (or level 1), it would be harder to be certain that the two groups are distinct, and an experimental solution of course might be to perform more measurements to get a more robust answer if this is considered important.

Similar results would be obtained using regression or the t-test, and we do not illustrate these calculations in detail. However, it is important to appreciate that when there are no interactions (for which see below when we discuss two-way ANOVA) there is no problem with unbalanced data, except that as usual, the more the measurements, the more confident we can be that there is a significant difference between the groups (or that the factor is significant or the variable is significant according to our terminology).

ANOVA and regression will provide the same p values as the t-test for all appropriate case studies presented in this book, as an example for the pre-arthritis case study 1 (see Table 5.8) which we will not repeat for brevity.

10.3.3   Multiple One-Way Design with Two Levels: Multilinear Regression

If we are not interested in interactions, we can treat factorial data as if each factor were independent. In this case, it is important that the factors are orthogonal and balanced. For brevity, we will not discuss unbalanced multifactorial data in this chapter, for more specialist designs consult the statistical literature. We recommend that where possible to stick to more straightforward and easily interpretable data and if some data are missing to estimate it where the proportion of missing data is small.

Case study 8 (Effect of environmental conditions on *Daphnia*) is a good example of an orthogonal factorial design. We will use coded values of the three factors salinity, temperature and hypoxia, throughout, see Chapter 6 for discussion.

- There are 24 experiments. For each of the three factors, 12 are at one level and 12 are at the other level.
- For each metabolite, we could then perform one-way two-level ANOVA separately.
- We will illustrate this by variable 1 (lactic acid).
- The first step is to centre the x variable (lactic acid), so for example, $x_{10} = 411.67$.
- We will now first consider factor 1 (salinity). If we use coded values, this is already centred.
- We will not go through the calculations in detail but summarise the main results

- $SSR = 7.942 \times 10^5$ and
- $SSE = 2.222 \times 10^6$.
- There are $\nu(R) = 1$ and $\nu(E) = 24 - 1 - 1 = 22$ degrees of freedom.
  - Hence, $MSSR = 7.942 \times 10^5$
  - and $MSSE = 1.010 \times 10^5$
  - Hence, $F = 7.862$
  - and p with $\nu_1 = 1$ and $\nu_2 = 22$ is equal to 0.0103
- This suggests lactic acid is likely to be influenced by salinity.

We could perform similar calculations for all factors and all metabolites. For lactic acid, we find for temperature, $p = 0.505$ and for hypoxia, $p = 0.533$, so only salinity significantly affects the concentration of lactic acid.

The calculated p values using ANOVA (or regression) are presented in Figure 10.5. Note that the same p values could have been obtained using the t-test in this case. The metabolites with $p < 0.05$ are listed in Table 10.5 for each of the three factors.

When performing one-way regression on a single factor and variable, we calculate models of the form $x = b_f c_f + e$ for factor f (assuming centred data so no intercept).

However, we could alternatively form a single multilinear model of the form $x = b_1 c_1 + b_2 c_2 + b_3 c_3 + e$. The values of the coefficients will be the same whether using one-way regression with a single factor or three factors.

- Now the c block is a 24×3 matrix $\boldsymbol{C}$ of coded values as in Table 6.3 with each unique set of conditions triplicated (we omit the full 24-row matrix for brevity as each unique condition is repeated 3 times). Sometimes this is called a design matrix.
- Since for each variable $\boldsymbol{x} = \boldsymbol{C}\,\boldsymbol{b} + \boldsymbol{e}$ where $\boldsymbol{b}$ is a 3×1 vector of coefficients, we can estimate $b = C^+ x = \left(C'C\right)^{-1} C'x$ using multilinear regression, so $\boldsymbol{b}' = [181.91 - 50.67\ 47.42]$ or $x = 181.91\,c_1 - 50.67\,c_2 + 47.42\,c_3 + e$ for variable 1.
- We can now calculate the errors for the full three-factor model.
  - The error for a three-factor model of the centred data is SSE $= \boldsymbol{x} - \boldsymbol{C}\,\boldsymbol{b} = 2.106 \times 10^6$.
  - and the regression sum of squares is $SSR = 9.098 \times 10^5$.
- There are $P = 3$ parameters in the model. Note that this does not equal the number of levels. Below we will discuss what to do when there are more than two levels and more than 1 parameter in the model.
- In this dataset, there are $I = 24$ observations. The total sum of squares of the original response data is characterised by $\nu(T) = I = 24$ degrees of freedom.
- Usually, we are interested in variance around the overall mean, which has $\nu(M) = M = 1$ degree of freedom.
- This leaves $\nu(A) = I - M = I - 1 = 23$ degrees of freedom for the mean adjusted total.

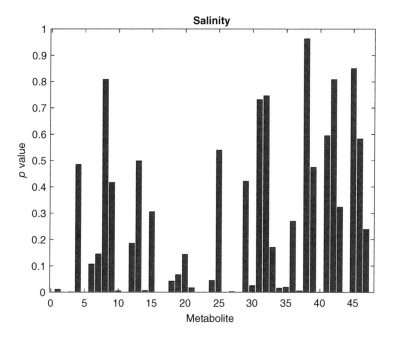

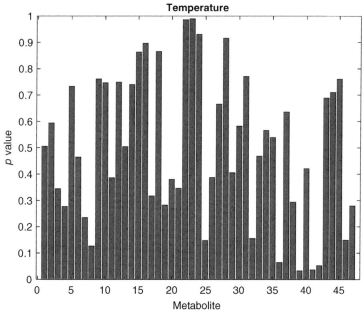

FIGURE 10.5 *P* values using one-way ANOVA for the 47 metabolites of case study 8 (environmental toxicology of Daphnia).

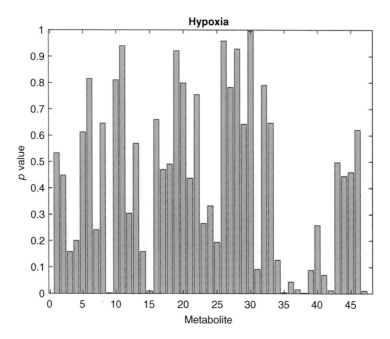

FIGURE 10.5   (Continued)

- In this example, the overall variance is then divided into two parts.
  - $\nu(R)$ is the degree of freedom for regression, and in this case $\nu(R) = P = 3$.
  - $\nu(E)$ is the remaining number of degrees of freedom used for estimating the error or within-group variance, in this case, $\nu(E) = I - M - P = 24 - 1 - 3 = 20$.
- This means that the relevant mean sums of squares are
  - $MSSR = 9.098 \times 10^5/3 = 3.032 \times 10^5$
  - The $MSSE = 2.106 \times 10^6/20 = 1.053 \times 10^5$
  - Hence, $F = 2.880$
- This results in a p value of 0.061 for the full three-factor model.

However, when performing multilinear regression we can additionally obtain p values for individual terms or factors. There are various approaches, all of which come to the same answer providing the data are balanced. For unbalanced data, there are a variety of approaches often called Type I, Type II and Type III errors that come to slightly different conclusions about p values when there are interactions or factors are not orthogonal. For interested readers, please consult the more specialist literature.

One simple approach to finding the p values of individual factors is to remove them one by one and compare the model with and without them.

- In order to determine the significance of the first factor we can calculate the model $x_{\{-1\}} = b_2 c_2 + b_3 c_3 + e_{\{-1\}}$ and compare it to the model $x = b_1 c_1 + b_2 c_2 + b_3 c_3 + e$ where $e_{\{-1\}}$ is the error when factor 1 is removed.

TABLE 10.5   Metabolites with $p < 0.05$ using one-way ANOVA or regression for the Daphnia case study.

Salinity	Temperature	Hypoxia
Lactic acid	Palmitic acid	Serine
Alanine	Stearic acid	Phosphoric acid
Norleucine		Methylgalactoside
1-Nonanol		Idose (Meox)
Glycerol		Myo-Inositol
Fumaric acid		N-acetyl-D-glucosamine (Meox)
Threonine		Oleic acid (Z)
N-acetylglutamic acid		Uridine-5'-monophosphate
L-Methionine		
D-(−)-Erythrose methoxyamine		
Aspartic acid		
L-5-oxoproline		
Proline		
Phenylalanine		
Glutamic acid		
Xylose methoxyamine		
L-Glycerol-3-phosphate		
Arginine [-Nh3]		
Galactose (Meox)		
Methylgalactoside		
Myo-Inositol		
Uridine		
Inosine		

- For each of the three models, we find that the sum of squares when each factor is removed in turn are given by
  - $SSR_{\{-1:-3\}} = [1.156\ 8.482\ 8.559] \times 10^5$
- However, for the overall model $SSR = 9.098 \times 10^5$ (see above).
- Therefore, the contribution to SSR for each term f in the model is $SSR_f = SSR - SSR_{\{-f\}} = [7.942\ 0.616\ 0.540] \times 10^5$.
- Note that these contributions are equal to the sums of squares for the corresponding one-factor models. This property depends on the factors being orthogonal and is not so otherwise.

- However, we use the error $SSE = 2.106 \times 10^6$ for the full three-factor model calculated above, as we are looking at how much removing each factor contributes to this overall error.
- The $MSSR_f = SSR_f/1 = [7.942 \ 0.616 \ 0.540] \times 10^5$ but $MSSE = SSE/20 = 1.053 \times 10^5$ as above. The degrees of freedom both for the overall model and the individual factors are illustrated in Figure 10.6.
- So, the F values are given by [7.542 0.585 0.512]. These are slightly different from the F values of the single-factor model, for example, for factor 1 (salinity) it was 7.862; this is due to differences in the number of degrees of freedom and $MSSE$.
- To obtain p values we calculate $1 - Fcdf(F,1,20)$ for each factor to obtain values of [0.012 0.453 0.482]. Note again that these are slightly different to those obtained for the one-factor models as above but still would result in similar conclusions, that salinity has a significant effect on lactic acid levels but temperature and hypoxia do not. Turning on its head we could say that lactic acid is a good marker for salinity, for example, if we were to analyse a group of *Daphnia* we could use this metabolite to determine whether they had lived in a saline environment or not.
- We could equivalently say that lactic acid is a potential marker for salinity but not for temperature or hypoxia. Chemometrics does not distinguish between cause and effect.

This information can often be presented in an ANOVA table as in Table 10.6. Sometimes t values rather than F values are listed, with t being the square root of F; however, the same p values should be obtained.

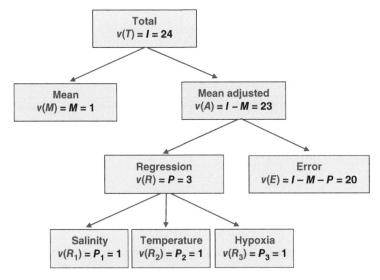

FIGURE 10.6   Degree of freedom tree for the three-parameter multilinear model for each variable for case study 8 (Daphnia) ignoring interactions.

TABLE 10.6   ANOVA table after multilinear three-factor regression for metabolite 1 (lactic acid) of case study 8 (Daphnia).

Source of variation	Sum of squares	Degrees of freedom	Mean sum of squares	F	p
Salinity (SSR_1)	7.942×10^5	1	7.942×10^5	7.542	0.012
Temperature (SSR_2)	0.616×10^5	1	0.616×10^5	0.585	0.453
Hypoxia (SSR_3)	0.540×10^5	1	0.540×10^5	0.512	0.482
Regression (SSR)	9.098×10^5	3	3.032×10^5	2.880	0.061
Error (SSE)	2.106×10^6	20	1.053×10^5		
Mean adjusted (SSA)	3.016×10^6	23			

10.3.4   Multilevel Designs

Multilevel designs are when there are three or more levels. For a two level, one-factor design, ANOVA is strictly not necessary, and the *t*-test will provide identical *p* values and so assessment of the significance of a factor or variable.

However, when there are more than two levels, the *t*-test is no longer adequate.

10.3.4.1   One-Way Multilevel ANOVA Test

The most straightforward and classical approach is multilevel ANOVA.

Instead of there being just two levels or groups, as in Sections 10.3.1 to 10.3.3, there are $K > 2$ levels or groups for at least one factor. Note the traditional terminology 'level' can be a little confusing, as there is no requirement the levels are quantitatively related as discussed above. For this example, we will treat the groups as unrelated in Sections 10.3.4.1 and 10.3.4.2, and related in Section 10.3.4.3.

- We will initially illustrate the method with variable 1 (glucose) of case study 5 NMR intensities in maize harvested at 4 different temperatures). Note that the same results are obtained whether the data are standardised or not; we will use the raw data below. The data consist of metabolite contents measured by NMR intensities in $\mu g\, g^{-1}$ dry weight except where the metabolites have not been identified in which case they are proportional to intensities.
- In this case study, there are $K = 4$ groups, each corresponding to a different harvesting temperature; for the analysis in this section, we will not take into account the different maize types.
- Each group therefore consists of $I_k = 54$ samples, which we will treat as replicates.
- The total sum of squares (sometimes omitted from ANOVA tables) is

$$SST = \sum_{i=1}^{I} x^2 = 4.083 \times 10^{10}.$$

- The mean is $\bar{x} = \sum_{i=1}^{I} x_i / I = 1.153 \times 10^3$ where I is the total number of samples (=216). Hence, the total sum of squares for the mean is $SSM = I\bar{x} = 216 \times (1.153 \times 10^3)^2 = 2.873 \times 10^{10}$.

- Hence, the total mean adjusted sum of squares $SSA = SST - SSM = \sum_{i=1}^{I}(x_i - \bar{x})^2 = (4.083 - 2.873) \times 10^{10} = 1.209 \times 10^{10}$.

- The between group sum of squares is given by $SSR = \sum_{k=1}^{K} I_k (\bar{\bar{x}}_k - \bar{x})^2$ where $\bar{\bar{x}}_k$ is the mean of data from level k, and $I_k = 54$ in this case, is the number of samples at level k; as the data are balanced, this will be the same at each level,

  - so $\bar{\bar{x}}_{k\{1:4\}} - \bar{x} = [0.513 \ -2.743 \ 0.096 \ 2.133] \times 10^3$
  - and $SSR = 54 \times [0.513^2 + -2.743^2 + 0.096^2 + 2.133^2] \times 10^6 = 6.666 \times 10^8$

- This leaves the within group sum of squares equal to $SSE = SSA - SSR = 1.209 \times 10^{10} - 6.666 \times 10^8 = 1.142 \times 10^{10}$ in this case.

- The degrees of freedom are illustrated in Figure 10.7. Hence,
  - $MSSR = 6.666 \times 10^8 / 3 = 2.222 \times 10^8$
  - $MSSE = 1.142 \times 10^{10} / 212 = 5.390 \times 10^7$

- Hence, in this case, $F = 4.123$ giving a p value $1 - Fcdf(F, 3, 212) = 0.0072$, so the difference between the means is highly significant for the metabolite glucose. However, as we will see this does not necessarily imply it is a good marker for temperature. If we return to PC plots (Figures 4.9 and 4.10) we see that PC1 appears most diagnostic of temperature effects, and the loading in PC1 of glucose is quite small in magnitude. The ANOVA test suggests there is a significant difference in some or all of the means between the four groups but not which or why. We will examine this limitation of the ANOVA test below.

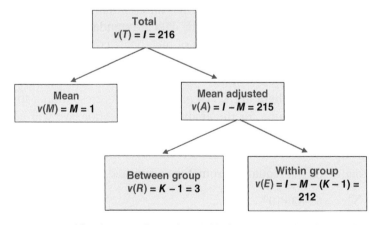

FIGURE 10.7 Degree of freedom tree for each variable from case study 5 (NMR of maize harvested at 4 temperatures).

10.3.4.2 One-Way Multilevel Regression with Dummy Variables: Unrelated Groups

An alternative to the ANOVA test is regression followed by ANOVA. The method described in this section assumes there is no relationship between the different groups. We will see that this approach provides equivalent results for the overall p value to the ANOVA test, but with additional insight. Section 10.3.4.3 describes the use of regression when the groups are modelled as quantitively related.

Regression in this situation requires the use of dummy variables as described below. These dummy variables can be defined as factors.

- For a K level experiment, we define $K-1$ dummy variables, so if there are four levels, we use three dummy variables. This is because the number of degrees of freedom is one less than the number of levels or groups.
- One level is designated as the reference group. It does not matter which one.
- The values of the $K-1$ dummy variables for all samples from the reference group are defined by $d_k = 0$
- Each other group g is denoted by a dummy variable $d_{g-1} = 1$ else $d_k = 0$.
- Hence, if there are $K = 4$ levels
  - for group (or level) 1 $\boldsymbol{d} = [\,0\,0\,0\,]$ which is the reference group
  - for group 2 $\boldsymbol{d} = [\,1\,0\,0\,]$
  - for group 3 $\boldsymbol{d} = [\,0\,1\,0\,]$
  - and for group 4 $\boldsymbol{d} = [\,0\,0\,1\,]$
- Hence, dummy variable or factor 1 represents the difference in means between levels 1 and 2, dummy variable 2 between levels 1 and 3, and dummy variable 3 between levels 1 and 4.
- To remove the intercept, we centre $\boldsymbol{x}$ and $\boldsymbol{d}$ and then form a model between the centred data $^{cen}\boldsymbol{X} = {}^{cen}\boldsymbol{D}\,\boldsymbol{b} + \boldsymbol{E}$ where $^{cen}\boldsymbol{X}$ is an $I \times 1$ vector, $^{cen}\boldsymbol{D}$ and $I \times (K-1)$ matrix and $\boldsymbol{b}$ a $(K-1)$ vector of coefficients.
- In the normal way, we can find $\boldsymbol{b}$ using the pseudoinverse by $^{cen}\boldsymbol{D}^+ \, {}^{cen}\boldsymbol{X}$ and so predict $^{cen}\hat{\boldsymbol{X}}$ and calculate SSR and SSE as described above. However, we can also calculate SSR_f for each factor or dummy variable, although the interpretation becomes a bit complex.

We will illustrate this for variable 1 (glucose) and case study 5 to compare with the calculations in 10.3.4.1.

- We define temperature 1 (20 °C) as the reference.
- Hence, the dummy variables
  - for maize harvested at 20 °C are [0 0 0]
  - at 16 °C [1 0 0]
  - at 13 °C [0 1 0]

- and at 8.5 °C [0 0 1].
- For each variable, we create a 216×1 $\boldsymbol{x}$ vector and a 216×3 $\boldsymbol{D}$ matrix. These do not need to be standardised, as each variable is handled separately.
- For variable 1 for the 59th sample (from group B), we find, for example, $x_{59} = 2.179 \times 10^4$ and $\boldsymbol{d}_{59} = [1\ 0\ 0]$.
- The next step is to centre both blocks, so after centring $x_{59} = 1.026 \times 10^4$ and $\boldsymbol{d}_{59} = [0.75\ -0.25\ -0.25]$ (we will omit the superscript centred below to simplify the notation).
- We then perform regression to obtain
  - $\boldsymbol{b}' = [-3.256\ -0.417\ 1.620] \times 10^3$
  - and for example $\hat{x}_{59} = -2.74 \times 10^3$ which is not a very good estimate.
- We can then calculate $SSR = 6.666 \times 10^8$ and $SSE = 1.142 \times 10^{10}$ resulting in the same F and p values as for the ANOVA test.

However, we now have additional information, which relates to the $\boldsymbol{b}$ coefficients for each of the dummy variables.

- We could write for variable 1 $x = b_1\ d_1 + b_2\ d_2 + b_3\ d_3 + e = -3.256\ d_1 - 0.417\ d_2 + 1.620\ d_3 + e$.
- This allows us to calculate the contribution of each of the dummy variables to the overall variances. We described one approach in Section 10.3.3, removing each variable at a time and for brevity just summarise the calculations.
- We can now calculate the F values for the three dummy variables and for glucose $F = [5.310\ 0.087\ 1.315]$ and corresponding p values $p = 1 - Fcdf(F,1,212)$ or $[0.022\ 0.768\ 0.253]$.
- How do we interpret this?
  - Dummy variable 1 corresponds to the difference in means between level 1 (20 °C) and level 2 (16 °C), and at $p = 0.022$ appears this is significant.
  - Dummy variable 2 corresponds to the difference in means between level 1 (20 °C) and level 3 (13 °C) and at $p = 0.768$ does not appear significant.
  - Dummy variable 3 corresponds to the difference in means between level 1 (20 °C) and level 4 (8.5 °C) at a $p = 0.253$ and is not very significant either.
- This suggests that although the mean of this variable does differ between temperatures, it is not a good marker for temperature and there could be other explanations for this difference. There is no room in this text to discuss this further; however, if we return to Figures 4.9 and 4.10, we see that glucose is quite close to the centre in the loadings of PC1 and therefore not necessarily a good marker for overall changes in temperature.
- Note that if we use a different reference temperature, for example, the lowest temperature and code, so that
  - Plants harvested at 20 °C are coded [1 0 0], at 16 °C as [0 1 0], at 13 °C as [0 0 1] and at 8.5 °C as [0 0 0],
  - we obtain $F = [1.315\ 11.909\ 2.078]$ and $p = [0.253\ 0.001\ 0.151]$, which provides p values of the difference in means between levels 1 and 4, 2 and 4 and 3

and 4. The p value for difference in means between levels 1 and 4 ($p = 0.253$), of course, will be the same no matter which of the two methods for coding are used, but this allows us to see how significant other differences between means should we wish.

We can now compare this behaviour to that of variable 9 (malate).

- We find that for the overall model $F = 91.242$ and $p = 0.0000$, suggesting there is definitely a difference between the means.
- We can then calculate F values for each of the dummy variables (using level 1 or temperature 20 °C as the reference).
- We find $F = [109.60\ 158.22\ 240.96]$. Although these are too large to calculate meaningful p values, we see that the larger the difference in temperature, the larger the F value, suggesting malate is potentially a good marker for harvesting temperature.
- If we return to the PC loadings plot in Figure 4.10, we see malate has a high positive score, suggesting its concentration is influenced by harvesting temperature.

Hence, the ANOVA test alone does not provide us with definitive evidence that the concentration of a metabolite is related to temperature, but if ANOVA is combined with regression, we obtain better insight.

10.3.4.3   One-Way Multilevel Multilinear Regression: Related Groups

Although the four levels are quantitatively related in case study 5, we have not yet used this information and in the analysis above only took into account that they are different but not their relationship. We can understand better the results by looking at the distributions of the measurements described above rather than just asking whether the differences in means are significant. Using dummy variables and regression can provide more insight compared to the ANOVA test if we interpret the coefficients.

We can alternatively regress c onto a single x vector whose values we will define $+1.5$ (level 1: 20 °C), $+0.5$ (level 2: 16 °C), $−0.5$ (level 3: 13 °C), $−1.5$ (level 4: 8.5 °C) for our example of the maize in case study 5. Note that we use regression to predict x from c and not the other way round in this chapter (for the one-factor one-variable or one-way two-level case it actually makes no difference to the p values), because in traditional statistics the c block (traditionally called x) can often be multivariate with several factors and interactions, whereas in traditional chemometrics it is the x block that is multivariate and in many cases, c consists of one or a small number of variables such as classifiers or calibrants.

- We will illustrate this again by variable 1 (glucose).
- We first centre the x vector as usual; the c vector is already centred using our method of coding and because the design is balanced – if using another approach or unbalanced data centre c first.
- Using centred data
  - if $x = c\,b + e$

- then $b = c^+ x$ as for univariate linear regression.
- Hence, we find
  - $b = -769.88$
  - so for example $\hat{x}_{18} = (1.5) \times (-769.88) = -1154.8$
  - which is quite a poor fit as $x_{18} = -313.33$.
- We can calculate for variable 1
  - $SSR = 1.600 \times 10^8$
  - $SSE = 1.193 \times 10^{10}$
  - $MSSR = SSR/\nu(R) = SSR/1 = 1.600 \times 10^8$
  - $MSSE = SSE/\nu(R) = SSE/(216-1-1) = 5.576 \times 10^7$
- This gives
  - $F = MSSR/MSSE = 2.87$ and $p = 1 - Fcdf(F,1,214) = 0.092$.
- This p value is higher than for the ANOVA test or dummy regression (0.0072) suggesting that once we take the quantitative relationship between the levels into account the distribution as related to temperature is less significant for glucose than had we not taken this into account.

If however, we consider variable 9 (malate), we obtain a value of $F = 203.84$ and $p = 0.000$ which corresponds to a very low p value obtained using the ANOVA test or regression using dummy variables and suggests malate is a good marker of temperature and that the metabolic pathways involving malate are influenced by harvesting temperature in this case.

10.3.4.4   Comparison and Interpretation

We have described three methods to determine p values of a four-level or group model on the maize NMR data. Of course, the first two methods can be used for any one-way multilevel experiment, and the third method can be used if the groups are quantitatively related.

- We find that both the ANOVA test (Section 10.3.4.1) and regression with dummy variables (Section 10.3.4.2) result in identical F ratios and p values for the overall model consisting of four groups.
- However, regression using dummy variables provides additional information, yielding insight into which of the four means are significantly different from each other, and would suggest that whereas both glucose and malate do have differences between means, the former would not be a good marker for temperature, which would not have been obvious from the ANOVA test. The ANOVA test states mainly that there is a difference in means, but not that this is monotonic.
- Hence, using regression with dummy variables is recommended over the ANOVA test providing the coefficients for the dummy variables can be interpreted. It is

possible to change which level is the reference to get some further insight. It is important to remember that differences between means are not independent. For example, if there are three levels if the means of level 1 and level 2 are similar, but level 1 and level 3 are significantly different, then levels 2 and 3 will be significantly different also by implication.

- If the levels are related quantitatively, as in many studies, an alternative is multilinear regression. We can get a clearer indication that malate is a good marker but glucose is not from these calculations, but this is only feasible if there is a relationship between the levels.

- The p values are lower than might normally be expected. p values are only indicative and are usually calculated and cited, but depend on a number of assumptions for example that of normality. These assumptions may not be well obeyed by the data.

- Nevertheless, variables can be ordered in terms of either F or p values, to estimate which are the most likely to be markers.

We can understand better the results by looking at the metabolite contents based on NMR intensities for each variable and each group of samples.

- The concentration distributions for both glucose and malate are presented in Figure 10.8. The vertical axis represents the number of samples within a small intensity window, which detail we will not discuss, but is of no direct significance to the interpretation.

- For glucose, we see that although the mean concentrations at the three lowest temperatures are related as expected, the mean at the highest temperature (20 °C) does not show this effect. A possible explanation could be of a metabolic pathway that kicks off as the temperature reduces (or this may relate to plant growth period in this particular case).

- However, this means that whereas the ANOVA test provides good evidence that there is some difference between means, regression with dummy variables and quantitative regression suggests this is not a straightforward temperature effect for glucose. Further analysis might be to investigate metabolism just over the three lowest temperatures, but we leave this to the interested reader.

- For malate, we can see that the content distributions are temperature related, suggesting it is a good marker for temperature.

- Although the assumptions of the tests described above are that the concentrations at each temperature are model by normal distributions, we see that at 20 °C this is not very well obeyed. More detailed investigation could ascertain whether the bimodal distribution is due to maize types having slightly different metabolisms at high temperature. However, the main preliminary conclusion is that the assumptions of normality and equal variance are not obeyed for the high-temperature data.

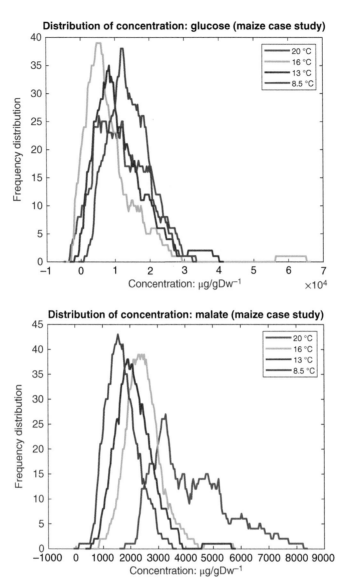

FIGURE 10.8   Concentration distributions of glucose and malate for case study 5 (NMR or maize harvested at 4 temperatures).

- Despite this, *p* values still provide guidance as to which metabolites are most related to growth temperature, and would be a good first step in identifying pathways and markers for further detailed study and can be used to rank variables as an alternative to other approaches discussed elsewhere in this chapter and Chapter 9.

10.4   MULTIPLE FACTOR (MULTIWAY) ANOVA TEST AND REGRESSION

Multiple factor experiments are when more than one factor is investigated. Examples are case study 5 (temperature and maize types) and case study 8 (salinity, temperature and hypoxia). Note that in case study 5, although the four temperatures are related quantitatively, the maize types are not. For case study 8, all three factors have a quantitative relationship, but the factors are studied at only two levels.

There are two definitions of multiway ANOVA. The simplest is when there is more than one factor. The more complicated one is when there are interactions between the factors.

In the former case, we could perform several one-way ANOVA tests or regressions. For two factors, this will lead to a model of the form (for centred data) $x = b_1 c_1 + b_2 c_2 + e$. The same values would be obtained for the coefficients and p values for each individual factor whether using two separate one-term models or a single two-term model, although the p value for the two-factor model has of course to be obtained using both terms. This principle has been illustrated in Section 10.3.3 for three factors.

In the latter case, we include interaction terms. When there are two factors there will be one such term so that $x = b_1 c_1 + b_2 c_2 + b_{12} c_1 \, c_2 + e$. We will look at interaction terms below but these are a key difference between one way and multiway ANOVA. In addition to calculating a p value for the interactions, the size of the error term will differ from the model without interactions as the interaction term will reduce the sum of square error, changing all the p values compared to the simpler model.

In addition to the usual assumptions, for example, normality and equal variances (which are approximate guidelines and can tolerate some deviation to obtain meaningful results if we view p values as guidelines rather than having an exact physical meaning), the factors should be independent, that is, uncorrelated, for two way ANOVA to be successful. Furthermore, to calculate interaction effects, it is preferred that designs are balanced. These restrictions mean more attention is required to the design of experiments (discussed in Chapter 6) than for one way or one-factor studies. We will not cover unbalanced multifactor designs in this chapter, and leave for more specialist texts and software, but suggest these are avoided where possible, with strategies such as estimating missing values or taking more replicates than necessary so as to discard any unusual measurements, where sampling and analytical problems may result in discarding some data points.

10.4.1   Simulated 2 × 3 Case Study: ANOVA Test and Regression

We will illustrate two-way ANOVA with a small 2×3 simulated dataset, with each level replicated $R = 5$ times as presented in Table 10.7.

- Rows 1–5 and 6–10 represent levels 1 and 2 of factor 1.
- Columns 1–3 represent levels 1–3 of factor 2.
- There are in total $I = 2 \times 3 \times 5 = K_1 \times K_2 \times R = 30$ elements.

TABLE 10.7 Simulated 2×3 dataset with 5 replicates at each level.

		Factor 2		
		Level 1	Level 2	Level 3
Factor 1	Level 1	31.256	19.655	10.744
		27.990	14.814	20.335
		27.336	20.193	12.807
		22.855	17.448	11.261
		28.314	7.402	13.382
	Level 2	20.865	17.821	18.203
		19.717	12.120	13.373
		15.859	11.579	18.675
		19.670	9.576	9.518
		20.062	9.532	14.768

10.4.1.1 ANOVA Test

We will describe the steps for an ANOVA test for the simulation of Table 10.7, both with and without interactions.

- We will annotate each element according to its level and replicate number.
  - Hence, $x_{2,1,4}$
  - is at level $k_1 = 2$ of factor 1
  - is at level $k_2 = 1$ of factor 2
  - and is replicate 4
  - so equals 19.670.
- The first step is to globally centre the data, so that we transform $x_{k_1 k_2}$ to $x_{k_1 k_2} - \overline{x}$ where we use the global mean of the data $\overline{x} = 17.238$ (not the column means).
- Hence, the transformed value of $x_{2,1,4} = 19.670 - 17.238 = 2.433$. We will use the terminology x for the centred data below, for simplicity. Note that the columns do not represent variables, but the entire data matrix represents a single variable.
- We can now calculate $SSA = \sum_{k_1=1}^{K_1} \sum_{k_2=1}^{K_2} \sum_{r=1}^{R} x_{k_1 k_2 r}^2 = 1102.92$ in the usual way.
- We can now calculate the sum of squares due to the group means or between group variance for each of the factors.
  - For the linear factors $SSR_f = \sum_{k_f=1}^{K_f} I_f \overline{\overline{x}}_{k_f}^2$ where I_f is the total number of elements in each group ($=15$ for factor 1 and 10 for factor 2).
  - so for factor 1 $SSR_1 = 15 \times ((1.185)^2 + (-1.185)^2) = 15 \times 2 \times 3.294 = 98.83$.

- and for factor 2 $SSR_2 = 10 \times ((6.155)^2 + (-3.225)^2 + (-2.931)^2) = 568.64$.
- If we want to include interactions, the calculations are a little more complicated.
  - For all $K_1 \times K_2 = 2 \times 3 = 6$ combination of factors first we calculate the group means $\bar{x}_{k_1 k_2} = \sum_{r=1}^{R} x_{k_1 k_2 r}/R$
  - or in our case $\begin{bmatrix} \bar{x}_{11} & \bar{x}_{12} & \bar{x}_{13} \\ \bar{x}_{21} & \bar{x}_{22} & \bar{x}_{23} \end{bmatrix} = \begin{bmatrix} 10.312 & -1.335 & -3.532 \\ 1.997 & -5.112 & -2.330 \end{bmatrix}$, where $\bar{x}_{11}$ is the group mean of measurements at level 1 for factor 1 and level 1 for factor 2 after global centring ($=(14.018 + 10.752 + 10.098 + 5.617 + 11.076)/5 = 10.312$).
  - so $SSR_{12} = R \sum_{k_1=1}^{K_1} \sum_{k_2=1}^{K_2} \left(\bar{x}_{k_1 k_2}\right)^2 - SSR_1 - SSR_2$

    $= 5 \times \left((10.312)^2 + (-1.335)^2 + (-3.532)^2 + (1.997)^2 + (-5.112)^2 + (-2.330)^2\right)$

    $-98.83 - 568.64 = 780.78 - 98.83 - 568.64 = 113.31$.
- To calculate the sum of squares for the error or within-group variance
  - for the linear model without interactions $SSE = SSA - SSR_1 - SSR_2 = 435.43$
  - and for the model including interactions $SSE = SSA - SSR_1 - SSR_2 - SSR_{12} = 322.14$.
- The degrees of freedom are:
  - $\nu(R_f) = (K_f - 1)$, so for factor 1: $\nu(R_1) = 1$, and for factor 2: $\nu(R_2) = 2$
  - for the interactions $\nu(R_{12}) = (K_1 - 1) \times (K_2 - 1) = 1 \times 2 = 2$.
- For the within group variance or error
  - without interactions $\nu(E) = I - 1 - \nu(R_1) - \nu(R_2) = 30 - 1 - 1 - 2 = 26$
  - and with interactions $\nu(E) = I - 1 - \nu(R_1) - \nu(R_2) - \nu(R_{12}) = 30 - 1 - 1 - 2 - 2 = 24$
  - where $I = K_1 \times K_2 \times R = 2 \times 3 \times 5 = 30$ is the total number of elements.
- The degree of freedom tree for the model with interactions is given in Figure 10.9.
- We can now calculate mean errors.
  - For the model without interactions, $MSSR_1 = SSR_1/1 = 98.84$, $MSSR_2 = SSR_2/1 = 284.31$ and $MSSE = MSSE/26 = 16.74$.
  - For the model with interactions, $MSSR_1 = SSR_1/1 = 98.84$, $MSSR_2 = SSR_2/1 = 284.31$ $MSSR_{12} = SSR_{12}/2 = 56.65$ and $MSSE = MSSE/24 = 13.42$.
- Hence, the F and p values for the model without interactions are:
  - $F_1 = MSSR_1/MSSE = 5.90$, so $p = 1 - Fcdf(F_1, \nu(R_1), \nu(E)) = 1 - Fcdf(5.90, 1, 26) = 0.022$
  - $F_2 = MSSR_2/MSSE = 16.97$, so $p = 1 - Fcdf(F_1, \nu(R_2), \nu(E)) = 1 - Fcdf(16.98, 2, 26) = 0.000$

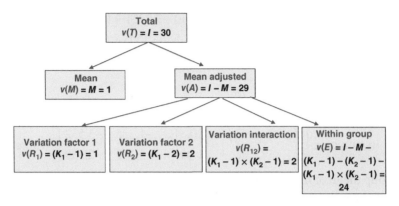

FIGURE 10.9   Degree of freedom tree for model with interaction for data of Table 10.7.

- The F and p values for the model with interactions are:
  - $F_1 = MSSR_1/MSSE = 7.36$, so $p = 1 - Fcdf(F_1, \nu(R_1), \nu(E)) = 1 - Fcdf(7.36, 1, 24)$
    $= 0.012$
  - $F_2 = MSSR_2/MSSE = 21.18$, so $p = 1 - Fcdf(F_2, \nu(R_2), \nu(E)) = 1 - Fcdf(21.18, 2, 24)$
    $= 0.000$
  - $F_{12} = MSSR_{12}/MSSE = 4.22$, so $p = 1 - Fcdf(F_{12}, \nu(R_{12}), \nu(E)) = 1 - Fcdf(4.22, 2, 24)$
    $= 0.027$
- Note that the p value for factor 1 differs quite a lot according to whether the interaction is included. This is because the F ratio differs and the number of degrees of freedom for the within-group variance. So although for the regression the sum of squares will be the same for factor 1 whether interactions are included or not, the resultant error will not be as an additional term is included in the model, also losing 2 degrees of freedom.
- The information can be arranged in an ANOVA table as in Table 10.8 for the model with interactions included. A slightly different table would be obtained without interactions. Note we can also calculate SSR and $MSSR$ if desired.

10.4.1.2   Regression with Dummy Variables

An alternative approach is regression, which involves using dummy variables as discussed before in the context of one-way ANOVA in Section 10.3.4.2.

We will continue to analyse the dataset of Table 10.7.

In order to perform regression, we have to unfold the dataset and define five dummy variables $\mathbf{d}$ where

- d_1 corresponds to whether factor 1 is at level 1 ($=0$) or level 2 ($=1$)
- d_2 corresponds to whether factor 2 is at level 2 ($=1$) or not
- d_3 corresponds to whether factor 2 is at level 3 ($=1$) or not
- $d_4 = d_1 d_2$ is the interaction between d_1 and d_2.
- $d_5 = d_1 d_3$ is the interaction between d_1 and d_3.

TABLE 10.8 ANOVA table for the 2×3 design of Table 10.7.

Source of variation	Sum of squares	Degrees of freedom	Mean sum of squares	F	p
(Factor 1) SSR_1	98.84	$(K_1 - 1) = 1$	98.84	7.36	0.0121
(Factor 2) SSR_2	568.64	$(K_2 - 1) = 2$	284.32	21.18	0.0000
(Interaction) SSR_{12}	113.30	$(K_1 - 1) \times (K_2 - 1) = 2$	56.65	4.22	0.0269
(Regression) SSR	780.78	$(K_1 - 1) + (K_2 - 1)$ $+ (K_1 - 1) \times (K_2 - 1) = 5$	156.16	11.64	0.0000
(Within group) SSE	322.14	$I - 1 - (K_1 - 1) - (K_2 - 1)$ $- (K_1 - 1) \times (K_2 - 1) = 24$	13.42		
(Total mean adjusted) SSA	1102.91	$I - 1 = K_1 \times K_2 \times R - 1 = 29$			

Hence, for factor 2,

- level 1 $[d_2\ d_3] = [0\ 0]$ is the reference level
- level 2 $[d_2\ d_3] = [1\ 0]$
- level 3 $[d_2\ d_3] = [0\ 1]$.

This is presented in Table 10.9, unfolding the x data to form a column vector. We will restrict the model including interactions in this section for brevity.

The 5 dummy variables, correspond to the $\nu(R_1) = (K_1 - 1) = 1$ degrees of freedom for factor 1, $\nu(R_2) = (K_2 - 1) = 2$ for factor 2 and $\nu(R_{12}) = (K_1 - 1)(K_2 - 1) = 2$ for the interactions. If we had a 3×4 design, there would be $2 + 3 + 2 \times 3 = 11$ dummy variables.

We can now perform regression as follows:

- Centre the columns both of the 30×1 x vector and 30×5 D matrix.
- Hence, the centred values of the 19th sample are $x_{19} = 2.433$ and with a corresponding row vector of $d_{19} = [0.500\ -0.333\ -0.333\ -0.167\ -0.167]$ as an example.
- We form a model $x = b_1 d_1 + b_2 d_2 + b_3 d_3 + b_4 d_4 + b_5 d_5 = D\ b$ with the d's this time being column vectors.
- We now calculate the coefficients b by $b = (D'\ D)^{-1} D'x = D^+ x$ to give
  $b' = [-8.315\ -11.648\ -13.845\ 4.539\ 9.517]$
- We find
  - $SSA = 1102.92$ as also calculated for the ANOVA test
  - $SSR = 780.77$; this equals $SSR_1 + SSR_2 + SSR_{12}$ as calculated for the ANOVA test with interactions but obtained in a different way
  - $SSE = 322.14$ as calculated by the ANOVA test.
- We can now remove each of the five dummy variables. We find
  - $SSR_{\{-1:-5\}} = [607.92\ 441.61\ 301.61\ 755.03\ 667.56]$
  - $SSR_{\{1:5\}} = [172.85\ 339.17\ 479.17\ 25.74\ 113.21]$

TABLE 10.9   Unfolded data of Table 10.7 together with dummy variables.

	Factor 1	Factor 2		Interactions	
x	d_1	d_2	d_3	$d_4 = d_1 d_2$	$d_5 = d_1 d_3$
31.256	0	0	0	0	0
27.990	0	0	0	0	0
27.336	0	0	0	0	0
22.855	0	0	0	0	0
28.314	0	0	0	0	0
19.655	0	1	0	0	0
14.814	0	1	0	0	0
20.193	0	1	0	0	0
17.448	0	1	0	0	0
7.402	0	1	0	0	0
10.744	0	0	1	0	0
20.335	0	0	1	0	0
12.807	0	0	1	0	0
11.261	0	0	1	0	0
13.382	0	0	1	0	0
20.865	1	0	0	0	0
19.717	1	0	0	0	0
15.859	1	0	0	0	0
19.670	1	0	0	0	0
20.062	1	0	0	0	0
17.821	1	1	0	1	0
12.120	1	1	0	1	0
11.579	1	1	0	1	0
9.576	1	1	0	1	0
9.532	1	1	0	1	0
18.203	1	0	1	0	1
13.373	1	0	1	0	1
18.675	1	0	1	0	1
9.518	1	0	1	0	1
14.768	1	0	1	0	1

TABLE 10.10   ANOVA table after regression with dummy variables for simulated data of Table 10.7.

Source of variation	Sum of squares	Degrees of freedom	Mean sum of squares	F	p
(Factor 1) d_1 SSR_1	172.85	1	172.85	12.87	0.0015
(Factor 2) d_2 SSR_2	339.17	1	339.17	25.27	0.0000
(Factor 2) d_3 SSR_3	479.17	1	479.17	35.70	0.0000
(Interaction) d_4 SSR_4	25.74	1	25.74	1.92	0.1788
(Interaction) d_5 SSR_5	113.21	1	113.21	8.44	0.0078
(Regression) SSR	780.77	$(K_1 - 1) + (K_2 - 1) +$ $(K_1 - 1) \times (K_2 - 1) = 5$	156.15	11.63	0.0269
(Error) SSE	322.14	$I - 1 - (K_1 - 1) - (K_2 - 1) -$ $(K_1 - 1) \times (K_2 - 1) = 24$	13.42		
(Total mean adjusted) SSA	1102.91	$I - 1 = K_1 \times K_2 \times R - 1 = 29$			

- Note that the sum $\sum_{d=1}^{D} SSR_d \neq SSR$ in this case because the dummy variables are not orthogonal; this contrasts with the previous examples.
- Using these values we find that $F = [12.88\ 25.27\ 35.70\ 1.92\ 8.43]$ and so $p = [0.001\ 0.000\ 0.000\ 0.179\ 0.008]$.
- This suggests all the main effects are important with $p \leq 0.001$ but the interactions are not as significant in this case.
- The corresponding ANOVA table is presented in Table 10.10.
- Interpreting these dummy variables is quite difficult especially if out of context and as this is a simulated case study we will not attempt to here.

10.4.2   Two-level Multiway Factorial Designs

A very common arrangement in metabolomics is a two-level full factorial design. These have been introduced in Section 6.2.3.2, and a good example is case study 8 (*Daphnia* in different environments as measured by GCMS) and these have been analysed as three one-way ANOVA or regression problems in Section 10.3.3. As these types of designs are very widespread we will look in further detail in this section. The emphasis is on the significance of the factors in this section, rather than the predicted values of the response *x*. Chapter 8 about multivariate calibration discusses the most widespread approach for prediction of a multivariate response. Of course, these aspects cannot be totally separated, but this text for reasons of brevity, is focussed on the most common usage of chemometric and statistical methods in metabolomics, rather than a broadly based comprehensive text on regression, of which there are many excellent ones available.

Factorial designs are orthogonal, and the simplest at two levels. The advantage of a two-level design is that multilinear regression, regression with dummy variables and the ANOVA test will give the same overall result. However, multilinear regression is computationally easier to include additional terms, if the number of factors and interactions are large.

For the *Daphnia*, for each variable when there are 3 (= F) factors we could build various models, a full model being of the form $x = b_0 + b_1 x_1 + b_2 x_2 + b_3 x_3 + b_{12} x_1 x_2 + b_{13} x_1 x_3 + b_{23} x_2 x_3 + b_{123} x_1 x_2 x_3 + e$ with

- $F!/[(F-0)!0!)] = 1$ intercept term
- $F!/[(F-1)!1!)] = F = 3$ linear terms
- $F!/[(F-2)!2!)] = 3$ two-factor interactions
- $F!/[(F-3)!3!)] = 1$ three-factor interactions

Note that there are 8 possible terms and the design without replicates in Table 6.3 consists of 8 unique experimental conditions.

Normally, higher level interactions are not observable, and if both blocks are centred we can remove the intercept term, so a practical model may be (for centred data)

$$x = b_1 x_1 + b_2 x_2 + b_3 x_3 + b_{12} x_1 x_2 + b_{13} x_1 x_3 + b_{23} x_2 x_3 + e$$

In general, for *F* factors studied at two levels each, there will be 2^F possible unique experimental conditions, and a similar number of terms, if all higher-order interactions are taken into account. Higher-order interactions can be removed from the model, as in practice we are unlikely to be interested in more than two-factor interactions, leaving additional degrees of freedom to assess the quality of the model, or alternatively can reduce the number of experiments by performing fractional factorial designs as described in Section 6.3.3.

We will restrict this section to analysis of the full factorial design of case study 8 for brevity, but similar principles can easily be extended to other two-level factorial data.

- We illustrate the method with variable 1 (lactic acid).
- There are 24 experiments, with 8 unique conditions and 3 replicates at each set of conditions.
- The first step is to code the factors as described above; note that it is normal to code these +1 and −1 rather than 0 and 1, but the latter would be possible if the data are centred afterwards. We will follow common conventions in this chapter.
- A 24×6 design matrix *D* can be set up, as in Table 10.11. Each column represents a variable. The columns are all centred using coding of ± but the *x* vector should also be centred so that there is no intercept term.
- The experimental data *x* is column centred by subtracting the column mean 1923.32 from each data point.
- Using centred data, we can now obtain the coefficients *b* for the six-term model via the pseudoinverse by $b = (D'D)^{-1} D' x = D^+ x$.

TABLE 10.11   Design matrix and peak intensities of lactic acid, for the 24 experiments of case study 8 (Daphnia) using a model with 3 linear and 3 interaction terms and coded values for the factor levels.

	x	D					
Experiment	Lactic acid	Sal	Temp	Hyp	Sal × Temp	Sal × Hyp	Temp × Hyp
1	1887	1	1	1	1	1	1
2	1893	1	1	1	1	1	1
3	2678	1	1	1	1	1	1
4	1863	1	1	−1	1	−1	−1
5	1610	1	1	−1	1	−1	−1
6	1783	1	1	−1	1	−1	−1
7	2533	1	−1	1	−1	1	−1
8	2050	1	−1	1	−1	1	−1
9	2157	1	−1	1	−1	1	−1
10	2335	1	−1	−1	−1	−1	1
11	2792	1	−1	−1	−1	−1	1
12	1682	1	−1	−1	−1	−1	1
13	1556	−1	1	1	−1	−1	1
14	1643	−1	1	1	−1	−1	1
15	2119	−1	1	1	−1	−1	1
16	1863	−1	1	−1	−1	1	−1
17	1967	−1	1	−1	−1	1	−1
18	1610	−1	1	−1	−1	1	−1
19	2044	−1	−1	1	1	−1	−1
20	1413	−1	−1	1	1	−1	−1
21	1676	−1	−1	1	1	−1	−1
22	1626	−1	−1	−1	1	1	1
23	1622	−1	−1	−1	1	1	1
24	1758	−1	−1	−1	1	1	1

- We find $\boldsymbol{b}' = [181.92\ -50.67\ 47.42\ -102.25\ 47.00\ 42.58]$.
- This means our prediction is $\hat{x} = 181.92c_1 - 50.67c_2 + 47.42c_3 - 102.25c_1c_2 + 47.00c_1c_3 + 42.58c_2c_3$ for centred data using coded values of the factors (the uncentred predictions are $\hat{x} + \bar{x} = \hat{x} + 1923.32$).
- So (for centred data) for the 16th sample, $x_{16} = -60.33$ whereas $\hat{x}_{16} = -173.33$ for example, the error is $e_{16} = -60.33 - (-173.33) = 113$.

- We now calculate various sums of squares
  - $SSA = 3.0162 \times 10^6$
  - $SSR = 1.2573 \times 10^6$
  - $SSE = 1.7589 \times 10^6$
- As there are $P = 6$ parameters in the model, we find, $\nu(R) = P = 6$, and $\nu(E) = I - 1 - P = 24 - 1 - 6 = 17$. The degree of freedom tree is similar to Figure 10.6 but with three extra degrees of freedom for the interactions (not illustrated for brevity).
- Therefore,
  - $MSSR = 1.2573 \times 10^6/6 = 2.0955 \times 10^5$
  - $MSSE = 1.7589 \times 10^6/17 = 1.0346 \times 10^5$
- This makes $F = 2.025$ and $p = 1 - Fcdf(2.025,6,17) = 0.118$ which suggests not a terribly good model. This is actually worse than the p value obtained in Section 10.3.3, which includes only the 3 linear terms. At first, this may seem strange, but the model with interactions contains six terms and suggests that although the fit will always be better as more terms are included, the improvement is not as large as would be expected. Note that the regression error for the full six-term model (1.7589×10^6) is, as expected, lower than the regression error for the three-term model of course (2.106×10^6) since there are additional terms, so a reduction in the absolute value of the error does not necessary correspond to a lower p value.

We can now look at each term individually.

- This we do by removing each variable one by one.
  - We find $SSR_{\{-1 \,:\, -6\}} = [0.463\ 1.196\ 1.203\ 1.006\ 1.204\ 1.214] \times 10^6$
  - $SSR_{\{1\,:\,6\}} = MSSR_{\{1\,:\,6\}} = [7.942\ 0.616\ 0.540\ 2.509\ 0.530\ 0.435] \times 10^5$; note that the sum of these elements equals SSR because the factors are orthogonal and is an advantage of factorial designs.
- Hence, we can calculate
  - $F = [7.677\ 0.596\ 0.522\ 2.425\ 0.512\ 0.421]$
  - so $p = 1 - cdf(F,1,17) = [0.013\ 0.451\ 0.480\ 0.138\ 0.484\ 0.525]$.
- An ANOVA table is presented in Table 10.12 and should be compared to Table 10.6. The p and F values differ slightly because the number of terms in the model changes. The contributions of the interaction terms are treated as error or noise in the simpler model, leading to higher SSE as less terms are included.
- This implies that for lactic acid, salinity appears quite significant, but the other factors and interactions are not very important. Obviously, this will differ according to metabolite, as we looked at the full metabolic profile in Section 10.3.3. We will not repeat this in this section, although for some metabolites, the interactions may provide further insight. In the section on ASCA, we will examine this in more detail.

TABLE 10.12 ANOVA table for data of Table 10.11.

Source of variation	Sum of squares	Degrees of freedom	Mean sum of squares	F	p
Salinity (SSR_1)	7.942×10^5	1	7.942×10^5	7.677	0.013
Temperature (SSR_2)	0.616×10^5	1	0.616×10^5	0.596	0.451
Hypoxia (SSR_3)	0.540×10^5	1	0.540×10^5	0.522	0.480
Sal × Temp (SSR_{12})	2.509×10^5	1	2.509×10^5	2.425	0.138
Sal × Hyp (SSR_{13})	0.530×10^5	1	0.530×10^5	0.512	0.482
Temp × Hyp (SSR_{23})	0.435×10^5	1	0.435×10^5	0.421	0.525
Regression (SSR)	1.2573×10^6	6	2.096×10^5	2.025	0.118
Error (SSE)	1.759×10^6	17	1.035×10^5		
Mean adjusted (SSA)	3.016×10^6	23			

We could alternatively have used dummy variables for this regression but would come to the same conclusions as using coded variables since there are two levels. The coded value for salinity could be defined as [1 0 0], for temperature [0 1 0] and for hypoxia [0 0 1] and so on for interactions and various combinations. The matrix D after centring would be identical except that +1 would be replaced by +0.5, and −1 by −0.5, so the coefficients b would be twice the magnitude. This would make no difference to the p values or predictions.

Alternatively, a three-way ANOVA test could be performed, but is not recommended, as the equations can be a bit complicated and there would be less interpretation. Hence, for multiple factorial designs at two levels, regression followed by ANOVA is recommended as the default.

10.5 ASCA

The methods reported above in this chapter are univariate, that is, they estimate the significance of each variable one at a time and do not take into account any correlations with other variables. The methods described in Chapter 9 although multivariate do not take into account interactions between the factors that influence the values of each variable, and are applied only to single-factor problems. If, of course, we want to study several factors, the experiments are designed so they are orthogonal, and interactions are not important or of interest, the methods in Chapter 9 are adequate.

However, there have been a variety of approaches developed that take into account both correlations or interactions between variables and between factors. Hence, for example, an experiment that investigates the effect of both temperature and light on plant growth, measuring the concentrations of 50 metabolites, may involve interactions/correlations in both the c and x blocks. ASCA is one of the most widespread approaches in chemometrics to investigate this situation, and we will restrict this section to ASCA. Other approaches such as MANOVA or ANOVA-PCA are available in the literature for interested readers.

We will initially illustrate the principles using small simulated data and then extend to case study 8 (influence of environmental factors on the metabolism of *Daphnia*).

We will restrict the discussion in this section to balanced designs. For unbalanced data, the calculations and interpretation can be a bit complex and if required readers are suggested to consult the specialist literature.

10.5.1 Simulated Dataset

- We will illustrate the principles by a 12×2 ($=I \times J$) dataset, whose response is affected by $F = 2$ factors.
- This dataset is presented in Table 10.13.
- Factor 1 is at three levels, which we code 1, 0, −1; factor 2 is at two levels which we code 1, −1.
- There are two replicates at each combination of unique conditions.

In order to study this we set up a matrix **D**, which involves dummy coding each factor/level combination.

- There are $K-1$ dummy variable for a K-level single variable factor.
- For factor 1, we define two dummy variables by $\delta_{1\{1\}} = [1\ 0]$ for level 1, $\delta_{1\{2\}} = [0\ 1]$ for level 2 and $\delta_{1\{3\}} = [-1\ -1]$ for level 3; note that there are other ways of defining the dummy variables but for brevity we will not compare these in detail and are following a common method for calculation within the ASCA literature.
- For factor 2, we will denote the dummy variable $\delta_{2\{1\}} = [1]$ for level 1 and $\delta_{2\{2\}} = [-1]$ for level 2.

TABLE 10.13   Simulated data for Section 10.5.1.

x_1	x_2	c_1	c_2
27.89	23.04	1	1
29.16	29.59	1	1
27.31	32.00	1	−1
25.82	32.57	1	−1
19.67	23.62	0	1
15.90	26.52	0	1
10.38	23.34	0	−1
15.20	19.23	0	−1
9.00	14.34	−1	1
4.07	17.09	−1	1
7.61	2.27	−1	−1
10.80	5.13	−1	−1

- One advantage of ANOVA is that we can study interactions. We can model $x_{ij} = b_{0j} + b_{1j}c_{1i} + b_{2j}c_{2i} + b_{12j}c_{1i}c_{2j} + e_{ij}$ where j represents the variable, i the sample and c the level (not the dummy variable).
- Hence, we can create an extra 2 dummy variables that correspond to the interactions between factors 1 and 2 which are the products of the single factor dummy variables. These have six levels, corresponding to all possible combinations of levels for factors 1 and 2, namely $\delta_{12\{1,1\}} = [1\ 0]$, $\delta_{12\{1,2\}} = [-1\ 0]$, $\delta_{12\{2,1\}} = [0\ 1]$, $\delta_{12\{2,2\}} = [0\ -1]$, $\delta_{12\{3,1\}} = [-1\ -1]$, and $\delta_{12\{3,2\}} = [1\ 1]$.
- Finally, we will not mean centre the data in this example. If mean centred the first dummy variable δ_0 is removed. For non-mean-centred data, the first dummy variable $\delta_0 = 1$. We could use other approaches which would involve mean centring the X matrix also.
- This results in a 12×6 dummy variable matrix D as presented in Table 10.14, the columns consisting of δ_0 (column 1), δ_1 (columns 2 and 3), δ_2 (column 4) and δ_{12} (columns 5 and 6). We will denote the jth column by d_j.

TABLE 10.14 Dummy variable matrix D corresponding to the two-factor model with interaction for the 12 unique conditions in Table 10.13 together with the factor matrix L.

Matrix D

Mean	Factor 1		Factor 2	Interactions	
1	1	0	1	1	0
1	1	0	1	1	0
1	1	0	−1	−1	0
1	1	0	−1	−1	0
1	0	1	1	0	1
1	0	1	1	0	1
1	0	1	−1	0	−1
1	0	1	−1	0	−1
1	−1	−1	1	−1	−1
1	−1	−1	1	−1	−1
1	−1	−1	−1	1	1
1	−1	−1	−1	1	1

Matrix L

Factor 1	0	1	1	0	0	0
Factor 2	0	0	0	1	0	0
Interaction	0	0	0	0	1	1

- The number of dummy variables will be denoted by N so in general D is an $I \times N$ matrix. If there are F factors each a K_f levels and we want to study single factors and all two-factor interactions, there will be $N = 1 + \sum_{f=1}^{F} (K_f - 1) + \sum_{f=2}^{F} \sum_{g=1}^{f} (K_f - 1)(K_g - 1)$ dummy variables.

In addition, we set up a matrix which we will denote L.

- Each row corresponds to a factor or interaction of interest.
- As there are three terms (or factors/interactions) we want to study, L has $T = 3$ rows.
- Each row has as many columns as D and each column corresponding to a dummy variable arising from the corresponding row has a value of 1, otherwise is equal to 0.
- So, L is a $T \times N$ matrix.
- This is also presented in Table 10.14 for our simulated example.
- The next step is the estimate the regression coefficients for each column j by $a_j = D^+ x_j$.
  - where j is the column (or variable) of X
  - D^+ is the pseudoinverse of D.
- These are presented in Table 10.15 for the model to give an $N \times J$ matrix A where $x_{ij} = a_{0j} d_1 + a_{1\{1\}j} d_2 + a_{1\{2\}j} d_3 + a_{2\{1\}j} d_4 + a_{12\{1,1\}j} d_5 + a_{12\{2,1\}j} d_6 + e_{ij}$

the first subscript of a referring to the original factor and the number in brackets to the corresponding dummy variable; '12{2,1}' implies the interaction (12) between factor 1 and 2, and {2,1} refers to the second dummy variable corresponding to factor 1 and the first (and only) to factor 2.

The next step is to determine a matrix for each term of interest.

- There are three terms of interest ignoring the intercept, each corresponding to a row of L.
  - Row m corresponds to a term we are interested in,
    - so in our case row 1 corresponds to factor 1, row 2 to factor 2 and row 3 to the interaction.
- Our question is whether each of these 3 terms is significant.
- We now create a new matrix M_m for each term m of dimensions $N \times N$ with all elements = 0 except diagonal elements $M_{m\{nn\}}$ where n equals the non-zero elements in row m of matrix L. This is easiest illustrated for M_1 in our example in Table 10.15 with equivalent matrices for M_2 and M_3.

TABLE 10.15 Steps in ASCA calculations for data of Table 10.13.

Regression coefficients A		
	x_1	x_2
a_0	16.90	20.73
$a_{1\{1\}}$	10.64	8.57
$a_{1\{2\}}$	−1.61	2.45
$a_{2\{1\}}$	0.71	1.64
$a_{12\{1,1\}}$	0.27	−4.62
$a_{12\{2,1\}}$	1.78	0.25

M_1					
0	0	0	0	0	0
0	1	0	0	0	0
0	0	1	0	0	0
0	0	0	0	0	0
0	0	0	0	0	0
0	0	0	0	0	0

X decomposed									
X_0		X_1		X_2		$X_3 = X_{12}$		E	
16.90	20.73	10.64	8.57	0.71	1.64	0.27	−4.62	−0.64	−3.28
16.90	20.73	10.64	8.57	0.71	1.64	0.27	−4.62	0.64	3.28
16.90	20.73	10.64	8.57	−0.71	−1.64	−0.27	4.62	0.75	−0.29
16.90	20.73	10.64	8.57	−0.71	−1.64	−0.27	4.62	−0.75	0.29
16.90	20.73	−1.61	2.45	0.71	1.64	1.78	0.25	1.89	−1.45
16.90	20.73	−1.61	2.45	0.71	1.64	1.78	0.25	−1.89	1.45
16.90	20.73	−1.61	2.45	−0.71	−1.64	−1.78	−0.25	−2.41	2.06
16.90	20.73	−1.61	2.45	−0.71	−1.64	−1.78	−0.25	2.41	−2.06
16.90	20.73	−9.03	−11.02	0.71	1.64	−2.05	4.37	2.47	−1.38
16.90	20.73	−9.03	−11.02	0.71	1.64	−2.05	4.37	−2.47	1.38
16.90	20.73	−9.03	−11.02	−0.71	−1.64	2.05	−4.37	−1.60	−1.43
16.90	20.73	−9.03	−11.02	−0.71	−1.64	2.05	−4.37	1.60	1.43

We are now set to go.

- The model can be decomposed into contributions from each term and the mean, in our case, $X = X_0 + X_1 + X_2 + X_3 + E$ where
  - X_0 is the mean
  - X_1 is the contribution due to factor 1
  - X_2 is the contribution due to factor 2
  - X_3 is the contribution due to the interaction of factors 1 and 2, alternatively denoted by X_{12}
  - E is the error or residual due to this model.

It is now easy to calculate the contribution of term t.

- $X_t = D M_m A$ for terms 1–3, and X_0 is just the mean.
- These are presented in Table 10.15.
- The error E is simply the residual $E = X - (X_0 + X_1 + X_2 + X_3)$

It is possible to calculate the sum of squares for each of the terms, including the total sum of squares.

- $SST = \sum_{j=1}^{J}\sum_{i=1}^{I} x_{ij}^2 = 1.048 \times 10^4, SSM = \sum_{j=1}^{J}\sum_{i=1}^{I} \bar{x}_j^2 = 0.858 \times 10^4,$

  $SSA = \sum_{j=1}^{J}\sum_{i=1}^{I} x_{ij}^2 - \sum_{j=1}^{J}\sum_{i=1}^{I} \bar{x}_j^2 = 0.190 \times 10^4$

- so $SSR_1 = \sum_{j=1}^{J}\sum_{i=1}^{I} x_{1ij}^2 = 1.593 \times 10^3, SSR_2 = \sum_{j=1}^{J}\sum_{i=1}^{I} x_{2ij}^2 = 38.33,$

  $SSR_{12} = \sum_{j=1}^{J}\sum_{i=1}^{I} x_{12ij}^2 = 191.92$

- and $SSE = \sum_{j=1}^{J}\sum_{i=1}^{I} e_{ij}^2 = 80.05$

As usual mean sum of squares and p values can be calculated, but we will look further at how PCA can be incorporated into the model and show the effects of each factor and interaction graphically.

We can perform PCA separately on each term (ignoring the intercept – the data could be centred as an alternative) to give a model of the form $X = X_0 + T_1 P_1 + T_2 P_2 + T_3 P_3 + E.$

- For each term, we can calculate the scores for the 12 rows (or samples) of the data matrix. However, there will only be K_f unique values of the scores for each term, falling into $K_f - 1$ dimensions. The data should be centred.
- We will not reproduce these calculations in full but for readers wishing to check their calculations $t_{1(3)} = [13.58\ 1.53]$ representing the scores for the estimated model for term 1, and sample 3.

- For a three-level model, there will be 3 different scores, each representing a single level.
- What is interesting is the errors for each sample using the contribution to the model from each of the three terms.
- For each sample and term m we can calculate $x_{me(ij)} = x_{m(ij)} + e_{ij}$.
- As an example for $m = 1$, $i = 7$ and $j = 2$, $x_{1(72)} = 2.45$ (see Table 10.15) and $e_{72} = 2.06$ so $x_{1e(72)} = 4.51$.
- We can then predict the scores using the corresponding model using $\hat{t}_{ti} = x_{te\{i\}} p_t'$ or for a two-component model and sample 7, $\hat{t}_{1(7)} = \begin{bmatrix} 0.368 & -6.028 \end{bmatrix}$.
- The mean of the predicted scores of samples at a single level will equal the scores of T_m.

This is best illustrated graphically and we can now visualise the data.

- We illustrate these in Figure 10.10 with the range or confidence limit indicated at each level.
- When there are more than 2 samples representing a single level, we enclose the data using a Mahalanobis distance of 2, otherwise, we just illustrate the two replicates using a straight line. The borders are indicative and should not be regarded as hard boundaries. It is beyond this introductory text to interpret these boundaries in statistical detail as we follow the normal steps in ASCA.
- Some software uses polygons to define the boundaries of each level or group as an alternative to ellipsoidal boundaries.
- The values of d for each level and factor are indicated in the figures.
- The number of samples in each group or level for each of the three terms is illustrated in these figures.
- It can be seen that the three levels of factor 1 are very well separated, suggesting it has a significant influence over the multivariate response.
- This is not so for factor 2 suggesting some overlap. The interactions are a little better separated than factor 2.

This approach allows a multivariate representation of how each significant each factor is. There are a variety of statistical interpretations that we leave to the reader and more specialist literature.

This analysis can be extended to individual variables; however, as there are only two variables in this example the graphs will not be very informative. We will look at case study 8 (environmental effects on *Daphnia*) in the next section, where there are 47 variables for better illustration.

10.5.2   Case Study: Environmental Effect on Daphnia

We will illustrate the application to case study 8. The data involves studying the influence of salinity, temperature and hypoxia on the metabolism of *Daphnia* as has been introduced elsewhere in the text, and in earlier sections of this chapter.

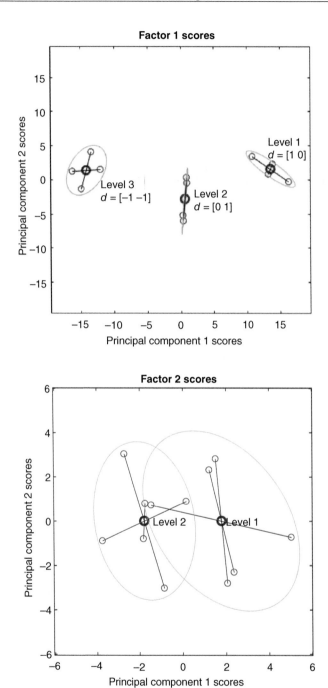

FIGURE 10.10   Scores using ASCA for the two factors and their interactions for the simulated case study of Table 10.13.

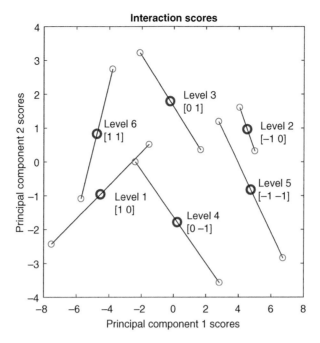

FIGURE 10.10 (*Continued*)

- The data are arranged as a three-factor full factorial design (see Chapter 6) with 3 replicates at each unique set of conditions.
- If we want to study all single-factor effects and all two-factor interactions we can model the data by $x_{ij} = b_{0j} + b_{1j}c_{1i} + b_{2j}c_{2i} + b_{3j}c_{3i} + b_{12j}c_{1i}c_{2j} + b_{13j}c_{1i}c_{3j} + b_{23j}c_{2i}c_{3j} + e_{ij}$
- Dummy variables for two levels directly correspond to a coded variable so that $x_{ij} = a_{0j} + a_{1j}d_{1i} + a_{2j}d_{2i} + a_{3j}d_{3i} + a_{12j}d_{12i} + a_{13j}d_{13i} + a_{23j}d_{23i} + e_{ij}$; if the dummy variables are coded ± 1, the coefficients a and b are the same in this case.
- In order to put the metabolites on the same scale we first standardise the X data; this mean centres each column so that $b_0 = a_0 = 0$. We will retain this term for compatibility, but it is irrelevant whether it is part of the model or not. Note that if analysing each metabolite separately as above standardisation is not necessary as each column is treated independently.
- Matrices D and L are presented in Table 10.16. As each factor is at two levels, these matrices are somewhat simplified compared to Table 10.14 as each factor corresponds to a single dummy variable, and the design matrix is identical to the matrix containing dummy variables.

For brevity, we will not describe each step in the analysis, as the detailed calculation procedures of ASCA have been described in Section 10.5.1 and primarily focus on the outcomes.

TABLE 10.16   Matrices D and L for case study 8 using a model with all single-factor terms and two-factor interactions.

Matrix D

Mean	Single factor			Interactions		
Mean	Salinity	Temp	Hypoxia	Sal × Temp	Sal × Hyp	Temp × Hyp
d_0	d_1	d_2	d_3	d_4	d_5	d_6
1	1	1	1	1	1	1
1	1	1	1	1	1	1
1	1	1	1	1	1	1
1	1	1	−1	1	−1	−1
1	1	1	−1	1	−1	−1
1	1	1	−1	1	−1	−1
1	1	−1	1	−1	1	−1
1	1	−1	1	−1	1	−1
1	1	−1	1	−1	1	−1
1	1	−1	−1	−1	−1	1
1	1	−1	−1	−1	−1	1
1	1	−1	−1	−1	−1	1
1	−1	1	1	−1	−1	1
1	−1	1	1	−1	−1	1
1	−1	1	1	−1	−1	1
1	−1	1	−1	−1	1	−1
1	−1	1	−1	−1	1	−1
1	−1	1	−1	−1	1	−1
1	−1	−1	1	1	−1	−1
1	−1	−1	1	1	−1	−1
1	−1	−1	1	1	−1	−1
1	−1	−1	−1	1	1	1
1	−1	−1	−1	1	1	1
1	−1	−1	−1	1	1	1

Matrix L

		Salinity	Temp	Hypoxia	Sal × Temp	Sal × Hyp	Temp × Hyp
Salinity	0	1	0	0	0	0	0
Temperature	0	0	1	0	0	0	0
Hypoxia	0	0	0	1	0	0	0
Sal × temp	0	0	0	0	1	0	0
Sal × hyp	0	0	0	0	0	1	0
Temp × hyp	0	0	0	0	0	0	1

We can calculate the sums of squares.

- As the data have been standardised, we do not need to calculate SST or SSM, there is little interest in uncentred data divided by a standard deviation.
- We find $SSA = 1128$. This is equal to 24 (samples)$\times$47 (variables) and is a consequence of standardising (using the population formula for standard deviation for scaling). Note that as the data have been standardised and so centred, $SSA = SST$. There is no statistical agreement whether to reduce the total numbers of degrees of freedom by 1 in this case, but as most software does this, we will follow the standard approach.

- For the single factors find $SSR_1 = \sum_{j=1}^{J}\sum_{i=1}^{I} x_{1ij}^2 = 244.01$, $SSR_2 = \sum_{j=1}^{J}\sum_{i=1}^{I} x_{2ij}^2 = 44.14$, $SSR_3 = \sum_{j=1}^{J}\sum_{i=1}^{I} x_{3ij}^2 = 85.81$

- For interactions $SSR_{12} = \sum_{j=1}^{J}\sum_{i=1}^{I} x_{12ij}^2 = 30.51$, $SSR_{13} = \sum_{j=1}^{J}\sum_{i=1}^{I} x_{13ij}^2 = 39.71$,

  $SSR_{23} = \sum_{j=1}^{J}\sum_{i=1}^{I} x_{23ij}^2 = 68.69$

- and $SSE = \sum_{j=1}^{J}\sum_{i=1}^{I} e_{ij}^2 = 615.13$.

- Check that $SSR_1 + SSR_2 + SSR_3 + SSR_{12} + SSR_{13} + SSR_{23} + SSE = SSA = I \times J = 1128$.

Compared to the simulated example, the relative error is much higher. This is because in real-world case studies, the effects we want to study may not be the main source of variability, so there will be many other factors.

Nevertheless, we can calculate the mean squared values of each factor and interaction and obtain an ANOVA table and p values as in Table 10.17. It can be seen salinity has a low p value of 0.019 and thus is likely to be significant. None of the other factors or interactions have $p < 0.05$. However, this is for all 47 metabolites or the entire metabolic profile, and individual metabolites may be more significant. The p value for regression using all 6 terms (0.076) suggests that the model might contain some terms that are significant.

The data can also be represented graphically as in Figure 10.11. The same conclusions can be drawn as for the sum of squares, for example, we see that the two levels for factor 1 (salinity) are well separated so this factor is likely to be significant.

In addition, though, we can gain greater insight into individual variables. Salinity is the most important factor overall, and we will look in more detail.

TABLE 10.17   ANOVA table using ASCA for case study 8.

Source of variation	Sum of squares	Degrees of freedom	Mean sum of squares	F	p
Salinity (Factor 1) d_1 SSR_1	244.01	1	244.01	6.74	0.019
Temperature (Factor 2) d_2 SSR_2	44.14	1	44.14	1.22	0.285
Hypoxia (Factor 3) d_3 SSR_3	85.81	1	85.81	2.37	0.142
Salinity × Temperature (Interaction) d_4 SSR_{12}	30.51	1	30.51	0.84	0.371
Salinity × Hypoxia (Interaction) d_5 SSR_{13}	39.71	1	39.71	1.10	0.309
Temperature × Hypoxia (Interaction) d_6 SSR_{23}	68.69	1	68.69	1.90	0.186
(Regression) SSR	512.87	6	85.48	2.36	0.076
(Error) SSE	615.13	17	36.18		
(Total mean adjusted) SSA	1128	23	49.04		

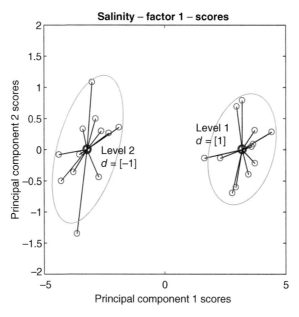

FIGURE 10.11   Scores of 2 PCs using ASCA for the single factor and interaction effects for case study 8.

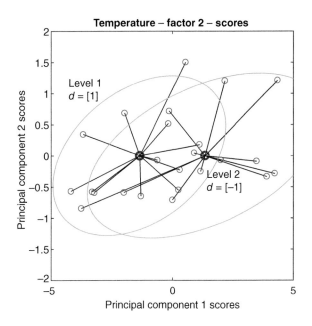

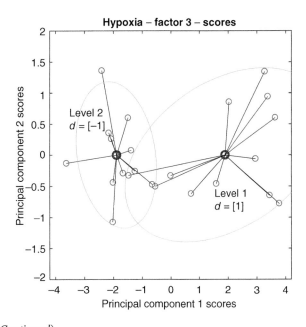

FIGURE 10.11   (*Continued*)

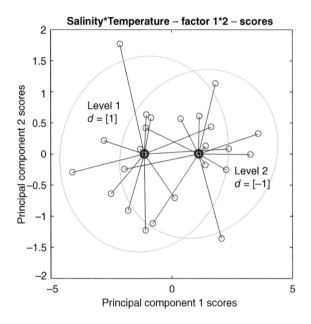

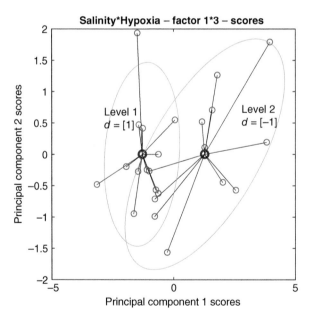

FIGURE 10.11   (*Continued*)

We can look at the loadings for each metabolite for individual factors or interactions. The loadings for salinity are presented in Figure 10.12, and we can highlight which variables have the loadings that are highest in magnitude to identify the most significant metabolites for this factor.

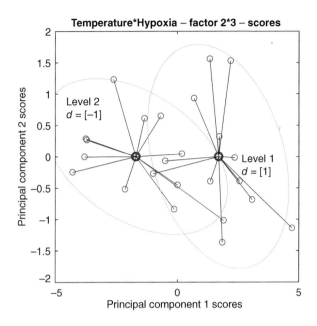

FIGURE 10.11   (Continued)

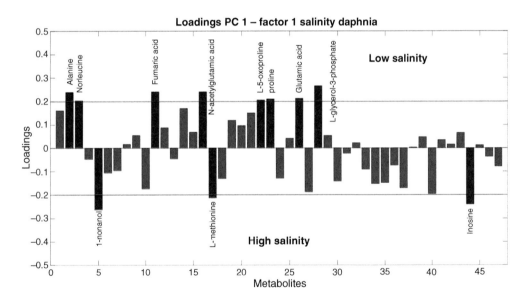

FIGURE 10.12   Loadings of PC 1 for salinity.

The sign of the loadings can provide insight as to whether an individual metabolite is a marker for high or low salinity. The metabolites with the loadings highest in magnitude are comparable to the loadings using a univariate regression model with the lowest p values and highest F values as discussed in Section 10.4.2, although the relationship is not exact as ASCA is a multivariate model. Figure 10.13 illustrates this

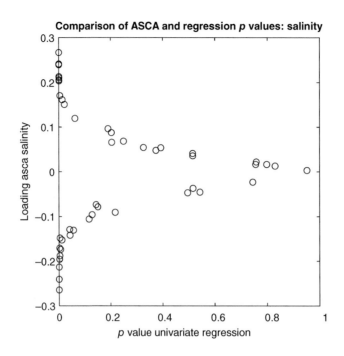

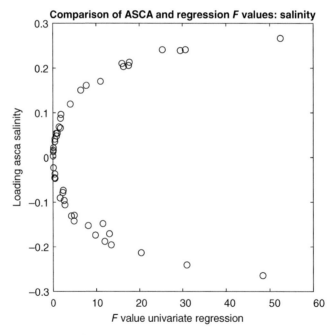

FIGURE 10.13   Comparison of p and F values from univariate regression to loadings from ASCA for factor 1 (salinity) of case study 8.

relationship. The more significant markers can be identified using either method, and a detailed comparison of the pros and cons is outside the scope of this text.

In summary, ASCA is a very flexible approach and can be used when we want a model that is multivariate both in the x block and c block, both to see whether factors and their interactions have an overall effect on the whole metabolic profile, or which individual metabolites and their interactions are influenced by individual factors. Whereas ASCA is a relatively recent method compared to some of the other more historic and primarily univariate approaches discussed in this chapter, it is nevertheless an important technique for many studies. It can be a bit complex when data are unbalanced, which extensions we will not discuss in this text but for balanced designs, it is becoming increasingly widespread at the time of writing this text.

Index

A

Alternating least squares (ALS), 35, 37
Analysis of variance (ANOVA), 19, 152, 201. *See also* Multiple factor (multiway) ANOVA test; Single factor (one-way-one-factor) ANOVA test
  calculation of, 358–361
  multivariate, 354
  N factor, 355
  one-way, 355
  principle of, 353
  two-way, 355
Analytical Chemistry journal, 5, 8
ANOVA. *See* Analysis of variance (ANOVA)
ANOVA simultaneous component analysis (ASCA) method, 201, 353, 389–390
  Daphnia, environmental effect on, 395, 397–405
  defined, 77
  simulated dataset, 390–397
Anti-Stokes scattering, 63
APCI. *See* Atmospheric pressure chemical ionisation (APCI)
apLCMS, 40
Arthur program, 18–19
Artificial intelligence (AI), 47
ASCA method. *See* ANOVA simultaneous component analysis (ASCA) method
Atmospheric pressure chemical ionisation (APCI), 30
Attenuated total reflectance (ATR), 65
Automated mass spectral deconvolution and identification system (AMDIS), 38
Auto-predictive model, 212, 306

B

Baseline correction, 55
BATMAN, 58
Bayesian statistics, 290–291
Bayesil, 58
Beginners' All-purpose Symbolic Instruction Code (BASIC), 16–17
Binning/bucketing data table, 57
Blood plasma
  malaria diagnosis in children, 69–70
  rheumatoid arthritis in humans study, 67–69

C

CAMO Software, 19
C and C++ software, 17
Case studies, 14–15
  blood plasma, rheumatoid arthritis using, 67–69
  characteristics of, 66, 67
  FTIR, nitrates on wheat, 64, 107–109
  GCMS
    Daphnia Magna metabolism by, 76–77
    malaria diagnosis using, 69–70
  HPLC, Chinese herbal medicine using, 77–78
  instrumental techniques, 66–67
  LCMS
    diabetes in mice by, 78–79
    of pre-arthritis, 98–100
    rheumatoid arthritis using, 67–69
  NMR
    glucose intolerance/diabetes using, 71–72, 100–103
    metabolic/temperature changes in maize by, 72–73, 92–97, 166–169, 187–190
    triglycerides measurement using, 70–71, 107, 109–112
  Raman spectroscopy
    of bacterial faecal isolates, 109, 112–114
    Enterococcal bacteria by, 75–76
CCR. *See* Correctly Classified Rate (CCR)
Central dogma, of biology, 9, 10
Chemical shift scale, 52
Chemometrics

Data Analysis and Chemometrics for Metabolomics, First Edition. Richard G. Brereton.
© 2024 John Wiley & Sons Ltd. Published 2024 by John Wiley & Sons Ltd.
Companion website: www.wiley.com/go/Brereton/ChemometricsforMetabolomics

analytical and physical chemistry, 4–6
applications, 8
applied statistics, 1–4
chemical manufacturing, 3
history of, 1
mainstream chemistry, 5
MATLAB, 7
multivariate methods, 3
quantitative chemistry, 4
quantum/statistical mechanics, 4
scientific computing, 6–7
substantial renaissance, 9
workshops, 8
ChemSpider, 46
Chenomx NMR Suite, 58
χ^2-distribution
    cumulative distribution function, 171–174
    F distribution relates to, 174–177
    probability density distributions, 170, 171
    p values calculated using, 182, 185, 186
    variance of, 169–170
Chinese herbal medicine, using HPLC, 77–78
Chromatography, 27–29
Cluster sampling, 209
Column centring, data transformation, 125–130
Column standardisation, 129
    maize raw data, 138, 139
    NMR standardised triglycerides, 140–141
    row-scaled data matrix, 134–136
    standard deviation, 130
    standardised data matrix, 130–137
Comprehensive R Network (CRAN), 17
Computational mass spectral and retention
        libraries, 46–47
Confusion matrix, 288
Conjoint PCA, 261
Contingency table, 288, 294
Conventional modern-day statistical
        methods, 151
Cooley–Tukey algorithm, 51
Correctly Classified Rate (CCR), 289–290
Correlation optimized warping (COW), 49, 57
Cosmic rays, 64
Coupled chromatography mass spectrometry
    data matrices, 30–33
    gas chromatography, 27–29
    ionisation and detection, 29–30
    liquid chromatography, 27, 29
    peak tables (*See* Peak tables)
    procedural steps, 27
    software approaches and databases, 26
COW. *See* Correlation optimized
        warping (COW)
Cross-validation, 301, 317–319

Cumulative distribution function, 156
    χ^2-distribution, 171–174
    F-distribution, 177–179
    T^2-distribution, 179, 180
    normal distribution, 144, 153, 154, 158,
        160–162, 169, 170, 185, 252, 257, 263,
        264, 294, 321, 322, 339, 341, 352, 377
    t-distribution, 156–158, 161–166, 175, 191,
        192, 323, 351, 353

D
Daphnia, 76–77, 199
    environmental effect on, 395, 397–405
    multilinear regression, 365–371
    two-level multiway factorial designs,
        385–387
Data matrices, 30–33
Data table, for chemometric analysis, 55
    alignment, 56–57
    binning/bucketing, 57
    deconvolution and identification, 58–59
    uninteresting regions removal, 57
Degrees of freedom, one-way ANOVA, 357–358
Deming, Stan, 6
DENDRAL program, 47
Design of experiments (DoE), 195
    factors, response and coding, 198–199
    replication, 199–200
    simulated case study, 196
    statistical designs, 200–201
        fractional factorial designs, 204–206
        fully crossed designs, 201–202
        two-level full factorial designs, 202–204
    systematic experimental design, 197
Diabetes
    in mice, by LCMS, 78–79, 103–107
        disjoint PCA, 262–264
        negative ion, 324–326
        normal distribution, 158–161
        OPLS, 241–244
        QDA, 253–257
        t-distribution, 164–166
    NMR study of, 71–72, 100–103
Discriminatory variables, 190–194
Disjoint PCA, 261–264
    conjoint *vs.*, 261
    diabetes in mice, LCMS, 262–264
    prediction error, 261, 262
Dispersion peak shapes, 52
DoE. *See* Design of experiments (DoE)
D-statistic, 264–265
    cut-off value of, 266–267
    10 PC disjoint model, 267, 270–271
    2 PC disjoint model, 267–269

Dummy variables
    multiway regression with, 382–385
    one-way multilevel regression with, 373–375
Dynamic Link Libraries (DLLs), 17
Dynamic time warping (DTW), 56

E
Eigenvalues, principal components analysis, 89, 94, 96, 126, 230
Electron impact (EI), 29
Electrospray ionisation (ESI), 30
Enterococcus, 75–76
Enterococcus faecium, 75
Environmental toxicology, 195
Error matrix, 90
Error Rate (ERR), 289
Escherichia coli, 11
Euclidean distance, 182–184
Expert systems, 47

F
Faecal bacterial strains, by Raman spectroscopy, 109, 112–114
Fast Fourier Transform (FFT), 51
F-distribution
    cumulative distribution function, 177–179
    probability density distribution, 178
    p values for, 175–177
    t and χ^2-distribution, 174–177
Ferric Reducing Antioxidant Power (FRAP), 78
FID. *See* Free induction decay (FID)
Fisher, Ronald, 2, 3, 151, 154
Fluorescence, 64
Fortran (Formula Translation), 7, 16
Fourier transform (FT), 51
    principles, 50–52
    resolution, 53–54
    signal-to-noise, 53–54
Fourier Transform infrared (FTIR)
        spectroscopy. *See* Mid-infrared (MIR)
        spectroscopy
Fractional factorial designs, 204–206
Free induction decay (FID), 50, 51, 53
Fully crossed designs, 201–202
Fuzzy Warping (FW), 57

G
Gas chromatography mass spectrometry
        (GCMS), 27–29, 90
    Daphnia Magna metabolism, 76–77
    data analysis, 40–41, 46
    data matrix, 30–32
    datasets, 45
    ionisation and detection, 29–30
    malaria diagnosis in children, 69–70

GCMS. *See* Gas chromatography mass
        spectrometry (GCMS)
Genomes, 10
Genomics, 10
Genotype, 10
Glucose intolerance assessment, using
        NMR, 71–72
Gossett, William, 2
Graphical user interface (GUI), 34, 41

H
Habilitation, 5
Haemophilus influenzae, 10
High-performance liquid chromatography
        (HPLC), 8, 300
    Chinese herbal medicine studies using, 77–78
    single-wavelength, 47–49
Hotellings T^2, 179–181, 247
    p values calculated using, 182, 183, 186
    in software packages, 187
Hotelling statistic. *See D*-statistic
HPLC. *See* High-performance liquid
        chromatography (HPLC)
Human metabolome database (HMDB),
        46, 58
Hypothesis testing, 151–152

I
Icoshift method, 56
Impaired glucose tolerance (IGT), 71, 72, 287
Inelastic scattering, 63
Iterative target transform factor analysis
        (ITTFA), 35

J
Jack-knife approach, 338–339, 341
Join Aligner, 39
Jurs, Peter, 7

K
k Nearest Neighbour (kNN), 147
Kovats Retention Index (RI), 28, 29
Kowalski, Bruce, 1, 7
Kyoto encyclopaedia of genes and genomes
        (KEGG), 46

L
LCMS. *See* Liquid chromatography tandem
        mass spectrometry (LCMS)
LDA. *See* Linear discriminant analysis (LDA)
Leave one out (LOO), 301, 317, 318
Linear discriminant analysis (LDA), 4
    defined, 220
    Mahalanobis distance, 220–226
    misassigned samples, 222–227
    as multiclass classifiers, 272, 274–276

vs. QDA, 245–247
variance-covariance matrix, 220
Lipidomics, 14
Liquid chromatography tandem mass
    spectrometry (LCMS), 27–29, 210
  data analysis, 40–41, 46
  data matrix, 30–32
  datasets, 45
  diabetes in mice, 78–79, 103–107
    disjoint PCA, 262–264
    negative ion, 324–331
    normal distribution, 158–161
    OPLS, 241–244
    QDA, 253–257
    t-distribution, 164–166
  ionisation and detection, 30
  MS technique, 45
  negative ion, 324, 325
  pre-arthritic and control donors, 294–300
  of pre-arthritis, 98–100
  rheumatoid arthritis in humans study, 67–69
Loadings, principal components analysis, 86–87
  column-centred matrix, 126
  effect of nitrate on wheat, 108–109
  low-temperature maize data, 96–97
  maize raw data, 138
  negative ion mice, 105–106
  positive ion mice, 104
  raw and column centred data, 129
  row scaling, 116, 117, 137
  scores *vs.*, 88–89
  standardised data, 99, 101, 132, 133, 137
  whole standardised maize data, 92–95
Logarithmic data transformation, 136, 141–144
LOO. *See* Leave one out (LOO)

M
Mahalanobis distance
  calculation of *p* values and, 182–190
  Euclidean and, 183, 184
  linear discriminant analysis, 220–226
  multiclass classifier, 272
  NMR of maize, 187–190
  Quadratic Discriminant Analysis,
    245–250, 259
Mainstream applied statistics, 3–4
Maize, NMR
  metabolic changes in, 72–73, 92–97
  PLSDA, one versus all, 279, 281–285
  p values, 187–190
  QDA, 256–260
  t distribution, 166–169
Malaria diagnosis, using GCMS, 69–70
Mann–Whitney *U* test, 321–324

Martens, Harald, 19, 228
MATLAB, 7, 16–18, 56
Matrices, principal components analysis, 80–81
Matthews correlation coefficient
  (MCC), 289, 292
MCR. *See* Multivariate curve resolution (MCR)
Mean adjusted sum of squares (SSA), 363,
  380, 383, 399
Mean sum of squares errors (MSSE), 303
Metabolites, identification of, 60–61
Metabolome, defined, 11
Metabolomics, 4
  central dogma, of biology, 9, 10
  genomics, 10
  lipidomics, 14
  metabonomics, 14
  phenotypic expression, 11
  proteomics, 11
  transcriptomics, 10–11
  workflow, 12–14
Metabonomics, 14
Metalign, 39–40
Metlin, 46
Mid-infrared (MIR) spectroscopy, 61, 64–65
  nitrates on wheat, 64, 107–109
MIR spectroscopy. *See* Mid-infrared (MIR)
  spectroscopy
Misclassification Rate (MCR), 289
Moler, Cleve, 7, 16
MSDial, 40
Multiclass classifiers, 272
  LDA, 272, 274–276
  PLSDA, 275
    multilevel, 287
    one versus all, 275, 277–285
    one versus one, 285–287
    PLS2DA, 285–287
  simulated case study, 272, 273
Multiclasses performance classifica-
  tion, 291–292
Multilevel designs, ANOVA
  comparison and interpretation, 376–378
  with dummy variables, 373–375
  multilinear regression, 375–376
  one-way multilevel, 371–372
Multilevel PLSDA, 287, 341–347
Multilinear regression
  one-way ANOVA, 365–371
  one-way multilevel, 375–376
Multiple factor (multiway) ANOVA test, 379
  simulated 2 x 3 case study, 379, 380
    regression with dummy variables, 382–385
    steps for, 380–383
  two-level multiway factorial designs, 385–389

Multi-stage sampling, 209
Multivariate calibration
  optimisation, 317–319
  overview of, 305
  partial least squares regression, 306–311
  training and test sets
    errors measurement, 313–314, 316
    5 PLS component model, 312–313
    former terminology, 310–311
    predicted *vs.* observed estimation, 313, 314
    10 component model, 313, 315
Multivariate curve resolution (MCR),
    35–38, 151
mzMatch, 40
MZmine, 38–39

N
National Institute of Standards Technology
    (NIST), 38
NGT. *See* Normal glucose tolerance (NGT)
NIPALS algorithm. *See* Nonlinear Iterative
    Partial Least Squares
    (NIPALS) algorithm
Nitrates on wheat, by FTIR, 64, 107–109
NMR. *See* Nuclear magnetic resonance (NMR)
NMRProcFlow, 56–57
Nonlinear Iterative Partial Least Squares
    (NIPALS) algorithm, 84, 87, 89, 90, 228,
    229, 285, 306
Non-obese Diabetic (NOD) mice, 78
Normal glucose tolerance (NGT), 71, 72, 287
Normalisation. *See* Column standardisation
Normal (*z*) distribution
  cumulative distribution function, 156
  diabetic mice, 158–161
  one-tailed distribution, 153–154
  Quadratic Discriminant Analysis, 245
  row scaled intensity, 158–161
  standard normal distribution, 155
  t-distribution, 156, 157
  two-tailed distribution, 153–154, 156, 157
Nuclear magnetic resonance (NMR), 11
  advantage/disadvantage, 50
  apodization/baseline correction, 26
  data table, preparing, 55
    chemometric approach, 56–57
    deconvolution and identification, 58–59
  Fourier transform, 50–54
  frequency domain intensities, 13
  glucose intolerance/diabetes assessment,
    71–72, 100–103
  intensities, 371, 377
  maize
    average intensity of, 142

metabolic/temperature changes in,
    72–73, 92–97
  PLSDA, one versus all, 279, 281–285
  p values, 187–190
  QDA, 256–260
  standard deviation of, 143
  t distribution, 166–169
metabolites identification, 60–61
metabonomics in, 14
transformed spectra, preparing, 54–55
triglycerides in serum, 70–71, 107, 109–112,
    128, 140–141
Null hypothesis, 154

O
One-class classifier, 244
  different situations for, 244–245
  QDA (*See* Quadratic Discriminant
    Analysis (QDA))
  SIMCA, 260
    D-and *Q*-statistics, 264–266
    disjoint PCA, 261–264
    limits and decisions, 266–272
One-class models, 293–294
One-tailed distribution, 153–154
One versus all, PLSDA model, 275, 277–285
One versus one, PLSDA model, 285–287
One-way multilevel ANOVA test, 371–372
One-way-one-factor ANOVA test. *See* Single
    factor (one-way-one-factor)
    ANOVA test
OPLS. *See* Orthogonal Partial Least
    Squares (OPLS)
Optimisation, 300–304
  multivariate calibration, 317–319
Orthogonal Partial Least Squares (OPLS)
  correlated variation, 240
  defined, 240
  diabetes in mice, LCMS, 241–244
  orthogonal variation, 240

P
Partial Least Squares Discriminant
    Analysis (PLSDA)
  advantage in, 227
  for equal class sizes
    advantages, 234
    calibration, principles of, 229
    components, 232–234
    one-component, 231
    procedural steps, 228–230
    two-component, 231, 233
  loadings and weights, 326, 328–334
  as multiclass classifier, 275

multilevel, 287
one versus all, 275, 277–285
one versus one, 285–287
PLS2DA, 285–287
multilevel, 341–347
OPLS, 240–244
selectivity ratios, 346, 348–350
for unequal class sizes, 234–240
components comparison, 236, 237
decision threshold, 234
metabolites, 239, 240
misclassified samples, 238
predicted values, 237–238
weighted mean centring, 235, 236
Partial least squares (PLS), 4
components, 231, 317–319
weight matrix, 229
Partial least squares regression (PLSR)
auto-predictive model, 309, 311
calculation of, 306–307
defined, 306
5 component model, 308, 310
mean chromatographic intensity,
307, 308
1-component model, 307–309
Pattern recognition (PR), 210–211
PCA. See Principal component analysis (PCA)
Peak detection algorithms, 43
Peak-picking algorithm, 32
Peak tables, 30–33
data alignment
AMDIS, 38
apLCMS, 40
appropriate approach, 41–42
factors, 41
LCMS/GCMS data analysis, 40–41
Metalign, 39–40
MSDial, 40
multivariate curve resolution, 35–38
mzMatch, 40
MZmine, 38–39
XCMS, 33–35
manual inspection, 42–43
metabolites and annotated, 43–45
computational mass spectral and retention
libraries, 46–47
experimental libraries, 45–46
expert systems, 47
principles of, 32, 33
Pearson, Karl, 2
Percentage variance, 89
Performance classification
contingency table, 288
multiclasses, 291–292

one-class models, 293–294
two-class models, 288–291
PLS. See Partial least squares (PLS)
PLS2DA, 285–287
PLSDA. See Partial Least Squares Discriminant
Analysis (PLSDA)
PLSR. See Partial least squares
regression (PLSR)
PLS Toolbox, 18
Pre-arthritis, LCMS, 98–100, 294–300
Precision (PRE), 289
Prediction errors, 261, 262
Preliminary processing
inspection and preparation, 212
principal components analysis, 211
procedural steps, 211
simulated case studies, 212–217
Principal component analysis (PCA),
2, 6, 7, 70, 192, 201, 301, 325–326
calibration, 151
case studies
FTIR, nitrates on wheat, 107–109
LCMS, of diabetes in mice, 103–107
LCMS, of pre-arthritis, 98–100
NMR, for triglycerides in serum,
107, 109–112
NMR, of human diabetes, 100–103
Raman, of bacterial faecal isolates,
109, 112–114
conjoint, 261
data direct, visualising, 81–83
data transformation, 114
column centring, 125–130
column standardisation, 129–141
logarithmic transformation, 136,
141–144
row scaling (See Row scaling)
disjoint, 261–264
eigenvalues, 89
experimental data matrix, 84
loadings, 86–89
matrices, vectors and scalars, 80–81
missing data, 144–148
number reduction, 89–91
preliminary processing, 211
quality control samples, 147–148
scores, 84–86, 88–89
temperature on maize, NMR study, 92
scores and loadings plots, 92–97
variable plots, 92, 93
variable reduction, 149–150
Principal component loadings, 324–327
Probabilistic Quotient Normalisation
(PQN), 123

Probability density distribution
χ^2-distribution, 170, 171
F-distribution, 178
T^2-distribution, 179, 180
Probability density function, 249, 252
Proteomics, 11
PubChem, 46
p values, 151–152, 338–341
calculating *z*- and *t*-distributions, 162, 164
discriminatory variables, 192, 193
for *F*-distribution, 175–177
Mahalanobis distance, 182–190
multivariate calculation of, 182–190
normal *(z)* distribution and, 153–161
Python, 17

Q
QDA. *See* Quadratic Discriminant
Analysis (QDA)
Q statistic, 265–266
cut-off value of, 266–267
10 PC disjoint model, 267, 270–271
2 PC disjoint model, 267–269
Quadratic Discriminant Analysis (QDA)
class boundaries, 293
diabetic mice, of LCMS, 253–254
vs. LDA, 245–247
Mahalanobis distance, 245–250, 259
misclassified samples, 256, 257, 259
NMR, of maize, 256–260
normal distribution, 252
predictions, unusually assigned samples,
249, 251–252
probability density function, 249, 252
rank plot, 253
Quadrupole time of flight (QTOF), 30
Quality control samples, 123, 147–148
Quantitative structure-activity
relationship (QSAR), 8
Quantum mechanics, 4

R
Raman spectroscopy, 61–64
of bacterial faecal isolates, 75–76, 109, 112–114
RANSAC aligner, 39
Rayleigh scattering, 63
R code, 41
Recall *(REC)*, 289
Recursive segment-wise peak alignment
(RSPA), 56
Reference standards, row scaling to, 123–125
Regions of interest (ROI), 37, 38
Regression, one-way ANOVA, 361–363

Replication, Design of Experiments, 199–200
Residual sum of squares prediction
(RMSEP), 313
Residual sum of squares (RSS), 307
Retention time (RT), 28, 29, 32, 34
Rheumatoid arthritis, 67–69
RMSE. *See* Residual sum of squares (RMSE)
RMSECV. *See* Root mean square error of
cross-validation (RMSECV)
RMSEP. *See* Residual sum of squares
prediction (RMSEP)
Root mean square error of cross-validation
(RMSECV), 317, 318
Row scaling, 114–116
to constant total, 115, 117–120
data matrix, 134–137
to reference standards, 123–125
standard normal variates, 121–122
R programming language, 17
RSS. *See* Residual sum of squares (RSS)

S
Saccharomyces cerevisiae, 10
Sample selection
design of experiments
factors, response and coding, 198–199
replication, 199–200
simulated case study, 196
statistical designs, 200–206
systematic experimental design, 197
motivation, 195–196
sampling designs, 206–209
Sampling designs
cluster sampling, 209
multi-stage sampling, 209
sample size, 208
simple random sampling, 207, 208
statistical sampling methods, 206–207
stratified sampling, 208–209
systematic sampling, 207, 208
SAS. *See* Statistical Analysis System (SAS)
Scalars, principal components analysis, 81
Scientific computing, 6–7
Score Distance. *See* D-statistic
Scores matrix, 229
Scores, principal components analysis, 84–86
column-centred matrix, 126
effect of nitrate on wheat, 108–109
faecal bacterial strains by Raman, 113
vs. loadings, 88–89
low-temperature maize data, 96–97
maize raw data, 138
negative ion mice, 105–106

NMR triglycerides, 128
positive ion mice, 104
row scaling, 116, 117, 134–137
standardised data, 99, 101, 132, 133, 137
of standards and standardised malaria, 124
whole standardised maize data, 92–95
Selectivity ratios, 346, 348–350
Signal-to-noise ratio (S : N), 51, 53–54
SIMCA. *See* Soft independent modelling of class analogy (SIMCA)
Simple random sampling, 207, 208
Single factor (one-way-one-factor)
ANOVA test, 357
balanced design, at two levels
calculation, of ANOVA, 358–361
degrees of freedom, 357–358
regression, 361–363
t-test, 363
multilevel designs
comparison and interpretation, 376–378
with dummy variables, 373–375
multilinear regression, 375–376
one-way multilevel ANOVA test, 371–372
multiple one-way design, with two levels, 365–371
unbalanced design, at two levels, 364–365
Single-wavelength HPLC, 47–49
Singular Value Decomposition (SVD) algorithm, 87
Sirius package, 19
Soft independent modelling of class analogy (SIMCA), 4, 18
chemometric approach, 260
D-and *Q*-statistics, 264–266
disjoint PCA, 261–264
limits and decisions, 266–272
Software, 15–20
SpecAlign, 57
Square Prediction Error (SPE). *See Q* statistic
SSA. *See* Mean adjusted sum of squares (SSA)
SSM. *See* Sum of squares for the mean (SSM)
SSR. *See* Sum of squares for regression (SSR)
SST. *See* Total sum of squares (SST)
Standard normal distribution, 152, 153, 155
Standard normal variates (SNV), 107, 121–122
Statistical Analysis System (SAS), 19
Statistical mechanics, 4
Statistical Methods for Research Workers (Fisher), 2
Statistical Package for the Social Sciences (SPSS), 19
Statistical sampling methods, 206–207

Statistics
χ^2-distribution, 169–174
discriminatory variables, 190–194
F-distribution, 174–178
Hotellings T^2, 177, 179–181
hypothesis testing, 151–152
Mahalanobis distance, 182–190
p values, 151–152
multivariate calculation of, 182–190
normal *(z)* distribution and, 153–161
standard normal distribution, 152, 153
t-distribution and degrees of freedom, 161–169
Stokes scattering, 63
Stratified sampling, 208–209
Sum of squares for regression (SSR), 363, 372, 380–381, 383, 394, 399
Sum of squares for the mean (SSM), 363, 372
Systematic sampling, 207, 208

T
T^2-distribution. *See* Hotellings T^2
Target projection (TP), 244, 348
t-distribution, 156, 157
and degrees of freedom, 161–169
diabetic mice, LCMS, 164–166
F distribution relates to, 174–177
NMR profiling, of maize, 161, 166–169
population standard deviation, 161
sample standard deviation, 161
vs. z-distribution, 161–164
Temperature changes in maize, NMR, 72–73, 92–97
Terminology, 353–356
Tetramethylsilane (TMS), 52
Time domain. *See* Free induction decay (FID)
Total sum of squares (SST), 362, 371, 394
Transcriptomics, 10–11
Transformed spectrum, preparing, 54–55
Triglycerides in serum, using NMR, 70–71, 107, 109–112, 128, 140–141
t-test, 363
Two class classifiers
hard boundary principle in, 218
linear discriminant analysis, 220–227
linearly inseparable classes, 218, 219
non-linear boundary, 218, 219
Partial Least Squares Discriminant Analysis, 227–228
for equal class sizes, 228–234
OPLS, 240–244
for unequal class sizes, 234–240
three different linear boundaries, 218, 219

Two-class models, performance
        classification, 288–291
    Bayesian statistics, 290–291
    confusion matrix for, 288
    Correctly Classified Rate, 289–290
    positives and negatives, 290
Two-level full factorial designs, 202–204
Two-level multiway factorial designs,
        385–389
Two-tailed distribution, 153–154, 156, 157
Type 2 Diabetes, 287
Type 2 Diabetes Mellitus (T2DM), 71, 72

U
Univariate approaches/models, 320–324,
        352–353
UNSCRAMBLER software, 19

V
Validation
    pre-arthritis, LCMS, 294–300
    training set/calibration set, 294
Variable importance in projection
        (VIP) scores
    calculation, 333, 335
    defined, 333
    5 component model, 335–336
    p values from, 339, 340
    10 PLS component model, 336–338
Variable plots, of temperature on maize, 92, 93
Variable reduction, principal components
        analysis, 149–150

Variance-covariance matrix, 220, 245–247,
        249, 295
Vectors, principal components analysis, 81
Vibrational spectroscopy
    carbon atom, symmetric/asymmetric
        stretches of, 61–62
    fingerprint region, 61
    MIR spectroscopy, 61, 64–65
    Raman spectroscopy, 61–64
VIP scores. *See* Variable importance in
        projection (VIP) scores
Visual Basic for Applications (VBA), 16
Volcano plots, 349–351

W
Wilcoxon rank sum test. *See*
        Mann-Whitney U test
Wold, Svante, 1, 18
World Health Organisation, 69

X
XCMS, 33–35

Y
Yates' algorithm, 3
Youden, W.J., 5

Z
z-distribution
    p value for two-tailed, 156
    t-distribution *vs.*, 161–164
Zero-filling, 53, 54

Printed and bound by CPI Group (UK) Ltd, Croydon, CR0 4YY

16/04/2025

14658552-0005